# Communications
# in Computer and Information Science 1539

More information about this series at https://link.springer.com/bookseries/7899

Victor Taratukhin · Mikhail Matveev ·
Jörg Becker · Yury Kupriyanov (Eds.)

# Information Systems and Design

Second International Conference, ICID 2021
Virtual Event, September 6–7, 2021
Revised Selected Papers

 Springer

*Editors*
Victor Taratukhin ⓘD
National Research University Higher School
of Economics
Moscow, Russia

Jörg Becker ⓘD
University of Münster
Münster, Germany

Mikhail Matveev ⓘD
Voronezh State University
Voronezh, Russia

Yury Kupriyanov ⓘD
National Research University Higher School
of Economics
Moscow, Russia

ISSN 1865-0929          ISSN 1865-0937 (electronic)
Communications in Computer and Information Science
ISBN 978-3-030-95493-2          ISBN 978-3-030-95494-9 (eBook)
https://doi.org/10.1007/978-3-030-95494-9

This Springer imprint is published by the registered company Springer Nature Switzerland AG
The registered company address is: Gewerbestrasse 11, 6330 Cham, Switzerland

# Preface

We are delighted and honored to introduce to you the proceedings of the Second International Conference for Information systems and Design (ICID 2021) which took place during September 6–7, 2021, in Divnomorskoe on the Black Sea coast of Russia.

As stated in the original mission of the conference, ICID is set to become a global cooperation network that promotes open industry innovations in the academic environment. We focus on the practical results-based studies prepared by academic researchers and industry experts in Information Systems' design, deployment, and adoption. The ICID community also welcomes students to present their research and participate in the experimental workshops (e.g., Ideathons, Hackathons and the like).

The main value of ICID community development is seen in building global connections between academia and business. The ICID conference is an active international scientific event with more than 100 participants and guests from Germany, Russia, Japan, Belarus, the UK, Finland, Belgium, the USA, and other countries.

In 2021 the organizational committee reviewed 51 submitted papers, of which 31 were accepted. Based on the clear distinction of submitted papers' research focus, depth, and results application a decision was made to split accepted submissions into three sections, and thus ensure the proper positioning of the authors within the ICID community.

The three sections of ICID 2021 materials are as follows:

- Methodological support of analysis and management tools: theoretical-focused research
- Digital transformation of enterprises based on analysis and management tools: practical-focused research
- Young scientists' research in the areas of enterprise digitalization

ICID 2021 was run primarily in an off-line/on-site format and was attended by 70 people in person, with 40 participants joining on-line via videoconferencing. In accordance with the ICID 2021 program, the conference took place over two days with a plenary session followed by 26 presentations on the first day and 10 presentations on the second day.

The on-site participants included leading experts and teams from key regional universities, research centers, leading IT companies, industrial companies of the oil and gas and metals and mining sectors, and institutes of the Russian Academy of Sciences.

September 2021

Victor Taratukhin
Mikhail Matveev
Jörg Becker
Yury Kupriyanov

# Organization

## International Chair

Jörg Becker          University of Muenster, Germany

## General Chairs

Victor Taratukhin       National Research University Higher School of Economics, Russia, and University of Muenster, Germany

Mikhail Matveev       Voronezh State University, Russia

## Industry Chair

Yuri Kupriyanov       National Research University Higher School of Economics, Russia

## International Program Committee

Anton Ambrazhey     Peter the Great St. Petersburg Polytechnic University, Russia

Natalia Aseeva       National Research University Higher School of Economics, Nizhny Novgorod, Russia

Eduard Babkin        National Research University Higher School of Economics, Russia

Tatiana Gavrilova     St. Petersburg University, Russia

Nikita Golovin        Peter the Great St. Petersburg Polytechnic University, Russia

Evgeny Gryazin       Nordea Bank Oyj, Finland

Jörg Becker           University of Muenster, Germany

Yuri Kupriyanov      National Research University Higher School of Economics, Russia

Pavel Malyzhenkov    National Research University Higher School of Economics, Nizhny Novgorod, Russia

Liudmila Massel      Irkutsk National Research Technical University, Russia

Mikhail Matveev      Voronezh State University, Russia

Natalia Pulyavina     Plekhanov Russian University of Economics, Russia

Vladimir Silkin V.      Shirshov Institute of Oceanology of the Russian Academy of Sciences, Russia

Victor Taratukhin       National Research University Higher School of Economics, Russia, and University of Muenster, Germany

Alexander Zatsarinny   Federal Research Center "Informatics and management" of the Russian Academy of Sciences, Russia

## Organization Committees

| | |
|---|---|
| Anton Ambrazhey | Peter the Great St. Petersburg Polytechnic University, Russia |
| Valentina Anikushina | University of Heidelberg, Germany |
| Daria Gaskova | Melentiev Energy Systems Institute, Russia |
| Nikita Golovin | Peter the Great St. Petersburg Polytechnic University, Russia |
| Yuri Kupriyanov | National Research University Higher School of Economics, Russia |
| Artem Levchenko | Voronezh State University, Russia |
| Valeria Skibina | Lomonosov Moscow State University, Russia |
| Yulia Skrupskaya | National Research University Higher School of Economics, Russia |
| Victor Taratukhin | National Research University Higher School of Economics, Russia, and University of Muenster, Germany |

# Contents

**Methodological Support of Analysis and Management Tools:
Theoretical-Focused Research**

# Digital Transformation of Enterprises Based on Analysis and Management Tools: Practical-Focused Research

# Disciplinary Routing as a Factor of Activization of Scientific Mobility

Alexander A. Zatsarinnyy⬛ and Alexander P. Shabanov⁽⊠⁾⬛

Federal Research Center "Computer Science and Control" of the Russian Academy of Sciences, Vavilova Street 44-2, 119333 Moscow, Russia

**Abstract.** The research relates to the tasks of the digital economy solved in the federal projects "Information Infrastructure", "Digital Technologies" and "Digital Public Administration" by implementing platform solutions in the collaboration of organizational systems. The scientific task of forming a new automated data routing process in the digital platform, taking into account the classification of scientific disciplines, has been set and solved. The goal is to activate scientific mobility. The analysis of relations in collaboration, the semantics of knowledge bases and retrospective information about platform solutions is carried out. As part of the work on the implementation of the Program of Fundamental Scientific Research for the long-term period and the project 18-29-03091 of the RFBR, a new model of the digital platform was created. The novelty of the model lies in the automatic formation of the data routing process directly during the data transmission process. At the same time, due to the intensification of interactions with economic entities, the set goal is achieved and an increase in the effectiveness, significance and relevance of the results of scientific research for the development of the national economy and society is ensured.

**Keywords:** Digital economy · Transposition structures · Interdisciplinary research · Innovative development · Information interaction

## 1 Introduction

The present study relates to the problem of increasing the effectiveness of the scientific industry by increasing the number of scientific works implemented in transposition structures – industry, regional, cross-border and other clusters, corporate and social systems, innovative and other associations of organizational systems, both conducting research and implementing them.

The basic elements that ensure the transfer of research data and other information between organizational systems in transposition structures are digital platforms.

A scientific task has been set and is being solved – the development of a new model of a digital platform placed in a transposition structure, with the possibility of reproducing an automated data routing process based on a classifier of standards.

The location of the routing process under consideration when transmitting research data is shown in Fig. 1.

© Springer Nature Switzerland AG 2022
V. Taratukhin et al. (Eds.): ICID 2021, CCIS 1539, pp. 3–16, 2022.
https://doi.org/10.1007/978-3-030-95494-9_1

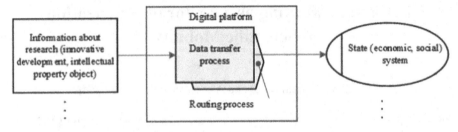

**Fig. 1.** The place of the routing process in the digital platform of the transposition structure.

The urgency of solving this problem is due to the creation of methodological and technical groundwork for increasing the level of scientific mobility through the development of platform solutions. The consequence of the application of this reserve is to increase the competitiveness of transposition structures in world markets.

Figure 2 shows the place of the task in the structure of regulatory and planning documents of the national program "Digital Economy" [1] and the project of the Russian Foundation for Basic Research (RFBR) #18-29-03091 [2].

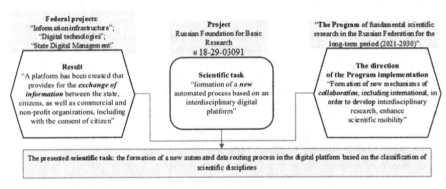

**Fig. 2.** The scheme of linking the scientific task to federal projects.

The analysis of relations in known transposition structures and in models of interconnections of data transmission route elements in the knowledge bases of well-known digital platforms led to the choice of the direction of development of a new routing process, namely, as a data transmission process. As an example, digital models of the elements involved in the routing process are shown (Fig. 3). The composition of the elements is determined depending on the delivery address. The address is determined depending on the coincidence of classification codes recorded by the subject-source and the subject-consumer of information about the research.

In this study, a digital platform with disciplinary routing is considered as a platform solution for increasing the level of scientific mobility. The article presents.

- Analysis of well-known platform solutions used for data transmission in networks of information interactions of transposition structures.

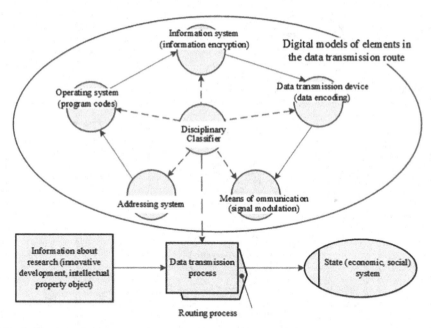

**Fig. 3.** Models of elements of the routing process embedded in the data transmission process.

- Development of a new digital platform model with disciplinary routing functions when transmitting data on scientific research, innovative developments and intellectual property objects.
- Development of a methodology for assessing the sufficiency of the performance of a digital platform while increasing the number of organizational systems supported in the transposition structure, both sources of information and its consumers.

The general novelty of the presented solutions lies in the centralized formation of the data routing process directly in the process of data transmission. The technical novelty lies in the automatic formation and execution of routing functions based on the data of the classifier of standards and data on the relationships of route elements linked to the addresses of data sources and consumers. The practical value of the presented solutions is to increase the effectiveness, significance and relevance of the results of scientific research for the development of the national economy and society.

## 2  Analysis of Well-Known Platform Solutions

### 2.1  The Look at the Digital Platform in Relation to the Documents of the Digital Economy

Figure 4 shows the place of the new digital platform model in the environment of the fundamental regulatory and planning documents of federal projects.

The fundamental documents include the passports of the federal projects "Information Infrastructure" [3], "Digital Technologies" [4], "Digital Public Administration" [5]

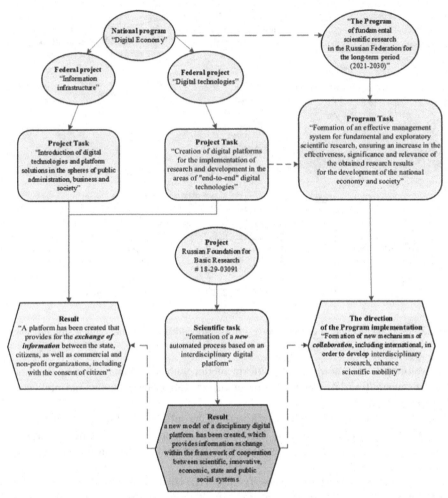

**Fig. 4.** The place of the presented digital platform in the environment of digital economy tasks (fragment).

and the Program of Fundamental Scientific Research in the Russian Federation for the long term (2021–2030) [6].

The emergence of a strategic direction to increase the effectiveness of the scientific industry [6] predetermined the trend of increasing the number of organizational systems both conducting research and implementing them. At the same time, along with departmental, industry and corporate organizational systems, small and medium–sized enterprises and innovation centers are increasingly appearing, which interact with large enterprises in transposition structures - federal collaborations, regional and industry clusters, urban conglomerations, and other associations. Some organizational systems conduct research and develop promising technical solutions, including intellectual property objects, while others develop projects based on them and create new technologies and products.

## 2.2 Prerequisites for the Creation of a Digital Platform with Disciplinary Routing

The growth in the number of transposition structures with technologies, for example, [7–21], which ensure the interaction of scientific, innovative, economic, educational and state actors, has led to the urgency of solving the problem of centralized reproduction of research data transmission processes. The following entities are at the center of solving this problem.

- Digital platforms as intermediaries in the transfer of research data between subjects of activity.
- Data on the classification needs of subjects-consumers of research information.
- Participation in research and in their provision of an increasing number of enterprises, authorities, investors, educational organizations and others.

To date, digital platforms for various purposes are used in economic and management sectors, for example, [22–31]. Structural and process tasks in these digital platforms are solved taking into account the world and Russian experience in the following areas.

1. Designing processes for collecting, processing, displaying and archiving information, processing and converting data in computer networks, data transmission and processing in data transmission networks [28, 30–32].
2. Creation of new business models [33], their transformation and routing [34].
3. Development of digital transport with global navigation satellite systems and protection of control channels [35, 36], emergency response modernization [36, 37].
4. Creation of platforms for research [38], innovative solutions for knowledge management [39], and research on the management of distributed [40] and cloud [41] resources of digital platforms.

At the same time, technical solutions for the predecessors of digital platforms – integrated control systems, control centers (centralized management), distributed control systems – have been most developed - up to the level of intellectual property objects. Based on this fact, the analysis of the possibility of using the following inventions as a digital platform in the direction under study was carried out.

- Integrated control system [42].
- The system of situational and analytical centers of the organizational system [43].
- Management Center of the organizational system [44].

These technical solutions relate to computing complexes and computer networks. The field of application is digital platforms for the implementation of research and development in the areas of "end-to-end" digital technologies [45, p. 52]. Due to the possibility of centralized reproduction of the data transmission process, the Management

Center of the organizational system is the closest, in comparison with technical solutions [42, 43], an analogue in the development of a new digital platform model. At the same time, the new model should automatically generate and reproduce the data transmission process with routing functions. The information core of the new model is the classifier of standards and the data transmission route.

## 3  Digital Platform Model with Disciplinary Routing

### 3.1  Block Diagram of a Digital Platform with Disciplinary Routing

Based on the above, a model of a digital platform with disciplinary routing has been developed – the Digital platform for supporting scientific research (Fig. 5).

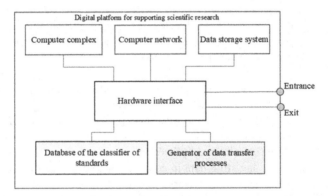

**Fig. 5.** Block diagram of a digital platform with disciplinary routing.

The technical result, which is achieved with the help of a new digital platform model, is to expand the functionality for reproducing the processes of transmitting research data. A new feature is the implementation of routing functions directly in the process of data transmission. The following new data are intended to perform the new function.

1. Data on the classification codes of studies.
2. Data on the relationships of transmission route elements.
3. Data on linking routes to multiple pairs of addresses of research sources and their consumers.

At the same time, the sources of research and their consumers are organizational systems in transpositional structures – subjects of scientific, innovative, economic, educational, financial and other activities, authorities.

The computing complex includes operating systems, including a hypervisor, and application programs of information systems that implement full software virtualization technology based on its hardware and software.

The interface of the equipment includes data transmission devices and means of communication.

The data storage system consists of data segments, each of which records and stores data on research conducted by one of the organizational systems in transposition structures, including data on their classification codes, and each segment is characterized by the address code of the corresponding transposition structure and organizational system.

## 3.2 Distinctive Features of the Digital Platform

Figure 6 shows a diagram of the application of a digital platform that has the following significant distinguishing features from its analogues [42–44].

1. Availability of a database of the classifier of standards and a generator of data transmission processes, the internal inputs and outputs of which are connected to the corresponding outputs and inputs of the equipment interface.
2. The database of the classifier of standards contains data on classification codes and on the interrelationships of elements of data transmission routes linked to pairs of addresses of the organizational system-the source of research data and the organizational system-the consumer of this data, which are linked to the same classification code.
3. In the computing complex, each virtual machine contains an operating system that differs in its characteristics from the operating systems in other virtual machines. At the same time, each operating system is equivalent in its characteristics to the operating systems that are equipped with one of the groups of organizational systems in the transposition structure. The number of such groups is equal to the number of virtual machines.
4. The equipment interface includes data transmission devices and means of communication, each of the data transmission devices corresponds in its characteristics to one of the data transmission devices in one of the groups of organizational systems in the transposition structure, and the number of such groups is equal to the number of data transmission devices in the equipment interface. Each of the means of communication by its characteristics corresponds to the means of communication in one of the groups of organizational systems in the transposition structure, and the number of such groups is equal to the number of means of communication in the equipment interface.
5. The generator of data transmission processes provides reproduction of the processes of transmission of research data, while each time new research data arrives in the data storage system, it performs the following actions.

- Determines in the data storage system the classification code of the received study and the address code of the organizational system-the source of research data.
- Selects in the data storage system the classification code of the address of the organizational system-the consumer of research data, whose classification code corresponds to the classification code of the received research.
- Selects in the database of the classifier of standards the models of route elements related to a pair of address codes of the organizational system-the source and the organizational system-the consumer of research data and alternately transmits the

received data to the route elements - devices and programs corresponding to their models in the database of the classifier of standards.

- If there are two or more consumers of research data with the same classification codes, the reproduction of the data transmission process is repeated for each route.

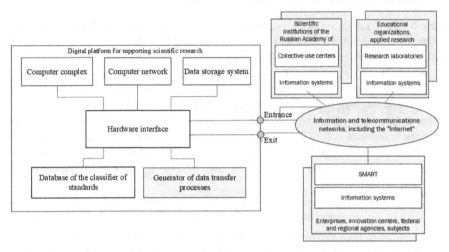

**Fig. 6.** Scheme of application of a digital platform with disciplinary routing

### 3.3 Semantic Database of the Classifier of Standards

The concept of IT Service Management [46] is proposed as the main methodological tool for structuring the database of the classifier of standards. The following terms and definitions related to this concept are used [47]:

- the Standards classifier database is a semantic database used to store configuration records throughout their life cycle, stores interrelated configuration elements and their attributes - routing elements and attributes, classification codes, addresses of organizational systems and other elements necessary for reproducing data transmission processes;
- a configuration management system is a set of tools, data and information used to support the management of service assets and configurations, to collect, store, manage, update, analyze and present data on all elements and their relationships;
- a configuration record is a record containing information about configuration items, each configuration record documents the lifecycle of an individual configuration item, configuration records are stored and maintained as part of a configuration management system;
- a configuration element (configuration element) is any component or other service asset that needs to be managed to reproduce data transmission processes.

The application of the IT Service Management concept ensures that the standards classifier database is filled with components relevant to new (innovative) production technologies for reproducing data transmission processes.

An example of configuration records in database routing segments are digital models of routing process elements embedded in the data transmission process, which are shown in Fig. 3.

An example of configuration records of classification codes in the database of the classifier of standards are configuration records of classification codes given in the All-Russian Classifier of Standards [48]. The All-Russian Classifier of Standards is part of the Unified System of Classification and Coding of Technical, Economic and Social Information of the Russian Federation. This classifier is harmonized with the International Classifiers of Standards and the Interstate Classifier of Standards.

## 4 Method of Evaluating the Performance of the Computing Complex

The performance of a computing complex in a digital platform with disciplinary routing is evaluated in order to control the sufficiency of computing resources assigned to each $i$-th virtual machine (Subsect. 3.2, paragraph 3) in accordance with the permissible values of the data processing waiting time $T_i$ and the probability $P_i$ ($\leq$T) of not exceeding this time, and $i = 1\dots V$, $V$ is the number of virtual machines.

When determining the resource performance for a virtual machine, follow these steps.

**Step 1.** Calculate the probabilities $P_N$ $(j)$ that the number of data packets waiting for service in the busy interval formed by $N$ data packets is on average no more than $j$.

$$P_N(j) = \frac{1}{N} \sum_{k=j+1}^{N} P_N^k(j), \text{ where} \tag{1}$$

$N$ – the number of data packets in the busy interval of the input buffer of the virtual machine, $N$ can take any integer values, and the processing time of one data packet is assumed to be constant;

$j$ – the number of data packets that can catch a data packet received by the $k$-th of $N$ received in the busy interval in the input buffer;

$P_N^k(j)$ – the probability that a data packet received by $k$-th from $N$ received in the busy interval in the input buffer will find o data packets in it, and $0 \leq j \leq k - 1$ и $1 \leq k \leq N$. It is determined using formulas (2)–(4):

$$P_N^k(j) = \frac{(N-1)!}{N^{N-2}}\{F1 + F2\}, \tag{2}$$

$$F1 = \frac{(k-j)^{k-j-2}}{(k-j-1)!} \times \left[ \frac{(N-k+j+1)^{N-k+j-1}}{(N-k+j)!} - \sum_{m=1}^{j-1} \frac{(N-k+j)^{N-k+j-m-1}}{(m-1)!(N-k+j-m)!} \right], \tag{3}$$

$$F2 = \sum_{y=0}^{x} \left[ \begin{array}{c} \dfrac{(-1)^y (x-y+1)^y}{y!(k-j+x-y-1)!} \\ \times (k-j+x-y)^{k-j+x-y-2} \end{array} \right] \times \left\{ \left[ \dfrac{(x+1)(N-k+j+1)^{N-k+j-x-1}}{(N-k+j-x)!} \right] - \left[ \sum_{z=0}^{j-x-1} \dfrac{(x+z)(N-k+j)^{N-k+j-x-z-1}}{z!(N-k+j-x-z)!} \right] \right\},$$

(4)

For $N = 3, 4...; k = 3... N; j = 1... k-1$.

Formulas (2)–(4) are based on the formula [49] for determining the probability $P_N^k(j)$ that a requirement serviced by the $k$-th in order in the busy interval with a duration of $N$ service intervals was waiting for service of $j$ service intervals, and the service interval is equal to the value of the service time of one data packet.

**Step 2.** Compile a table of probabilities $P_N(j)$ for different parameters $j$ and $N$.

Set acceptable values for each virtual machine for the duration of the busy interval, expressed by the maximum number of $N_{max}$ data packets.

Figure 7 shows a sample of probability values.

| | $N=10$ | $N=11$ | $N=12$ | $N=13$ | $N=14$ | $N=15$ |
|---|---|---|---|---|---|---|
| $j=0$ | 0,1 | 0,09091 | 0,08333 | 0,07692 | 0,07143 | 0,06667 |
| $j=1$ | 0,42872 | 0,39786 | 0,37108 | 0,34763 | 0,32693 | 0,30854 |
| $j=2$ | 0,76379 | 0,72922 | 0,69692 | 0,66684 | 0,63887 | 0,61288 |
| $j=3$ | 0,93814 | 0,91896 | 0,8989 | 0,87839 | 0,85776 | 0,83725 |
| $j=4$ | 0,99016 | 0,98431 | 0,97711 | 0,96872 | 0,95929 | 0,949 |
| $j=5$ | 0,99912 | 0,99813 | 0,99661 | 0,99448 | 0,9917 | 0,98827 |
| $j=6$ | 0,99996 | 0,99988 | 0,99969 | 0,99936 | 0,99883 | 0,99806 |
| $j=7$ | 1 | 1 | 0,99998 | 0,99995 | 0,99989 | 0,99978 |
| $j=8$ | 1 | 1 | 1 | 1 | 0,99999 | 0,99998 |
| $j=9$ | 1 | 1 | 1 | 1 | 1 | 1 |

**Fig. 7.** Sampling of probability values $P_N(j)$.

As an example, let us take:

- the allowed waiting time $T_i$ for data processing in terms of the number of data packets is 4;
- the permissible probability $P_i (\leq T)$ of not exceeding this time is 0.94.

Then, using the calculated data (Fig. 7), we obtain an acceptable value for the duration of the busy interval. The number of data packets indicates this value. In our example, $N_{max} = 15$.

**Step 3.** During the project on the development of the transposition structure, in connection with the introduction of a new organizational system into it, which will require the resources of the digital platform, the actual indicators of the duration of the intervals of employment of the input buffers of virtual machines, expressed by the number of data packets, are measured.

**Step 4.** Based on the obtained values and using the data of the probability tables PN (j), a comparative analysis is carried out and a decision is made on the optimal or rational, depending on the availability of material and other means, for example:

- Redistribution of computing resources between virtual machines, taking into account the resources allocated for the new organizational system,
- Alternatively, the creation of a new virtual machine with the placement of new computing resources, due to the prospect of further scaling.

A new, distinctive feature of the presented method for evaluating the performance of a computing complex is the simplification of the process of measuring performance indicators by reducing measurements to the number of busy intervals and control points - to the number of input buffers of virtual machines.

## 5 Conclusion

The research relates to the scientific directions of the digital economy in terms of increasing the efficiency of the scientific industry by increasing the number of research implementations, innovative developments and intellectual property objects in economic, managerial and social organizational systems.

The initiation of such a study is due to the factor of activation of the scientific industry, which is in demand in connection with the administrative support of state bodies, expressed in the formulation and development of federal projects of the national program on the digital economy.

The research was preceded by a number of works in the field of building integrated control systems, control centers and digital platforms of various scales and purposes. These works were carried out with the participation of the authors of this article. These are works on the creation of methods and models for the construction of integrated control systems, control centers and digital platforms of various scales and purposes. Based on the results of these works, a scientific task has been set for conducting this research – the creation of a new model of a digital platform. The digital platform is introduced into a transposition structure for the purpose of automatic reproduction of data transmission processes. The condition for such reproduction of processes is the coincidence of classification codes of research and classification codes that are in demand by consumers of research. Such consumers are subjects of economic, managerial and social activities.

The following specific tasks have been set and solved:

- the purpose of solving the scientific problem has been determined – the development of methodological and technical groundwork for increasing the level of scientific mobility through the development of platform solutions;
- the analysis of well-known platform solutions is carried out, prerequisites for the creation of a digital platform with disciplinary routing are identified and the choice of the direction of development of a new routing process in a digital platform as a process as part of the data transmission process is justified (Sect. 2);

- the new model of a digital platform has been developed, the core of which is a semantic database with digital models of interconnected elements of data transmission routes between organizational systems-sources of information about research and organizational systems-consumers of this information, the latter implement scientific research in their technologies and in their products (Sect. 3);
- the method for assessing the sufficiency of the digital platform performance has been developed.

The novelty of the developed model and methodology lies in the centralized structuring of the semantic database, the formation and reproduction of data transmission processes based on it with the ability to control and, if necessary, modernize the system of virtual machine resources in the digital platform.

The practical significance of the research results lies in ensuring the intensification of interactions between producers of scientific research, innovations, inventions and their consumers. As a result, the activation of scientific mobility and the improvement of the effectiveness of research.

## References

1. Program the Digital Economy of the Russian Federation. https://www.researchgate.net/pub lication/331993372. Accessed 15 Oct 2021
2. The Russian Foundation for Basic Research. Project No. 18-29-03091. https://kias.rfbr.ru/ind ex.php. Accessed 15 Oct 2021
3. Passport of the Federal project "Information Infrastructure". https://digital.ac.gov.ru/. Accessed 15 Oct 2021
4. Passport of the Federal project "Digital Public Administration". https://digital.ac.gov.ru/. Accessed 15 Oct 2021
5. Passport of the Federal project "Digital technologies". https://digital.ac.gov.ru/. Accessed 15 Oct 2021
6. The program of fundamental scientific research in the Russian Federation. http://static.govern ment.ru/media/files/skzO0DEvyFOIBtXobzPA3zTyC71cRAOi.pdf. Accessed 15 Oct 2021
7. Andreeva, T., Astanina, L.: Analysis of the industrial structure of clusters included into the register of the Ministry of Industry and Trade of Russia. Trends Manage. **4**, 111–127 (2018). https://doi.org/10.7256/2454-0730.2018.4.28159
8. Rezanov, V.K., Rezanov, K.V., Zvereva, E.V., Yuylyan, T.: Cross-border umbrella structures as the basis of a cross-border international cluster. Power Adm. East Russ. **4**(85), 90–99 (2018). https://doi.org/10.22394/1818-4049-2018-85-4-90-99
9. Tashenova, L.V., Babkin, A.B.: Typology and structure of industrial clusters. Manage. Russ. Abroad **1**, 4–14 (2019)
10. Balash, O.S., Manukyan, M.M., Yuklasova, A.V.: Development of innovative activity of integrated industrial structures. Vestnik Samara Univ. Econ. Manage. **3**(9), 7–11 (2018)
11. Rezanov, V.K., Zhang, J.: The umbrella structure as a basis for sustainable development of international cooperation and cross-border oil and gas cluster. Far East Probl. Dev. Archit. Constr. Complex **1**(1), 334–338 (2019)
12. Mrochkovsky, N.S.: Developing models of organization management on the basis of their integration in digital clusters. Vestnik Plekhanov Russ. Univ. Econ. **1**(1), 159–163 (2020). https://doi.org/10.21686/2413-2829-2020-1-159-163

13. Suboch, F.: Prospects for development and features of the associative concept in the construction of the latest transpositional structures, including clusters. Agric. Econ. **3**(298), 20–40 (2020)
14. Matyukin, S.V.: Models of transformation of the holding companies structures into clusters: scenarios under the context of the Russian economy. Vlast' (Authority) **5**(28), 143–151 (2020)https://doi.org/10.31171/vlast.v28i5.7590
15. Serikova, N.V.: Clusters as a way to implement network communications of business and other structures in the context of digital transformation. Econ. Entrep. **12**(125), 739–744 (2020). https://doi.org/10.34925/EIP.2021.125.12.147
16. Novikova, I.V., Sanko, G.G., Timofeeva, Y.A.: Cluster as a network structure and factor of economic growth of national economy. Trudi BGTU. Ceria 5: Ekonomika i upravleniya **2**(214), 22–27 (2018)
17. Kudryashov, V.S.: The structure of the formation and functioning of territorial cluster at the regional level. Uchenii zapiski Tambovskogo otdelenia RoSMU **9**, 56– 68 (2018)
18. Khalilov, N.R.: The structure of the unified information platform of interaction in machine-building cluster. Problemi sovremennoj ekonomiki **2**(66), 207–210 (2018)
19. Bezpalov, V.V., Skripnik, O.B., Lochan, S.A., Petrosyan, D.S.: Regional clusters: concept, structure and tendencies of development. Audit i finansovij analiz **3**,153–161 (2018)
20. Napolskich, D.L.: Analysis of the dynamics of changes in the organizational structure of innovation clusters in the Russian Federation. Napolskich. Innovatsii technologii upravleniya i prava **1**(24), 18–26 (2019)
21. Romanova, A.T., Popova, M.V.: Cluster as a large-scale project to increase the competitiveness of production and economic structures. Vestnik Moskovskogo gumanitanogo-ekonomicheskogo instituta **1**, 80–86 (2018)
22. Kuznetsova, S., Markova, V.: The problems of formation a business ecosystem based on a digital platform: using the example of 1C company platform. Innovations **2**(232), 55–60 (2018)
23. Raunio, M., Nordling, N., Kautonen, M., Rasanen, P.: Open innovation platforms as a knowledge triangle policy tool – evidence from Finland. Foresight STI Gov. **2**(12), 62–76 (2018). https://doi.org/10.17323/2500-2597.2018.2.62.76
24. Doroshenko, M., Miles, I., Vinogradov, D.: Knowledge intensive business services: the Russian experience. Foresight Russ. **4**(8), 24–39 (2014)
25. Zatsarinnyy, A., Kozlov, S., Shabanov, A.: Interoperability of organizational systems in addressing common challenges. Manage. Issues **6**, 43–49 (2017)
26. Ivanov, D.A., Ivanova, M.A., Sokolov, B.V.: Analysis of transformation trends in enterprise management principles in the era of Industry 4.0 technology. SPIIRAS Proc. **5**(60), 97–127 (2018). https://doi.org/10.15622/sp.60.4
27. Zatsarinnyy, A., Shabanov, A.: Models and methods of cognitive management of digital platform resources. Syst. Control Commun. Secur. **1**, 100–122 (2019). doi: 1024411/2410-9916-2019-10106
28. Ognivtsev, S.: The concept of the digital platform of the agro-industrial complex. Int. Agric. J. **2**(362), 16–22 (2018)
29. Korovin, Y., Tkachenko, M.: Software and hardware platform for building a digital field system. Oil Econ. **1**, 84–87 (2017)
30. Grigoriev, M., Maksimtsev, I., Uvarov, S.: Digital platforms as a resource for improving the competitiveness of supply chains. News St. Petersburg State Univ. Econ. **2**(110), 7–11 (2018)
31. Maximtsev, I.: Digital platforms and digital finance: problems and development prospects. News St. Petersburg State Univ. Econ. **1**(109), 7–9 (2018)
32. Zatsarinnyy, A., Shabanov, A.: Method of centralized reproduction of information transmission processes in the digital platform control loop. Procedia Comput. Sci. **186**, 63–69 (2021). https://doi.org/10.1016/J.PROCS.2021.04.125

33. Potapov, S.: Research of process of information transfer on virtual routes in the radio network of the communication system with mobile objects. Radio Commun. Theor. Equip. **3**, 11–23 (2019)
34. Danilova, S., Mikhail, I., Rakitskiy, S., Rakitskiy, D.: Variant of robotic system with protected control channel. I-methods **4**(11), 1–10 (2019)
35. Arendt, C., Nötzel, J., Boche, H.: Reliable Communication under the influence of a state-constrained jammer: an information-theoretic perspective on receive diversity. Probl. Inf. Transm. **2**(55), 3–27 (2019). https://doi.org/10.1134/S0555292319020013
36. Kozlov, S., Kubankov, A., Shabanov, A.: Innovations in control systems of actions of robotic objects in the field of emergency response. In: Wave Electronics and its Application in Information and Telecommunication Systems, WECONF 2019 (2019). https://doi.org/10.1109/WECONF.2019.8840139
37. Kozlov, S., Kubankov, A., Shabanov, A.: On the transformation of research data transmission processes in the digital platform. In: Wave Electronics and its Application in Information and Telecommunication Systems, WECONF 2021 (2021). https://doi.org/10.1109/WECONF 51603.2021.9470757
38. Mantsivoda, A.V., Ponomaryov, D.K.: A Formalization of document models with semantic modelling. Bull. Irkutsk State Univ. Ser. Math. **27**, 36–54 (2019). https://doi.org/10.26516/1997-7670.2019.27.36
39. Kozlov, S., Kubankov, A., Shabanov, A.: On the role of the semantic knowledge model in ensuring the stability of reproduction of data transmission processes. In: Wave Electronics and its Application in Information and Telecommunication Systems, WECONF 2020 (2020). https://doi.org/10.1109/WECONF48837.2020.9131521
40. Zatsarinnyy, A., Shabanov, A.: Methods of computer simulation based on shared digital platform. In: CEUR Workshop Proceedings, vol. 2426, pp. 17–23 (2019). http://ceur-ws.org/Vol-2426/paper3.pdf
41. Volovich, K., Zatsarinnyy, A., Kondrashev, V., Shabanov, A.: Scientific research as a cloud service. Syst. Means Inf. **1**(27), 73–84 (2017). https://doi.org/10.14357/08696527170105
42. Patent No. 2630393 C1 Russian Federation, MPC G06F 7/76 (2017)
43. Patent No. 2533090 C2 Russian Federation, MPC G06F 17/00 (2014)
44. Patent No. 127493 U1 Russian Federation, MPC G05B 19/00 (2013)
45. Passport of the national program Digital Economy of the Russian Federation. http://government.ru/info/35568/. Accessed 15 Oct 2021
46. Results of the all-Russian research IT Service Management. Information Management (2019)
47. ITIL® Glossary of Terms English v.1.0. https://www.axelos.com/corporate/media/files/glossaries/itil_2011_glossary_gb-v1-0.pdf. Accessed 15 Oct 2021
48. The All-Russian Classifier of Standards. https://elib.kuzstu.ru/method2/mks.pdf. Accessed 15 Oct 2021
49. Shabanov, A.P.: The basic model of the state of the productive resource of the information system. In: Proceedings of the 52nd Scientific Conference: Part XI. Business Information Systems, Moscow, MFTI, pp. 47–49 (2009)

# Challenges and Prospects for Cloud-Based Enterprise Systems in Tradition-Focused Cultures: A Design Thinking Case Study

Artem Levchenko[1]([⊠]) [iD] and Victor Taratukhin[2] [iD]

[1] Voronezh State University, Voronezh, Russia
artem.levchenko@sap.com

[2] Germany and Higher School of Economics, University of Muenster, Muenster, Russia

**Abstract.** Cloud-based enterprise systems, a growing segment in the global market for enterprise IT solutions, have met with strong resistance in traditions-focused cultures such as Japan's. Low cloud concept utilization has caused difficulties during the coronavirus pandemic, as remote work was not always technically possible due to the limitations of on-premises solutions. Our work applied Design Thinking methodology to form a foundation stone upon which cloud providers may build a Cloud Mindset in Japan and similar markets. This article considers both the endogenous agility that SaaS can engender with organisations and the exogenous possibilities that it brings about in terms of connections with other organisations. It describes the case study and covers six phases, from understanding the problem to testing a prototype of the solution. The research aims to act as a practical inquiry into the appliance of Design Thinking methodology in the new, challenging field of building Cloud Mindsets in multi-language environments. Consideration of the cultural semiosphere forms a necessary element of the research. The scientific achievement of this work is to make useful modifications to the classical methodology in this field. The practical value of the work lies in its applications for cloud providers in countries with similar cultural aspects to Japan.

**Keywords:** Cloud-bases enterprise systems · Cloud mindset · SaaS · Design thinking · Creativity

## 1 Introduction

Enterprise systems are essential for business in the twenty-first century. Ongoing technological progress provides new IT models such as the Internet of Things, blockchain, on-live analytics, machine learning, and cloud. All of these have proven their value for business over recent years.

Cloud technologies offer an opportunity for enterprise systems to use IaaS (Infrastructure as a Service), PaaS (Platform as a Service), and SaaS (Software as a Service) models. On the one hand, it reduces time-to-value metrics, avoids infrastructural expenses, operational and maintenance costs, and simplifies system architecture [1]. Cloud-based apps are enriched with intrinsic cloud features, including multi-tenancy, scalability, agility, and elasticity [2]. Some research even uses the "cloud-friendly" term to define companies relating to cloud technology or saying the sector has become

© Springer Nature Switzerland AG 2022
V. Taratukhin et al. (Eds.): ICID 2021, CCIS 1539, pp. 17–30, 2022.
https://doi.org/10.1007/978-3-030-95494-9_2

"cloudy" [3]. On the other hand, these models place greater demands on IT-business alignment and the endogenous agility and flexibility of change management processes. The latter can be especially crucial and challenging in countries with a tradition-focused culture and low rate of business adaptation to changing environments, such as Japan. In response, Cloud-services providers, meeting with strong resistance, had to adapt their businesses in these countries. They initiated robust business development initiatives to establish clarity around Cloud technology, and pushed digital transformation projects for traditional customers, who prefer on-Premises concepts and complete control of IT infrastructure management processes. The great advantage of Cloud technologies is that they promote the opportunity to delegate maintenance processes outside of the customer organisation. But they require giving third-party companies access secure customer data, as well as to some extent dictating how a company's IT environment and business processes must be organised.

This research aims to define a systematic approach for describing business challenge resolution by using available scientific knowledge in the area of cloud-based enterprise systems, and determining its applicability by practical realization in the business environment. It tests the endogenous agility of SaaS provider companies and their expanded exogenous opportunities in the form of modified business connections with customers. This research takes a people-process-technology approach to determining how to break down market resistance in the Cloud-adaptation area. The Design Thinking approach was selected to build a system analysis of the creation of a Cloud Mindset in the Japanese market.

## 2 Related Works

Following areas of science articles were analysed: Cloud technology adaptation challenges and solutions, Design Thinking appliance for software engineering and business processes building, Enterprise systems evolution with Japan's cultural specifics.

Key concepts, architectural principles, state-of-the-art implementation, and research challenges of Cloud computing, such as Automated service provisioning and Software frameworks, are highlighted in [4] and are still relevant now. Service-oriented architecture and microservices and challenges resulting from multi-tiered, distributed, and heterogeneous cloud architectures cause uncertainty that has not been sufficiently addressed [5]. The area of new actual research challenges also relates to the self-adaptation process of Cloud applications from control engineers' perspective. These research challenges are due to the nature of software applications [6]. A different area of challenges is architecture-related challenges for cloud-based software systems driven by IT systems' complexity growth. Unique challenges exist in this area as the systems to be designed range from pervasive embedded systems and enterprise applications to smart devices with the Internet of Things [7]. Researchers provided a classification of the state-of-the-art of cloud solutions to sort different challenges. It's argued there is the need for model-driven engineering techniques and methods facilitating the specification of provisioning, deployment, monitoring, and adaptation concerns of multi-cloud systems at design-time and their enactment at run-time [8].

Several practical solutions were developed for particular industries and countries. New challenges were raised and resolved by the unique requirements of the e-Healthcare industry for using cloud services by e-Healthcare providers. These challenges are regulatory, security, access adaptation, inter-cloud connectivity, resource distribution [9]. From a regional perspective, local challenges and acceptance of cloud-based services, where the organisational context is based on local governments in Australia, are highlighted in [10].

Analysis of studies of the existing literature on design thinking for innovation and accounts of using design thinking for innovation in practice was done in the perspective of organisational culture [11]. It is argued that the power of design thinking is in the tension between seemingly opposite ways of thinking, such as analytic thinking versus intuitive thinking and linear thinking versus thinking in iterative processes. For design thinking to flourish, it needs to be embedded in an organisational culture capable of maintaining a dynamic balance on a number of fundamental tensions in innovation processes.

Especially, Design Thinking is in wider use within software companies in the context of multinational organisations. Research in Design Thinking suggests that being exposed to Design Thinking changes the mindset of employees [12]. Evaluation of how the Design Thinking approach is integrated with Agile Software Development methodologies was done in the past. The results show that most of the integrated models are applied throughout the software life cycle [13]. The integrated models resulted in a better approximation of end-users and the development team, improving the software's quality and usability. Design Thinking is used to gather customer requirements during product development. Industry-specific case study research describes the need to facilitate an in-depth understanding of healthcare stakeholders' realities in the context of their day-to-day experiences, identifying the need to introduce a pre-software requirements phase [14]. Thus, Design Thinking techniques were used to inform healthcare innovation.

Besides corporate use, Design Thinking is used as a methodological approach for the instruction of Software Engineering at the undergraduate level [15]. It aims to create innovative software products from scratch and go beyond the typical "analysis – design – implementation – testing" process to reinterpret it with the "empathize – define – ideate – prototype – testing" proposed by Design Thinking. A case study on how universities plan to implement Design thinking strategies to support graduate students' project-based education is presented in [16]. Design thinking helps not only in software engineering but also in organisation adaptation – moving from a product-centric focus into an organisational focus. A case study from the aged care field on developing environmental sustainability strategies using design thinking is presented in [17]. Changing existing methods are quite typically made for particular Design Thinking appliance use-cases. Another industry example of Design Thinking use for sustainable smart energy system design requires extended and modified methods to suit the content of the study [18]. For Cloud topics, there is a small number of articles. One study proposed a method of evaluating cloud platform service usability by implementing design thinking for integration analysis [19]. These results show the combination of using grounded theory for persona creation and the subsequent tasks for evaluation was effective in capturing the perspectives of tenants seeking information on the cloud platform service.

Questionnaire-based research was done into Japanese companies' awareness of Cloud Computing, main concerns, and challenges, and into what Cloud computing lacks to become primarily accepted and adopted [20]. It is pointed out that more cloud computing services are yet to be created and improved to appear attractive and increase the adoption rate among the Japanese market. In that respect, security issues are the main obstacles to overcome, not only for fortifying the underlying IT infrastructure in the provider's possession but also in terms of how to deal with disruptions in the continuity of the services and how to guarantee service availability. 13.2% of the respondents are still unaware of the term Cloud Computing. It is worth mentioning that none of them think of it as a trend that will disappear soon.

Existing studies widely cover Design Thinking's appliances for software engineering, but not for cloud-based enterprise systems. Cloud adaptation challenges' correlation with selected countries' cultural specifics were discussed in several articles, but no defined solution was found. This knowledge gap highlights the scientific value of the current research.

## 3   Design Thinking Workshop Preparation and Conducting

### 3.1   Workshop Preparation

The workshop preparation included goal definition, preliminary research, workshop plan preparation, and Design Thinking methodic adaptation according to the defined objectives.

Two main objectives were defined – science research and practical workshop goals that are correlating. First is a Design Thinking methodology knowledge area enhancement, and second is commercial benefits gain for the cloud provider in the highly competitive market. The workshop goal was to define the strategic and change management initiatives, including action plan definition for the market development by generating Cloud Mindset. Purely Cloud-driven market growth is calculated as a Compound Annual Growth Rate (CAGR) for the selected procurement and business network market. For instance, according to Gartner, with a CAGR of 15%, Cloud is the main driver for growth. By 2024, cloud subscription revenues will likely account for 99% of the APJ market. As known, Japan does not have the reputation of a fast-changing environment [20]. For Japan's market, this proportion is relatively lower, driven by a lack of local trust in modern emerging technologies and firm intention to have complete control of the end-to-end business processes that contradicts approaches based on IaaS, PaaS, and SaaS models.

In Japan, there is a proverb called "sanpo-yoshi". "Sanpo" indicates three parties, and 'yoshi' represents something good in Japanese. The proverb means "good for everyone" [21]. Since these three parties have conflicting areas of interest, it is necessary to admit that while the three parties seek their benefits, they also need to make concessions to achieve the collectively beneficial state. Japan's semiosphere was used to consider national specifics, such as respect, safety, and stability perspectives (see Fig. 1). These national qualities are crucial for the initiative's creation for cloud-based systems acceptance and adaptation for the market. Design Thinking as an approach to getting that agreement between "sanpo" is well suited for this culture because it's also traditionally based on different opinions, group discussion, and communication [22].

Hasso-Plattner Institute (HPI) defined six phases of design thinking taken as a basis for the workshop planning: Understand, Observe, Point of View, Ideate, Prototype, Test (see Fig. 2) [23]. Phase selection is similar to the intuitive workflow process of a designer. All the phases are qualitatively essential and can't be skipped, according to the methodology.

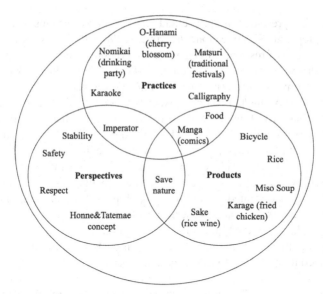

**Fig. 1.** Japan's semiosphere

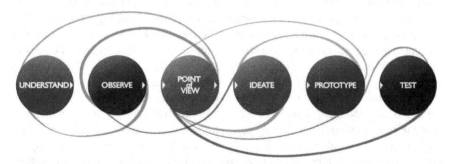

**Fig. 2.** The design thinking process model at HPI

The Understand phase was done beforehand and relies on integral marketing research and previous customer experience with SaaS business solutions. It was identified that more than half of negative customer escalations were caused by a misunderstanding of cloud-based enterprise systems specifics, such as limited options to change the by-design functionality, listed ways and points of system enchantments via the APIs, scheduled events for the releases to fix the system bags, and new functionality utilization. Internal SaaS provider corporate data was used for problem area preliminary research, and additional questionary was prepared, distributed to the subject-matter experts, collected during the interview sessions, and analysed for the knowledge gaps. Following the methodology, the preparation of a "Personas" technique has a special place. Therefore, the preparation is described in more detail below.

Personas are a representation of the cloud services' end-user; thus, a detailed questionnaire was prepared. The questionnaire covered the following aspects: name and age, position and area of responsibility, background (hobby, family, personality), quote (something he or she says often), stakeholders (who does he or she work closely with, who are his/her "army," which resists against him/her), known business priorities (short and long term goals, how his/her company exceeds over competitors), current initiatives (central theme, what drives him/her, what does he or she care, what goes him/her to lead change), success measurement (known KPIs), and others motivators. The questionnaire was built in a format of the persona's description template. It was used during the interviews with experts in cloud provider companies who were interacting with the customers a lot and can represent customers' voices. After the consultations, three customer personas were prepared and printed for the workshop. It provides a necessary input for the "Meet your customer" exercise at the observation phase conducted during the workshop. This exercise was based on the "Empathy Map" technique.

The workshop plan also includes the following Design thinking methods and techniques related to the point of view definition phase:

- "Magazine cover creation" exercise. This exercise was based on the "Context Map" technique that was transformed. Considering that person for the map is being created beforehand, the outcomes should be used on the next steps and focus on the prospects. It covers following dimensions: customers, competitors, technology, company, situation, economic or social trends, industry. The outcome for this exercise is a visual document that enables you to start a conversation about core stakeholders. Discussing the points of disagreement lead to new thoughts and understanding of the company and will set a singular base for further design thinking activities.
- "Current/Future/Barriers" technique [24] for current customer experience and value definition and "the ideal future" barriers definition. In the current stage, participants state their assumptions (may include perceptions, requirements, or constraints) about a situation or the problem. Then they try to reverse the premises to see if new opportunities are revealed. It creates a clearer picture of the vision, outcomes, or problem (it's crucial that it's a shared view of the problems). In the original technique, four vertical flip chart paper pieces, taped in a row, are used: Current, Barriers, How Might We and Future. We decided to reduce it to three by moving the "How Might We" part to the following stages for a deeper discussion. Input data for the "Future" section was taken from the previous "Magazine cover creation" exercise.

After the break in the schedule, the Ideate phase starts. This phase includes two main exercises:

- The first exercise is based on a combination of two techniques: "Dreamer/ Realist/Critic" and "Affinity Mapping". The purpose is to go through a complete idea generation-exploration-evaluation cycle. To have at the end an expansive idea, practical ideas, and evaluated ideas. The core concept is to wish as a dreamer and play the realist to work it into a practical idea, then put on the critic hat to poke holes in the idea. Originally this technique was developed by the Walt Disney Company [25, 26]. As a next stage, the final ideas go through the "Affinity Mapping" technique to sort and prioritize ideas or data quickly and visually for the next exercise. We asked the group to organise the collection of ideas or data into clusters based on similar characteristics.

- The combined ideas were carried to the extended version of the "Who + Do" technique [27], which is quite similar to the "create a project plan" method. The final exercise with the name "Who-Do-Through-When" was for the clear RACI model defining. We use results of this exercise for the further prototyping phase.

Final results are presented to the whole audience for their awareness and feedback collection. Prototyping and Testing phases were planned for the out of the workshop discussion because most of the ideas prototyping requires deeper dialogue with a broader audience. The reason for this is the nature of ideas – it's a business operation of a global IT company. Some of them were pretty clear for immediate implementation. Some of them require careful study with extensive data analysis. Some of them relate to the stakeholders out of the participant's list. We expected these outcomes, and the goal was strategic and change management initiatives, including action plan definition for the market development for the next following-up activities.

## 3.2 Workshops

The results of practical innovation are three essential components: design thinking process, variable space, and multidisciplinary teams. As the design thinking process was described above, the spaces and the team's characteristics are listed below.

The workshop's environment (see Fig. 3) was specially designed and prepared for the brainstorming and Design Thinking workshops-like conduction. It includes a spatial room, whiteboards, flipcharts, a sufficient number of colored markers, stickers, and other stationery. Interior is also designed differently from the classical corporate-style meeting rooms to activate the right mood in participants. Considering that the sessions were conducted during the pandemic time, additional working environment requirements were requested, such as enough space for social distance and regular airing of the room.

The working group was primarily formed from the department's management representatives, who were empowered to make decisions regarding the company processes change or strongly influenced these changes. Some specialists were also invited to the workshop to get feedback from different corporate hierarchy levels. The following SaaS organisations' departments participated: sales, sales support experts, functional and technical consulting, post-deployment customer management organisation, value advisory, and product engineering. Three groups were defined with four participants each, excluding one facilitator for each group (one leading facilitator and two supportive) – fifteen participants in total. Two groups were Japanese-speaking, one English-speaking. The whole explanation and instructions were done in English with additional Japanese instructions for the two groups individually. The company's top management representatives performed a session opening to motivate the participants.

The Observe Point of View definition phase was conducted successfully with high engagement active involvement from all the participants. Persona definition created by one team is presented in Fig. 4 as an output example.

**Fig. 3.** A special environment for conducting the workshop

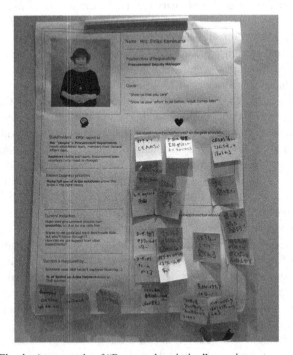

**Fig. 4.** An example of "Persona description" exercise outcomes

The Point of View definition phase was done more actively by warmed up groups and made a clear focus on the target state. The "Magazine cover creation" exercise's outcomes (see Fig. 5) were painted by hand real and imagined magazines covers with the core stakeholder (CPO, CFO, CIO) on the cower.

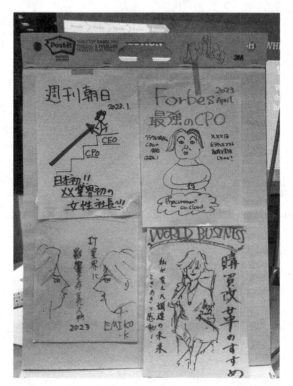

**Fig. 5.** An example of "Magazine cover creation" exercise outcomes

The ideate phase was the most challenging period for the teams. It triggered the deep cross-team collaboration around internal business processes, market positioning, cooperation with partners, and other themes crucial for the Cloud business. An example of one "Dreamer/Realist/Critic" and "Affinity Mapping" exercise outcomes is shown in Fig. 6.

A strategic and change management initiative, including an action plan definition for the market development for the subsequent following-up activities, as defined on the last stage (see Fig. 7). The team managers' participation created the unique environment when each team can see the other's team contributed to the cross-teams initiatives and ready to help and contribute itself.

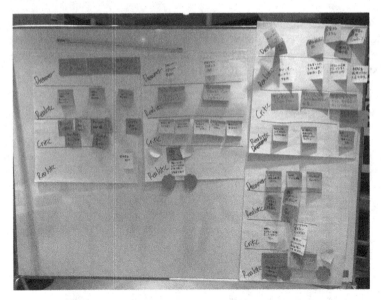

**Fig. 6.** An example of "Dreamer/Realist/Critic" and "Affinity Mapping" exercise outcomes

**Fig. 7.** An example of one "Who-Do-Through-When" exercise outcomes

A final presentation of the results was done in the native languages of each team – Japanese for the two groups and English for the one unit (the facilitator of the English-speaking team provided simultaneous translation). It was important in Japanese culture to use the native language to understand each responsibility and detail of the further collaboration. Top management representatives did the closing words to appreciate the productive engagement and highlight that prototyping, testing, and implementation will be done regularly.

### 3.3  After Workshop Activities

After workshop activities included two workstreams: results formalization and regular meetings for the idea's realization tracking.

The final results were packed into the PPT file and consisted of final ideas with a timeline and responsible persons and photos from the session, including photos of all exercise outcomes, translated into English.

Regular meetings were scheduled bi-weekly to track the execution statuses and drive the organisations' change management initiatives. Prototyping and testing activities were done selectively for some ideas.

## 4  Results and Discussions

During the workshop, seventy-six ideas were generated. Nine of them were grouped, prioritized and defined as a strategic initiative for execution.

The leading nine ideas relate to the following two main groups:

1. Endogenous initiatives and processes, primarily strategic initiatives that create new strategies or change existing SaaS provider company processes.
2. Exogenous initiatives and processes that affect interactions with customers.

To deploy ideas demanded by the needs of change management, we analysed existing ongoing strategic initiatives in the company. Internal strategic initiatives are typically hard to initiate from scratch as they require a considerable time investment, budget approval, and other formal procedures, especially in big international companies. Luckily, we were able to initiate a three-year market unit strategy, so all the change management ideas were merged with this project. It allows the delegating of execution while receiving regular status updates on realization progress. For instance, a new type of cloud-consumption-based customer contract requires various teams' involvement. For this idea, the decision was to proceed with prototyping and testing phases according to the Design Thinking methodology.

Other internal initiatives that did not have interaction with existing strategic initiatives were built as new initiatives. Ideas such as recent sales play run and rearrangement of technical architecture positioning also require prototyping and testing phases – something which was achieved, and all ideas relating to which were agreed for production use. Simple ideas, such as increasing internal awareness of particular topics, extension of the participants' list for the Quality Business Review with strategic customers – were implemented without prototyping and testing because their benefits are quite obvious.

Ideas related to interactions with SaaS customers mainly fall under the responsibility of marketing departments. They relate to the organisation of business development customer events, creation of specific digital transformation customer reference stories, and generation of different market-specific media to build market awareness around cloud-driven business transformation projects.

Before the Design Thinking workshop, potential resolution areas for Cloud Mindset generation related only to marketing activities. The workshop facilitated the discovery of new improvement areas and defined change management opportunities. The Design Thinking approach activated deep cross-team interactions and generated creative solutions. The positive feedback received from the top management team demonstrates the high practical value of the study.

## 5   Conclusion and Future Work

In Japan, as in other tradition-focused cultures, Cloud-based enterprise systems have met with strong resistance – even as they have grown rapidly in importance in the global IT market. Our case study shows the practical application of Design Thinking methodology in resolving the challenge that SaaS providers face in breaking down culturally-ingrained resistance to the technologies they offer. Our scientific research looked at possible improvements to Design Thinking methodology, with classical techniques adjusted systematically to be applied to the field of Cloud study by SaaS providers. One structural element added to the standard methodology is a consideration of the cultural semiosphere as a way to refine classical Design Thinking techniques. Design Thinking methodology allowed us to define a foundation stone upon which Cloud operators in the Japanese market may build a Cloud Mindset. The ability to first identify which initiatives to pursue and then deploy them demonstrates SaaS providers' endogenous agility, while exogenous agility is demonstrated by increased possibilities for connections with customers.

Six phases of the Design Thinking approach, from understanding the problem to testing prototype solutions, were described and covered in explanations of required changes in the standard technique applied in the IT industry. Design Thinking workshops enabled us to define clear next steps to processes, which then led to the realization of ideas capable of achieving commercial success. Three essential components of the innovation's practical results – Design Thinking process, variable space, and a multidisciplinary team – were combined to achieve the study's aims. The Design Thinking approach showed promising results when several teams' results pointed toward one solution appropriate to a particular organisation. The methods developed can now be used by other SaaS providers in countries with similar semiospheric elements – especially those with tradition-focused cultures and low cloud-based enterprise system adoption in the market.

Our next work will look at applying the results gained from our market research in Japan to other markets. We expect new Design Thinking modifications and different workshop outcomes to be influenced by cultural aspects of the acceptance of Cloud-based enterprise systems.

## References

1. Euripides, G.M.P., et al.: Internet of Things as a Service (iTaaS): challenges and solutions for management of sensor data on the cloud and the fog. Internet Things 3–4, 156–174 (2018)
2. Khan, M.A., Salah, K.: Cloud adoption for e-learning: survey and future challenges. Educ. Inf. Technol. 25(2), 1417–1438 (2019). https://doi.org/10.1007/s10639-019-10021-5
3. Avram, M.G.: Advantages and challenges of adopting cloud computing from an enterprise perspective. Procedia Technol. 12, 529–534 (2014)

4. Zhang, Q., Cheng, L., Boutaba, R.: Cloud computing: state-of-the-art and research challenges. J. Internet Serv. Appl. 1(1), 7–18 (2010). https://doi.org/10.1007/s13174-010-0007-6

5. Pahl, C.: Architectural principles for cloud software. ACM Trans. Internet Technol. 18–2, 1–23 (2018)

6. Farokhi, S., Jamshidi, P., Brandic, I., Elmroth, E.: Self-adaptation challenges for cloud-based applications: a control theoretic perspective. In: 10th International Workshop on Feedback Computing 2015 (2015)

7. Chauhan, M.A.: Architecting cloud-enabled systems: a systematic survey of challenges and solutions. Softw. Pract. Expertise 47, 599–644 (2017)

8. Ferry, N.: Towards model-driven provisioning, deployment, monitoring, and adaptation of multi-cloud systems. In: 2013 IEEE Sixth International Conference on Cloud Computing, pp. 887–894. IEEE, Santa Clara, CA, USA (2013)

9. Liu, W.: e-Healthcare cloud computing application solutions: Cloud-enabling characteristics, challenges and adaptations. In: 2013 International Conference on Computing, Networking and Communications (ICNC), pp. 437–443. IEEE Computer Society, San Diego, CA, USA (2013)

10. Ali, O.: Cloud computing technology adoption: an evaluation of key factors in local governments. Inf. Technol. People 34–2, 666–703 (2020)

11. Prud'homme van Reine, P.: The culture of design thinking for innovation. J. Innovation Manage. 5–2, 56–80 (2017)

12. Dobrigkeit, F.: Design thinking in practice: understanding manifestations of design thinking in software engineering. In: Proceedings of the 2019 27th ACM Joint Meeting on European Software Engineering Conference and Symposium on the Foundations of Software Engineering (ESEC/FSE 2019), pp. 1059–1069. Association for Computing Machinery, New York, USA (2019)

13. Cesar Pereira, J.: Design thinking integrated in Agile software development: a systematic literature review. Procedia Comput. Sci. 138, 775–782 (2018)

14. Carroll, N.: Aligning healthcare innovation and software requirements through design Thinking. In: 2016 IEEE/ACM International Workshop on Software Engineering in Healthcare Systems (SEHS), pp. 1–7. Association for Computing Machinery, Austin, TX, USA (2016)

15. Corral, L.: Design thinking and agile practices for software engineering: an opportunity for innovation. In: Proceedings of the 19th Annual SIG Conference on Information Technology Education (SIGITE 18), pp. 26–31. Association for Computing Machinery, New York, NY, USA (2018)

16. Taratukhin, V.: Next-Gen design thinking for management education. Project-based and game-oriented methods are critical ingredients of success. Dev. Bus. Simul. Experiential Learn. 47, 261–265 (2020)

17. Clune, S.J.: Developing environmental sustainability strategies, the Double Diamond method of LCA and design thinking: a case study from aged care. J. Cleaner Prod. 85, 67–82 (2014)

18. Tushar, W.: Exploiting design thinking to improve energy efficiency of buildings. Energy 197, 1–19 (2020)

19. Ng, K.H.: Design thinking for usability evaluation of cloud platform service-case study on 591 house rental web service. In: 2018 IEEE International Conference on Applied System Invention (ICASI), pp. 247–250. IEEE, Chiba, Japan (2018)

20. Khare, A., Khare, K., Baber, W.W.: Why Japan's digital transformation is inevitable. In: Khare, A., Ishikura, H., Baber, W.W. (eds.) Transforming Japanese Business. FBF, pp. 3–14. Springer, Singapore (2020). https://doi.org/10.1007/978-981-15-0327-6_1

21. Hosono, S.: Innovation drivers of ICT toward service evolution: a study of service generations in Japan. In: Kosaka M., Wu J., Xing K., Zhang S.Y. (eds.) Business Innovation with New ICT in the Asia-Pacific: Case Studies, pp. 37–55. Springer, Singapore (2021)

22. Taratukhin, V.: The future of project-based learning for engineering and management students: towards an advanced design thinking approach. In: The American Society for Engineering Education Annual Conference & Exposition Proceedings, pp. 1–15. ASEE, Salt Lake City, Utah, USA (2018)
23. Plattner, H., Meinel, C., Weinberg, U.: Design Thinking. Mi-wirtschaftsbuch, Munich, Germany (2009)
24. Michalko, M.: Cracking Creativity: The Secrets of Creative Genius, Revised ed. edn. Ten Speed Press, Berkeley, CA, USA (2001)
25. Dilts, R.: Strategies of Genius: Sigmund Freud, Leonardo da Vinci, Nikola Tesla, vol. 3. Meta Publications, Aptos, CA, USA (1994)
26. Tausch, S., Steinberger, F., Hußmann, H.: Thinking like Disney: supporting the Disney method using ambient feedback based on group performance. In: Abascal, J., Barbosa, S., Fetter, M., Gross, T., Palanque, P., Winckler, M. (eds.) INTERACT 2015. LNCS, vol. 9298, pp. 614–621. Springer, Cham (2015). https://doi.org/10.1007/978-3-319-22698-9_42
27. Gray, D., Brown, S., Macanufo, J.: Gamestorming: A Playbook for Innovators, 1st edn. O'Reilly Media, Sebastopol, CA, USA (2010)

# Modeling Antenna Arrays on Cylindrical Surfaces for Aircraft Control from Ground Objects

S. N. Razinkov[1] , D. N. Borisov[2]([✉]) , and A. V. Bogoslovsky[1]

[1] Military Educational and Scientific Center of the Air Force "Air Force Academy from Professor N. Ye. Zhukovsky and Yu. A. Gagarin", 394064 Voronezh, Russia
[2] Voronezh State University, University pl. 1, 394018 Voronezh, Russia
borisov@sc.vsu.ru

**Abstract.** Using electrodynamic models for systems of elementary electric dipoles located on cylindrical surfaces obtained using the method of induced currents, the polarization components of the electric field for a cylinder excited by a system of elementary electric vibrators are estimated. The surface currents of the cylinder are found taking into account the electromagnetic couplings in the array under the boundary conditions for the superposition of the vibrator fields. When choosing the calculated distributions of currents and fields of the carrying surface as an initial approximation for solving the problem of exciting vibrators interacting through secondary radiation using the software for three-dimensional electrodynamic modeling CST MWS - Computer Simulation Technology Microwave Studio, we calculated the directional pattern and directional action of the antenna array. The boundary value problem for an array of electric vibrators located on the lateral surface of the cylinder is solved by the method of moments when the complex amplitudes of surface currents are represented by a set of discrete values at grid nodes with rectangular and tetrahedral cells. Regularities of changes in directional patterns and coefficients of directional action for antenna arrays when placed on ideally conducting cylinders and cylinders with thin dielectric coatings, depending on the number of elements for different electrical dimensions and electrophysical properties of bearing screens, have been investigated.

**Keywords:** Array of vibrators · Electrodynamic model of the system elementary electric dipoles · Software for electrodynamic modeling of antennas

## 1 Introduction

To analyze the electromagnetic environment and conditions for the propagation of radio waves in the interests of deploying communication networks in remote areas and on terrain with difficult profile, electronic systems are used, which are located on the sides of unmanned aerial vehicles (UAV) and are designed to communicate with ground-controlled approach for information exchange [1–3].

The advantages of unmanned systems include: a significant reduction in overall dimensions; high mobility and speed of deployment, the use of mobile control stations;

© Springer Nature Switzerland AG 2022
V. Taratukhin et al. (Eds.): ICID 2021, CCIS 1539, pp. 31–42, 2022.
https://doi.org/10.1007/978-3-030-95494-9_3

the ability to service with a minimum number of operators; lack of strict requirements for the launch pad; high economic performance [4, 5].

There are several classification features of unmanned vehicles (Table 1) [6].

**Table 1.** UAV classification.

| UAV classification signs | | | |
|---|---|---|---|
| Distance of application | Takeoff weight | Constructive execution | Special purpose |
| Long distance over 500 km | Heavy over 500 kg | Aircraft type<br>Helicopter type<br>Balloons<br>Airships<br>Multi rotary<br>Hybrid<br>Reactive | Optoelectronic systems<br>Communication, relaying and warning systems<br>Radar systems<br>Cargo containers |
| Medium distance up to 500 km | Average up to 500 kg | | |
| Short distance up to 250 km | Lungs up to 200 kg | | |
| Short-haul up to 100 km | Small up to 30 kg | | |

The key components of radio-electronic complexes are antenna systems, the characteristics of which affect the energy availability of objects in the monitoring area [7, 8] and indicators of the spatial-frequency selectivity of signal and interference reception [7–9]. When creating antenna systems that meet the requirements for functional efficiency and placement conditions on carriers [7, 10, 11], the greatest financial and time costs are associated with their development, and not the manufacture of prototypes.

Antennas of aircraft are characterized by small mass and size characteristics and shapes of structures that provide a slight deterioration in aerodynamic characteristics and preservation of the mechanical strength of the carrier when placing target loads on board [12, 13].

The development of antenna systems of radio-electronic complexes placed on aircraft implies the analysis of receiving-emitting structures on the bearing surfaces corresponding to the sections of the carrier bodies [5, 14]. The essence of the analysis is determine the electromagnetic field at any point in space surrounding the antenna system. The sources of the field are currents and charges distributed over the surface of the receiving-emitting structure [10, 15]. Based on the found distribution of surface currents and charges, rational shapes and sizes of antennas are established in accordance with the requirements for directivity of radiation (reception) of electromagnetic field [7, 10].

In accordance with the principles of discrete synthesis of radio engineering devices, the development of the design of aircraft antennas includes the construction of topological schemes of antenna arrays in accordance with the intended purpose and conditions of use; determination of the tactical and technical characteristics of antennas that ensure the required efficiency of the onboard radio-electronic complex [15–17].

The design options are determined using the decomposition rule, which consists in specifying the antenna structure, finding the design parameters, and comparing the radiation patterns and directive gain with the required characteristics.

As a rule, when designing antenna systems, it becomes difficult to choose a method and program for solving electrodynamic problems. The choice of software is reduced to finding a compromise between the speed of calculation, its accuracy, the required computing resources [18, 19]. From this point of view, the CST MWS – Computer Simulation Technology Microwave Studio [20], Matlab Antenna Toolbox [21, 22] and some other programs [23, 24] are effective tools for designing antenna systems for aircraft.

Moreover, the computational core CST MWS provides analysis of models by the methods of moments, finite elements, finite differences in the time domain, geometric and physical optics, generalized diffraction theory, as well as the multilevel method of fast multipoles. The electrodynamic modeling software implements the functions of hybrid calculation of characteristics and full-scale multiphysics computational experiments to determine the characteristics of the efficiency of signal transmission (reception) [20].

However, due to the closed internal content of the electrodynamic modeling programs and the use of general computational algorithms that are not optimized for calculating the characteristics of a particular antenna system, it is difficult to control the calculated errors and identify the reasons for their occurrence [18].

In the interests of overcoming these difficulties in the development of unmanned electronic systems in the proposed work, electrodynamic models of systems of elementary electric dipoles on cylindrical screens are constructed and an analysis of the directional patterns and directivity of the vibrator arrays from the composition of radio electronic systems of aircraft is carried out. According to [10, 17, 18, 25–28], it is advisable to use a round cylinder in the model representation of fuselage fragments of small unmanned aerial vehicles of an aircraft type.

The purpose of the article is to substantiate the technology for the design and modeling of complex antenna systems, as well as investigation the regularities of changing their characteristics depending on the number of elements, electrical dimensions and electrophysical properties of bearing surfaces.

## 2 Modeling a Vibrator Antenna Array on a Round Perfectly Conducting Cylinder

Modeling an antenna array of electric vibrators located on the lateral surface of a circular ideally conducting cylinder is based on finding asymptotic estimates of the electric field in the far zone of the bearing surface with the initial representation of surface currents by the distribution of currents of a cylinder of infinite length. To calculate the polarization components of the electric field of the array, it is advisable to use the CST MWS electrodynamic modeling software when finding the initial distribution of surface currents of the bearing ground surface [17, 18] from the results of solving the problem of exciting an infinite ideally conducting cylinder [29].

To set the location of the antenna elements on the carrier surface and calculate the cylinder currents, in the work was used a cylindrical coordinate system $(\rho, \phi, z)$; to calculate the field, radiation patterns and directive gain of the array, was set a spherical coordinate system $(r, \phi, \theta)$, the center of which was aligned with the origin of the

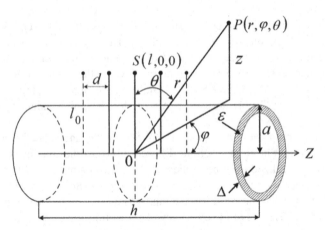

**Fig. 1.** Antenna system schematic

cylindrical coordinate system; the angle $\theta$ was measured from the axis $Oz$ as shown in Fig. 1.

The array consists of $N$ electric vibrators, characterized by height $l_o$ and spaced $d$; the first element is placed at the coordinate point $(a, 0, l)$. The edges of the cylinder pass through parallel planes with coordinates $z = \mp h/2$. The dimensions of the cylinder and array meet the conditions: $(N - 1)d \le h - |l|$, $|l| \le h/2$.

The unnormalized complex radiation pattern of the array is determined by the expression [9]

$$\dot{F}(\theta, \phi) = \sum_{n=1}^{N} i_n^* f_n(\theta, \phi) \tag{1}$$

where $i_n$ – the complex amplitude of the current for the $N$-th antenna element, $n = 1, \ldots, N$,

$$f_n(\theta, \phi) = D_n(\theta, \phi) \exp\left\{ -j\frac{2\pi}{\lambda}(a\cos\phi \sin\theta + (l + (n - 1)d)\cos\theta) \right\} \tag{2}$$

partial diagram of the $n$-th elementary electric vibrator [7], $D_n(\theta, \phi)$ – the diffraction factor of the bearing ground surface [17, 18], $\lambda$ – the wavelength, $*$ – the sign of the complex conjugation.

Using the accepted designations in (1), we represent the expression for the directive gain of the array in quadrature forms [9]

$$G(\theta, \phi) = 4\pi \left( \sum_{n=1}^{N} \sum_{p=1}^{N} i_n^* C_{np}(\theta, \phi) i_p \right) \cdot \left( \sum_{n=1}^{N} \sum_{p=1}^{N} i_n^* S_{np} i_p \right)^{-1}, \tag{3}$$

where

$$C_{np}(\theta, \phi) = f_n(\theta, \phi) f_p(\theta, \phi), \quad n, p = 1, \ldots, N \tag{4}$$

elements of the matrix characterizing the electromagnetic interaction of the $n$-th and $p$-th antenna elements [7],

$$S_{np} = \int_0^{2\pi} \int_0^{\pi} f_n(\theta, \phi) f_p(\theta, \phi) \sin\theta \, d\theta \, d\phi, \quad n, p = 1, \ldots, N, \tag{5}$$

– elements of the matrix characterizing the power of isotropic radiation of a hypothetical antenna system, according to the level of which the power emitted by the array is estimated [7].

We find the diffraction factors $D_n(\theta, \phi)$, $n = 1, \ldots, N$, assuming, in accordance with the conditions of application of the method of induced currents [9, 18], that with the electrical dimensions of the generatrix of the cylinder $h/\lambda \gg 1$, the amplitudes of the currents at points with coordinates $z = \mp h/2$ are small and waves of surface currents reflected from the edges of the cylinder and flowing beyond the edges, can be neglected.

Representing the electric field of the array in the form of a superposition fields of the antenna elements and the cylinder, we represent polarization components in the form [17, 19]:

$$\dot{E}_\theta(r, \theta, \phi) = -j\frac{\exp(-j\frac{2\pi r}{\lambda})}{2r\lambda} \left[ \mu_0 c \sum_{n=1}^{N} M_{\theta n}(\theta, \phi) + \sum_{n=1}^{N} L_{\phi n}(\theta, \phi) \right] \tag{6}$$

$$\dot{E}_\phi(r, \theta, \phi) = -j\frac{\exp(-j\frac{2\pi r}{\lambda})}{2r\lambda} \left[ \mu_0 c \sum_{n=1}^{N} M_{\phi n}(\theta, \phi) + \sum_{n=1}^{N} L_{\theta n}(\theta, \phi) \right], \tag{7}$$

where

$$\begin{bmatrix} M_{\theta[\phi]n}(\theta, \phi) \\ L_{\theta[\phi]n}(\theta, \phi) \end{bmatrix} = \begin{bmatrix} M_{\theta[\phi]}(\theta, \phi) \\ L_{\theta[\phi]}(\theta, \phi) \end{bmatrix} \exp\left\{ -j\frac{2\pi}{\lambda}(a\cos(\phi - \phi_0)\sin\theta + (l + (n-1)d)\cos\theta \right\} \tag{8}$$

– diffraction functions of the $n$-th element, $n = 1, \ldots, N$,

$$\begin{bmatrix} M_\theta(\theta, \phi) \\ L_\theta(\theta, \phi) \end{bmatrix}$$

$$= a\sin\theta \int_0^{2\pi} \int_{-h/2}^{h/2} \begin{bmatrix} H_\phi(a, \phi', z') \\ -E_\phi(a, \phi', z') \end{bmatrix} \exp\left[ j\frac{2\pi}{\lambda}(a\cos(\phi - \phi')\sin\theta + z'\cos\theta) \right] dz' \, d\phi', \tag{9}$$

$$\begin{bmatrix} M_\phi(\theta, \phi) \\ L_\phi(\theta, \phi) \end{bmatrix} = a \int_0^{2\pi} \int_{-h/2}^{h/2} \begin{bmatrix} -H_z(a, \phi', z') \\ E_z(a, \phi', z') \end{bmatrix} \tag{10}$$

$$\times \exp\left[ j\frac{2\pi}{\lambda}(a\cos(\phi - \phi')\sin\theta + z'\cos\theta) \right] dz' \, d\phi'$$

– distribution functions of magnetic and electric field components on the surface of finite length cylinder [30], $H_{\phi[z]}(r', \phi', z')$ and $E_{\phi[z]}(r', \phi', z')$ – azimuthal and longitudinal components of magnetic and electric fields on the surface $r' = a$ an infinitely extended cylinder, $\mu_0$ – magnetic permeability of free space, $c$ – speed of light.

By analogy with [30], using the method proposed in [26] for approximating fields by series of cylindrical functions and representing the components of magnetic and electric fields by equivalent currents, we calculate the integrals using the antiderivatives included in (9), (10). We find the diffraction functions (8) for the components of the array fields on the lateral surface of the cylinder by summing the series of trigonometric functions for azimuthal harmonics [29, 30] with the weight coefficients found in an analytical form.

From the definition of the partial diagram of the $n$-th element of the array (2) and the expression (6)–(10) for calculating the fields (6), (7), we obtain the rule for calculating the diffraction factor of the lateral surface of an ideally conducting cylinder.

$$D_n(\theta, \phi) = \sqrt{D_{n\theta}^2(\theta, \phi) + D_{n\phi}^2(\theta, \phi)}, \quad n = 1, \dots N, \tag{11}$$

where

$$D_{n\theta}(\theta, \phi) = \frac{2\pi^2 \mu_0 l_0 a}{\lambda^2} \left[ 2\cos\theta \cos\phi \exp\left( j \frac{2\pi}{\lambda} \sin\theta \cos\phi \right) - \sum_{m=0}^{\infty} \varepsilon_m j^m \cos(m\phi) \gamma_{nm}^{\theta} \right], \tag{12}$$

$$D_{n\theta}(\theta, \phi) = \frac{2\pi^2 \mu_0 l_0 a}{\lambda^2} \left[ 2\sin\phi \exp\left( j \frac{2\pi}{\lambda} \sin\theta \cos\phi \right) - \sum_{m=0}^{\infty} j^m \sin(m\phi) \gamma_{nm}^{\phi} \right], \tag{13}$$

$l_o$ – length of electric vibrator, $\varepsilon_m = \begin{cases} 1, & m = 0 \\ 2, & m \neq 0 \end{cases}$.

Expressions for calculating the factors $\gamma_{nm}^{\theta}, m \geq 0$ and $\gamma_{nm}^{\phi}, m \geq 1$ in (12) and (13), obtained in [30] for axisymmetric surfaces and for a perfectly conducting cylinder have the form [31]:

$$\gamma_{nm}^{\theta} = \int_{-\infty}^{\infty} \frac{k}{\alpha} \left[ \frac{\dot{H}_m^{(2)}\left(\frac{2\pi a}{\lambda}\alpha\right)}{H_m^{(2)}\left(\frac{2\pi a}{\lambda}\alpha\right)} - \left(\frac{m\lambda}{2\pi a}\right)^2 \frac{H_m^{(2)}\left(\frac{2\pi a}{\lambda}\alpha\right)}{\dot{H}_m^{(2)}\left(\frac{2\pi a}{\lambda}\alpha\right)} \right]$$

$$\times \frac{\exp\left[j\frac{2\pi}{\lambda}(h - (l + (n-1)d))(\cos\theta - k)\right] - \exp\left[-j\frac{2\pi}{\lambda}\left(\frac{h}{2} - (l + (n-1)d)\right)(\cos\theta - k)\right]}{\cos\theta - k} dk \tag{14}$$

$$\gamma_{nm}^{\phi} = \int_{-\infty}^{\infty} \frac{H_m^{(2)}\left(\frac{2\pi a}{\lambda}\alpha\right)}{\alpha \dot{H}_m^{(2)}\left(\frac{2\pi a}{\lambda}\alpha\right)}$$

$$\times \frac{\exp\left[j\frac{2\pi}{\lambda}(h - (l + (n-1)d))(\cos\theta - k)\right] - \exp\left[j\frac{2\pi}{\lambda}\left(\frac{h}{2} - (l + (n-1)d)\right)(\cos\theta - k)\right]}{\cos\theta - k} dk \tag{15}$$

where $\alpha = \sqrt{1-k^2}$, $H_m^{(2)}(\ldots)$ – the Hankel function of the second kind of the $m$-th order, $\dot{H}_m^{(2)}\left(\frac{2\pi a}{\lambda}\alpha\right)$ – its derivative with respect to the variable $r$ at the point $r = a$.

Using (1), (2), (11)–(15) was calculated the normalized radiation pattern $F(\theta, \phi)\frac{|\dot{F}(\theta,\phi)|}{\max\limits_{\theta,\phi}|F(\theta,\phi)|}$, and on the basis of (3)–(5), subject to expressions (2), (11)–(15), the coefficients of directive gain for $N = 3..5$ with step $d/\lambda = 0.25..1$, located on a perfectly conducting cylinder with an electric length $h/\lambda = 5$ and an electric cross-sectional radius $a/\lambda = 0.3$. The electric length of the vibrators $l_0/\lambda = 0.1 \div 0.2$, the relative distance for the first antenna element from the end of the cylinder $l/\lambda = 0.3$.

In Fig. 2 shows the directive gain of array for $N = 3$ elements with $d/\lambda = 0.5$ and $l_0/\lambda = 0.1$.

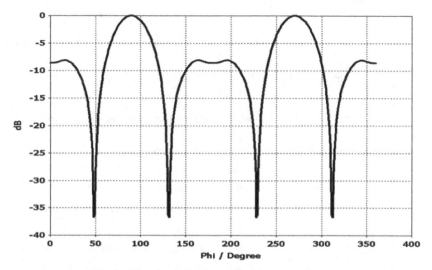

**Fig. 2.** Directive gain of an array for $N = 3$ vibrators.

In Fig. 3 shows the directive gain of the array for $N = 5$ elements with $d/\lambda = 0.25$ and $l_0/\lambda = 0.2$.

Table 2 shows the directive gain of the calculated antenna arrays.

Thus, based on the distribution of surface currents of an infinitely long ideally conducting surface with axial symmetry, using the CST MWS and Matlab Antenna Toolbox electrodynamic modeling software's, the design of the antenna array and the calculation of the electric field and directivity of the considered antenna systems were carried out.

Including, the calculation was carried out by the method of moments when the complex amplitudes of the surface currents of the cylinder are represented by a set of discrete values at grid nodes with rectangular and tetrahedral cell shapes. To reduce computational costs in the areas of coverage of the cylinder surface with a grid with rectangular cells, the Multilevel Subgridding Scheme method is used, which forms conformal layers with reduced sampling step. The polarization components of the field and the characteristics of the array were determined by activating the built-in calculator Frequency

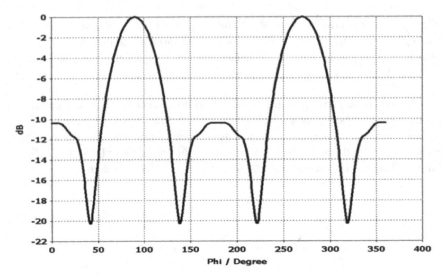

**Fig. 3.** Directive gain of an array for $N = 5$ vibrators.

**Table 2.** Maximum values of directive gain for vibrator array on the cylinder.

| Relative length of vibrators, $l_0/\lambda$ | Operating frequency, GHz | Number of antenna elements | | | | | | | |
|---|---|---|---|---|---|---|---|---|---|
| | | $N = 3$ | | | | $N = 5$ | | | |
| | | Relative inter-element distance, $d/\lambda$ | | | | | | | |
| | | 0,25 | 0,5 | 0,75 | 1 | 0,25 | 0,5 | 0,75 | 1 |
| | | Directive gain, dB | | | | | | | |
| 0,1 | 3 | 3,82 | 6,07 | 3,94 | 4,09 | 6,68 | 8,38 | 10,8 | 8,96 |
| | 6 | 4,57 | 4,54 | 4,29 | 7,88 | 7,56 | 10,3 | 11,6 | 9,3 |
| 0,2 | 3 | 7,43 | 7,15 | 5,34 | 6,0 | 6,96 | 9,69 | 11,4 | 10,3 |
| | 6 | 8,75 | 5,6 | 6,1 | 9,02 | 7,55 | 10,2 | 11,8 | 10,9 |

Domain Solver, which automatically generates the coordinates of the grid points to represent the search area for solving the problem of exciting a cylinder with an array of discrete fragments. The results of the analytical calculation of the currents and fields for cylinder in numerical format were imported from the computer algebra system by the Combine Calculation Results module.

Comparison of the table results with Fig. 2 and Fig. 3 allows us to conclude that the dependences are almost completely identical. The angular distribution of the array field, found using the electrodynamic modeling software, contains deeper (by 3... 5.3 dB) local extrema.

# 3  Modeling a Vibrator Array on a Dielectric Coated Cylinder

The distance between the antenna elements and the characteristic dimensions of the bearing surface are chosen the same as for the array located on the lateral surface of an ideally conducting cylinder.

Electrodynamic modeling of array on a cylinder with a dielectric coating is implemented under boundary conditions on an ideally conducting surface $(a, \phi, z)$ and the outer surface of a dielectric coating $(a + \Delta, \phi, z)$, an analytical calculation of the polarization components of the electric and magnetic fields of an infinitely long ideally conducting cylinder with a dielectric coating, excited by a hypothetical elementary source, is performed.

Using the CST MWS and Matlab Antenna Toolbox electrodynamic modeling software's, the field and directivity indices of an array of electric vibrators on a cylinder of finite length are calculated by discretizing the bearing ground surface from the conditions of the most accurate representation of surface currents with the minimum possible number of counts.

The boundary conditions on an ideally conducting surface are determined by the equality of the tangential components of the total electric field [15]

$$\dot{E}_{z[\phi]}^{(m)}(a, \phi, z) = 0, \tag{16}$$

where the superscript "$m$" in parentheses corresponds to the field on a perfectly conducting surface.

At the interface between two media (at $a + \Delta$), the conditions of continuity of the tangential components of the electric and magnetic fields are controlled [31]:

$$\dot{E}_{z[\phi]}^{(d)}(a + \Delta, \phi, z) = \dot{E}_{z[\phi]}^{(m)}(a + \Delta, \phi, z), \quad \dot{H}_{z[\phi]}^{(d)}(a + \Delta, \phi, z) = \dot{H}_{z[\phi]}^{(m)}(a + \Delta, \phi, z), \tag{17}$$

where the superscript "$d$" in parentheses denotes the field in the dielectric medium.

Taking into account the axial symmetry of the cylinder, we represent the components of the electric and magnetic fields in the form of infinite Fourier series in azimuthal harmonics:

$$\dot{E}_{z[\phi]}^{(m,d)}(a, \phi, z) = \sum_{s=-\infty}^{\infty} \exp(-js\phi) \int_{-\infty}^{\infty} \dot{E}_{z[\phi]\tilde{\alpha}s}^{(m,d)} \exp(-j\tilde{\alpha}z)d\tilde{\alpha}, \tag{18}$$

$$\dot{H}_{z[\phi]}^{(d)}(a + \Delta, \phi, z) = \sum_{s=-\infty}^{\infty} \exp(-js\phi) \int_{-\infty}^{\infty} \dot{H}_{z[\phi]\tilde{\alpha}s}^{(d)} \exp(-j\tilde{\alpha}z)d\tilde{\alpha}, \tag{19}$$

where $\dot{E}_{z[\phi]\tilde{\alpha}s}^{(m,d)}$ and $\dot{H}_{z[\phi]\tilde{\alpha}s}^{(d)}$ – are the weight coefficients determined through the functions of the sources and the reflection coefficients of the electric and magnetic fields, $\tilde{\alpha}$ – is the longitudinal wave number [31].

As a result of applying the boundary conditions (16) and (17) to represent the electric and magnetic fields (18) and (19), a closed system of equations for the coefficients $\dot{E}_{z[\phi]\tilde{\alpha}s}^{(m,d)}$ and $\dot{H}_{z[\phi]\tilde{\alpha}s}^{(d)}$ is formed.

Expressions for calculating the functions of sources for electric and magnetic types, establishing the parametric dependences of the weight coefficients in (18) and (19) on the dielectric constant of the medium $\varepsilon$, are given in [15, 29, 30]; the reflection coefficients of the fields were found in [31].

In Fig. 4 shows the radiation power of an antenna array of $N = 3$ vibrators located on a cylinder with a dielectric coating, characterized by a dielectric constant of $\varepsilon = 3.6$ and a thickness of $\Delta/\lambda = 0.1$.

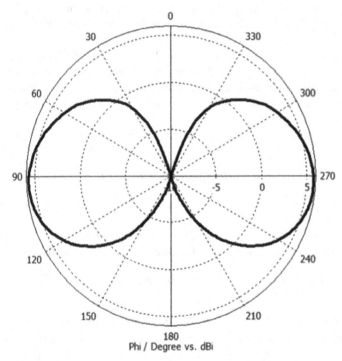

**Fig. 4.** Radiation power of an array of 3 vibrators on the side surface of a perfectly conducting dielectric coated cylinder.

The directional gain of the grating is 5.72 dB. It was found that due to the preliminary calculation of the currents of the bearing ground surface based on the method of induced currents, it becomes possible, when using a computing tool based on an Intel Core i7-10700K processor with 32 GB of RAM, to perform multiple calculations of the characteristics of array in order to determine the rational parameters of their structures.

## 4   Conclusion

In modern conditions, for the design of complex antenna systems of aircraft, it is necessary to use a set of analytical and numerical design (aggregate of models). Using analytical expressions for the polarization components of the superposition of electric fields for systems of elementary electric dipoles and cylindrical bearing surfaces obtained using the method of induced currents and the CST MWS electrodynamic modeling software, an electrodynamic analysis of vibrator arrays developed for radio-electronic complexes of aircraft control systems was carried out. The directional patterns and directivity of arrays for cases of placing antenna arrays on ideally conducting cylinders and cylinders with thin dielectric coatings are investigated.

It is shown that with the combined use of analytical models and the electrodynamic modeling software, the possibility of multiple calculations of the characteristics of the arrays is provided to establish the rational parameters of their structures during design.

# References

1. Zhang, Z.: Antenna Design for Mobile Devices. Wiley, Boston (2011)
2. Zhi Ning, C.: Antennas for Portable Devices. Wiley, Singapore (2007)
3. Volakis, J.L., Chi-Chih, C., Fujimoto, K.: Small Antennas: Miniaturization Techniques & Applications. Mc. Graw Hill, Milan (2010)
4. Sharma, S., Chieh, J.: Multifunctional Antennas and Arrays for Wireless Communication Systems. Wiley-IEEE Press, Hoboken (2021)
5. Mittra, R.: Developments in Antenna Analysis and Design, vol. 2. SciTech Publishing, Croydon (2018)
6. Verba, V.S., Tatarsky B.G.: Unmanned aerial vehicle complexes. In: 2 Books: UAV-Based Robotic Systems. Radio Engineering, Moscow (2016)
7. Ashihmin, A.V.: Design and Optimization of Ultra-Wideband Antenna Devices and Systems for Radio Control Equipment. Radio and Communications, Moscow (2005)
8. Avdeev, V.B., Ashikhmin, A.V.: Modeling of Small-Sized Ultra-Wideband Antennas. Voronezh State University, Voronezh (2005)
9. Neganov, V.A., Tabakov, D.P., Yarovoy, G.P.: Modern Theory and Practical Application of Antennas. Radio Engineering, Moscow (2009)
10. Reznikov, G.B.: Aircraft Antennas. Soviet Radio, Moscow (1967)
11. Hallas, J.R.: Basic Antennas: Understanding Practical Antennas and Design. The National Association for Amateur Radio, Newington (2009)
12. Guha, D., Antar, Y.M.M.: Microstrip and Printed Antennas: New Trends, Techniques and Applications. Wiley, Noida (2011)
13. Kishk, A.: Advancement in Microstrip Antennas with Recent Applications. InTech, Rijeka (2013)
14. Bakhrakh, L.D., Kremenetsky, S.D.: Synthesis of Radiating Systems (Theory and Calculation Methods). Radio and Communication, Moscow (1974)
15. Neganov, V.A., Osipov, O.V., Raevskiy, S.B., Yarovoy, G.P.: Electro-Dynamics and Distribution of Radio Waves. Radio and Communication, Moscow (2004)
16. L'vova, L.A.: Radar Detectability of Aircraft. Publisher of All-Russian Research Institute of Technical Physics named after Academician E.I. Zababakhin, Snezhinsk (2003)
17. Bogoslovsky, A.V., Borisov, D.N., Razinkov, S.N., Razinkova, O.E.: Electro-dynamic analysis and synthesis of the array of elementary electric vibrators on the side surface of a circular cylinder of final length. In: Proceeding XXVI International Scientific and Technical Conference "Radar, Navigation, Communication". Voronezh State University Publishing House, Voronezh (2020)
18. Shatrakov, Yu.G., Rivkin, M.I., Tsymbayev, B.G.: Aircraft Antenna Systems. Mechanical Engineering, Moscow (1979)

19. Bogoslovsky, A.V., Borisov, D.N., Razinkov, S.N., Razinkova, O.E.: Modeling and analysis of directional arrays of elementary electric vibrators on the side surface of a cylinder of final length. In: Proceeding XX International Scientific Conference "Informatics: Problems, Methods, Technologies", Wellborn, Voronezh (2020)
20. Kurushin, A.A., Plastikov, A.N.: Designing Microwave Devices in the CST Microwave Studio Environment. Publish of Moscow Power Engineering Institute, Moscow (2011)
21. Makarov, S., Iyer, V., Kulkarni, S., Best, S.: Antenna and EM Modeling with MATLAB Antenna Toolbox, 2nd edn. Wiley, Hoboken (2021)
22. Gross, F.B.: Smart Antennas with MATLAB, 2nd edn. McGraw-Hill Education, Sydney (2015)
23. Kogure, H., Kogure, Y., Rautio, J.C.: Introduction to Antenna Analysis Using EM Simulators. Artech House, Boston (2011)
24. Visser, H.J.: Approximate Antenna Modeling for CAD. Wiley, Chichester (2009)
25. Kashin, V.A.: Methods of phase synthesis of antenna arrays. Foreign radioelectronics. Successes Modern Radioelectron. 1, 47–60 (1997)
26. Markov, G.T., Vasiliev, E.N.: Mathematical Methods of Applied Electrodynamics. Soviet Radio, Moscow (1974)
27. Ananin, E.V., Vaksman, R.P., Patrakov, Yu.M.: Methods for reducing radar signature. Foreign Radioelectron. 4(5), 5–21 (1994)
28. Pontryagin, L.S., Boltyansky, V.G., Gamkrelidze, R.V., Mishchenko, E.F.: Mathematical Theory of Optimal Processes. Science, Moscow (1983)
29. Vasiliev, E.N.: Excitement of Bodies of Rotation. Radio and Communication, Moscow (1987)
30. Kuehl, H.H.: Radiation from a radial electric dipole near a long finite circular cylinder. IRE Trans. Antennas Propag. AP-9, 546–553 (1961)
31. Shorokhova, E.A.: Radiation and diffraction of electromagnetic waves in natural and artificial inhomogeneous material media. Dissertation Dr. Phys.-Math. Research Institute of Measuring Systems named after Yu. Sedakova, Nizhny Novgorod (2010)

# Project-Based Education in COVID-19 Era. Disseminating Design Thinking in New Reality

Natalia Pulyavina[1]([✉]) [iD], Anastasija Ritter[1], Nadezhda Sedova[1] [iD],
and Victor Taratukhin[2] [iD]

[1] Plekhanov Russian University of Economics, Moscow, Russia
{pulyavina.ns,sedova.nv}@rea.ru
[2] WWU, Muenster, Germany
victor.taratukhin@ercis.uni-muenster.de

**Abstract.** This paper is devoted to Design Thinking as a method of project-based education in COVID-19 Era. COVID-19 has dramatically changed all spheres of our life, including education. Traditional teaching methods have been replaced by online and distance ones. The study analyses the advantages of Design Thinking in the New Reality, including in the online format, and some of its disadvantages. Real examples of remote project work formulated according to the principles of design thinking include participation of a team of students in the "Stanford Rebuild" global eight-week innovation sprint held by Stanford Graduate School of Business. In the course of that sprint they developed a comprehensive socially oriented online platform "COHELP-19". Another example is the successful participation of Plekhanov University students in the first international research "Ideathon" organized by the University of Munster, "Higher School of Economics" National Research University and SAP that resulted in the development of a prototype application for people experiencing psychological, environmental, and physiological discomfort due to the coronavirus pandemic. Based on the conducted surveys, the authors have shown that the remote project work of students based on the Design Thinking method allowed them to get a useful experience of interaction in a team, taught them to think "outside the box", brightened up their days spent at home, brought new emotions, and reduced the level of fatigue from video conferences.

**Keywords:** Design thinking · Project-based education · On-line projects · Prototyping · Innovation

## 1 Background of Design Thinking Using in Remote Mode

The coronavirus pandemic that broke out at the beginning of 2020 turned our lives upside down. It forced us to limit contacts with other people and abandon the traditional formats of work and study.

Before COVID-19 we hardly imagined that education could be completely transferred to the online format. However, it turned out that this is quite real. Thus, design thinking as a project-based learning method aimed at developing creative and cognitive skills, critical thinking, the ability to integrate knowledge from different fields and apply

V. Taratukhin et al. (Eds.): ICID 2021, CCIS 1539, pp. 43–51, 2022.
https://doi.org/10.1007/978-3-030-95494-9_4

them to solve emerging problems, has shown its applicability and efficiency in a remote format.

The idea of using design as a way of thinking was generated by Herbert Simon in his book "The Sciences of the Artificial", published in 1969 [14]. Later it was developed by William Hannon, who founded the Institute for Design Management in 1975 and wrote "Design Thinking" in 1987 based on his own experience of combining design and business. Only in the early 2000s the method became popular thanks to Hasso Plattner and David Kelly - researchers and businessmen who created the Hasso Plattner Institute Design school and began to teach design thinking as a business approach [10].

Currently, the methodology developed by them is one of the most frequently used in project activities employed by business, management, medicine, and education, especially, considering our new reality. The basic rules of design thinking state that those who use it:

• shall be curious and involved in the process as much as possible and think "outside the box";
• shall not be afraid to make mistakes because in most cases the negative experience helps finding the right solutions;
• shall trust and listen to the opinions of other team members;
shall never give up;
• shall not criticize other people's suggestions and ideas;
shall be confident in the success of teamwork and aim to achieve great results [5].

## 2   Design Thinking as a Project-Based Education Method in New Reality

Design thinking is used to find non-obvious solutions to a wide range of problems [3, 4, 11, 13], as well as to create innovative products and services that can make the lives of their users easier [2, 7, 8]. This is proved by successful participation of Plekhanov Russian University of Economics students in two international sprints held under coronavirus restrictions and resulted in Plekhanov University version of Design Thinking designed as part of collaboration with University of Muenster, Germany [1, 15].

The first sprint was the eight-week project "Stanford Rebuild" organized by the Stanford Graduate School of Business. Its goal was to develop products and services for solving public problems caused by COVID-19. "Stanford Rebuild" was the competition where the Plekhanov Russian University of Economics students' team applied the methodology of design thinking for the first time. They used design thinking as a team in order to develop an unprecedented aggregated Internet platform aimed at overcoming the negative consequences of the pandemic (job loss, mental and cognitive overload, burnout due to constant being at home and studying or working online). The simplicity and accessibility of the methodology, the enthusiasm and creativity of the students helped them create a unique prototype and improve their professional and soft skills. According to the results of the sprint, the Plekhanov team was recognized as one of the two hundred finalists of the "Stanford Rebuild" project (Fig. 1).

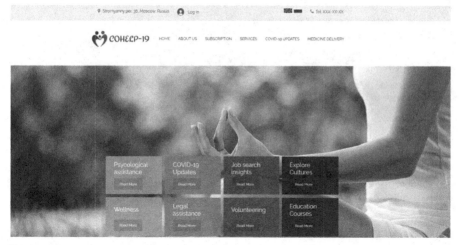

**Fig. 1.** The "COHELP-19" platform developed by Plekhanov University Team during "Stanford Rebuild" project.

Another example of the successful application of design thinking by students of Plekhanov University is the victory in the first international research "Ideathon" organized by the University of Munster, the "Higher School of Economics" National Research University and SAP. In less than two days, the students team became a single "organism" and developed a prototype of a mobile application for people experiencing psychological, environmental, and physical discomfort in self-isolation and wanting to move to the countryside for these reasons. Thanks to the use of design thinking methodology, team spirit, a high level of self-organization, compliance with deadlines and constant communication, the students at the Plekhanov Russian University of Economics proved themselves at the presentation of their product, received positive feedback from the jury and earned the first place among other nineteen bright, strong, and worthy competitors from twelve countries (Fig. 2).

A comprehensive analysis of the teamwork of students of Plekhanov Russian University of Economics showed the following advantages of using design thinking both as the traditional and remote mode (Table 1):

Because of COVID-19, video conferences have become an integral part of work and study. According to the Stanford University research, the number of users of the Zoom platform has increased from 10 million people (December 2019, when the coronavirus pandemic was not yet known) to 300 (April 2020 – booming spread of coronavirus). People who were "locked" in their homes and constantly used "Zoom" felt physical discomfort from lack of movement and prolonged use of the monitor, experienced cognitive overload and "Zoom fatigue" – depression, mood swings, emotional and professional burnout, the desire to retire after video conferences, etc. The experience of the Plekhanov University team showed that remote project work carried out using the design thinking methodology did not cause such "side effects", but on the contrary allowed students to get a useful experience of interaction in a team, taught them to think "outside the box", made their days at home brighter and caused to experience new emotions when faced

**Fig. 2.** The mobile application for relocation created by Plekhanov University Team during "Ideathon" project.

**Table 1.** The design thinking advantages

| For the traditional mode | For the remote mode |
| --- | --- |
| • Human orientation, i.e. understanding and feeling consumers' problems, finding effective solutions;<br>• Simplicity of used tools;<br>• Wide application area;<br>• Innovative products that make people lives easier and more convenient as a result;<br>• Creativity and thinking "outside the box";<br>• Development of teamwork skills, i.e. learning to find a compromise, improve decision making, defend your point of view, listen to the opinions of other people and participate in effective communication;<br>• Development of personal, professional, and soft skills;<br>• Turning challenges into opportunities | All the design thinking advantages for the traditional format<br>• Saving time and resources;<br>• Low labor intensity;<br>• Communication with team members and potential consumers regardless of where they are at the moment;<br>• Great opportunity for survey;<br>• Easy data collection and systematization;<br>• Visualization via computer programs;<br>• Cloud data storage |

with interesting people. This was proven by the results of a survey conducted among the participants of innovative sprints.

It turns out that it is easier for most students to communicate with unfamiliar people in the online format. As part of teamwork, it contributes to a rapid transition from the "formation" stage to the "activity" stage and increases the effectiveness of interaction, a high degree of which is noted by about 58% of respondents. Project activity of students using design thinking in a remote format does not cause cognitive overload but contributes to the development of hard and soft skills (57% and 79% of respondents, respectively, think so), improves teamwork skills (according to 86% of students' opinions) and does not lead to "Zoom fatigue".

However, there are two sides of a coin. So, using design thinking in the traditional format has some disadvantages:

- no guarantees of getting a fast, high-quality and relevant result;
- the non-linearity of the process necessitates going through the stages of design thinking again and again to perform the same iterations until a certain result is achieved;
- risks of obtaining biased data in public opinion polls due to the homogeneity of the respondents' group;
- the inability to develop a unique product in the absence of team cohesion;
- the need for training in the methodology of design thinking or inviting specialists;
- resource intensity.

In the case of online format, design thinking methodology disadvantages are supplemented by:

- the need to have computer skills, special equipment, software and Internet access;
- the risk of technical problems;
- the lack of communication with other team members;
- the inability to express emotions fully, manually create a prototype and touch it during testing;
- low physical activity and the lack of movement.

These disadvantages may be attributed to the fact that in the conditions of the new reality [6], when applying the design thinking methodology "Zoom", "Skype", "Microsoft Teams" and "Discord" applications have been used for discussion and live communication; e-mail, social networks and messengers ("Google Mail", "Facebook", "Twitter", "Instagram", "WhatsApp", "Viber" etc.) have been used for written communication, and cloud data storage ("Google Drive", "iCloud", etc.) have been used for joint document management. At the same time, the content of the stages of design thinking carried out via the Internet remained unchanged, and effective virtual alternatives were found to the classically applied methods, technologies, and tools. This resulted in developing the Plekhanov University version of Design Thinking. Its stages shall be described below.

## 3 Plekhanov University Version of Design Thinking

The first step of the Design Thinking – "empathize" technique – involves putting team members in the user's place, feeling his/her problems, understanding their physical and emotional needs, i.e. empathizing with them. The goal of this stage is understanding what is important for a person, as well as finding so-called "pain points". In the process of traditional design thinking it can be done via interviews and surveys, in Plekhanov

University version these methods are replaced by remote conferencing, questionnaires in social networks and Google Forms. Their advantages include: saving the time that the interviewer and interviewee may spend on getting to the meeting place; the ability of users to complete the survey at a convenient time for them; automatic storing of answers; and the possibility to reach a wide range of people and learn the opinions of those who live in other regions and countries.

The use of computer technologies at the second ("define") stage of design thinking makes it easier to collect, sort, organize and group the results of Internet surveys. Video communication and mind mapping desks are further used to discuss these results as a team and highlight a narrowly focused problem that bothers users. In order to create "personas" in the Plekhanov University version of Design Thinking, image editors such as "Photoshop", "Paint", etc. are used to visualize typical representatives of the target audience of a future innovative product or service. It is also important to come up with their names, fields of activity, hobbies, life goals and dreams in the process of online communication. The tools described above reduce the complexity and duration of the design thinking procedures and allow team members to realize their creative potential and inspire others.

The next stage of the project methodology is called "ideate". It consists of group work aimed at developing unusual solutions to a previously identified users' problem. This stage involves the absence of criticism and stereotypes in the team, its positive attitude, focus on innovation, unlimited imagination, flexibility of thinking and the ability to listen to others. During their activities, members of the group have to put aside their fears and put forward all the solutions that come to their mind for public discussion, regardless of their originality and atypical nature. In the process of "brainstorming", it is important to write down all the ideas without exception and not stop at the first one that seems to be the most advantageous. To go through these steps, team members can use traditional text editors such as "Microsoft Word", "Notepad" etc., screen demonstration, mind mapping desks and drawing tools.

When the list of solutions to the users' problem is formed, project developers can choose the idea. It is necessary to evaluate each of the team members' proposals based on the developed criteria, select one or more of the most innovative and progressive by voting, or synthesize all their ideas into one. The team must not hesitate in choosing a very broad direction of their activity, because the process of design thinking implies the possibility of correcting mistakes at all stages of the project. Plekhanov University students' practice shows that the choice of ideas can be made both in one conference session and in several, since group participants may need additional time to make a final decision. The discussion, visualization, representation, and selection of the most successful ideas are usually performed with the help of above-mentioned software and applications.

The fourth stage of the method ("prototype") involves checking viability of the chosen idea and creating a product or service model, i.e. a simplified version that can solve the users' problem. A prototype may be simple during an early stage of development, but it must be completed and made more complex during testing. This step of design thinking is used to adjust and develop a product in details or to understand that the product is useless, stop the process and save resources. In the traditional mode, a prototype can be drawn on paper, made from scrap materials, or even built from "Lego". In Plekhanov University version of Design Thinking, there are extra options for creating it: graphic and video editors, games ("The Sims", "Minecraft", and etc.), services that allow to animate static models and develop mobile applications, websites, and so on.

**Table 2.** Steps of Plekhanov University version of design thinking.

| The design thinking stage | Tools used during this stage | The advantages of used tools |
| --- | --- | --- |
| "Empathize" | Social media polls and Google Forms | Saving time, convenience of survey, automation of recording responses and the ability to interview a wide range of people all over the world |
| "Define" | Graphic design software ("Adobe Photoshop", "Sketch", "Paint", and etc.), mind mapping desks | Low labor intensity and saving time during creating personas, the most plausible visualization |
| "Ideate" | Word processing programs ("Microsoft Word", "Note", and etc.), screen demonstration, painting tools, mind mapping desks | Visualization of the ideas of the whole team, ease of perception and visibility |
| "Prototype" | Graphic design and video editing software; services for animation and website creation, games | Detailed creation of prototypes, the possibility of animation |
| "Test" | Social media polls and Google Forms | Automation of recording responses, their rapid subsequent processing, coverage of a wide range of people all over the world |

The benefits of an invented product or service must be evaluated not by team members, but by ordinary users who can check how effective and easy to use the prototype is. It is essential to get feedback from them that will help project developers to know consumers better, more accurately formulate their problem and eliminate the disadvantages of the developed product or service. All this constitute the fifth and final stage of design thinking – "test" – that involves observing and communicating with persons who evaluate the prototype. Both steps can be implemented in an online format via video or audio communication or feedback forms that give team members the opportunity to get the opinion of any person, regardless of their location. During the testing process, it is important to ask users what they are not satisfied with and try to understand the reasons. Testing a prototype allows seeing the product problems in due time, eliminate them and thereby save a great amount of time, as it is much better to improve the project at initial stages than to invest a huge number of resources during its launch and fail (Table 2).

Although "test" is the final stage of design thinking, it does not mean that the work on the project ends there. The inability to produce a perfect product or service, constantly changing reality, globalization, trends, the political and economic situation in the world as well as other factors make the team of entrepreneurs systematically return to certain stages of project activity and even start the process again. As the practice shows, the more iterations a product goes through, the more useful, convenient, and effective it becomes for the user. The transfer of the design thinking methodology to a remote format opens more opportunities for repeating iterations since the use of computer tools allows to move from one stage to another in a shorter time and achieve results faster (Fig. 3).

**Fig. 3.** Communication between Plekhanov University team members in "Zoom".

## 4  Conclusion

Thus, the coronavirus pandemic has proved the applicability of design thinking as a method of project-based learning in the online format. It turned out that sometimes this method of team interaction is more applicable, efficient, and successful compared to the traditional one. An example of this was the participation of Plekhanov Russian University of Economics students in international innovative projects, that allowed them to show their creativity, cohesion and efficiency and create unique prototypes that can solve social problems in the conditions of a new post-COVID reality. Moreover, they created Plekhanov University version of Design Thinking that can be adjusted and applied not only to education, but to the other aspects of our life. As future work, more theoretical research is needed, for instance on computer-supported group decision making [12] in order to create a comprehensive framework and to validate such approach at experimental level.

# References

1. Becker, J., Taratukhin, V., Pulyavina, N.: Next-gen design thinking. Using project-based and Game-oriented approaches to support creativity and innovation. In: CEUR Workshop Proceedings: ICID 2019 - Proceedings of the 1st International Conference of Information Systems and Design, Moscow (2020)
2. Brown, T., Katz, B.: Change by Design, Revised and Updated: How Design Thinking Transforms Organizations and Inspires Innovation. Harper Business (2019). 304 p.
3. Buchanan, R.: Systems thinking and design thinking: the search for principles in the world we are making. She Ji J. Des. Econ. Innov. **5**, 85–104 (2019)
4. Buhl, A., Schmidt-Keilich, M., Muster, V., Blazejewski, S., Schrader, U., Harrach, C.D., et al.: Design thinking for sustainability: why and how design thinking can foster sustainability-oriented innovation development. J. Clean. Prod. **231**, 1248–1257 (2019)
5. Cross, N.: Design Thinking: Understanding How Designers Think and Work. Berg, Oxford UK and New York (2011). 140 p.
6. Fauville, G., Luo, M., Queiroz, A.C.M., Bailenson, J.N., Jeff, H.: Zoom Exhaustion & Fatigue Scale, Social Science Research Network (2021). https://ssrn.com/abstract=3786329. Accessed 06 Aug 2021
7. Hoolohan, C., Browne, A.L.: Design thinking for practice-based intervention: co-producing the change points toolkit to unlock (un)sustainable practices. Des. Stud. **67**, 102–132 (2020)
8. Dorst, K.: The core of 'design thinking' and its application. Des. Stud. **32**(6), pp. 521–532 (2011). https://doi.org/10.1016/j.destud.2011.07.006
9. Kelley, T.: Creative Unleashing the Creative Potential Within Us All – Currency (2013). 304 p.
10. Martin, R.: The Design of Business: Why Design Thinking is the Next Competitive Advantage. Harvard Business Review Press (2009). 208 p.
11. Van der Bijl-Brouwer, M., Dorst, K.: Advancing the strategic impact of human-centered design. Des. Stud. **53**, 1–23 (2017). https://doi.org/10.1016/j.destud.2017.06.003
12. Poole, M.S., DeSanctis, G.: Microlevel structuration in computer-supported group decision making. Hum. Commun. Res. **19**(1), 5–49 (1992)
13. Pourdehnad, J., Wexler, E., Wilson, D.: Systems & design thinking: a conceptual framework for their intergration. In: 55th Annual Meeting of the International Society for the Systems Sciences, pp. 807–821 (2011)
14. Simon, H.: The Science of the Artificial – The Massachusetts Institute of Technology Press (1996). 231 p.
15. Taratukhin, V., Pulyavina, N., Becker, J.: The Future of Design Thinking for Management Education. Project-Based and Game-Oriented Methods are Critical Ingredients of Success. 2020, Developments in Business Simulation and Experiential Learning: Proceedings of the Annual ABSEL Conference, Pittsburgh, USA, pp. 261–265. ABSEL (2020)
16. Wilkerson, B., Trellevik, L.-K.: Sustainability-oriented innovation: Improving problem definition through combined design thinking and systems mapping approaches. Think. Skills Creat. **42** (2021). https://doi.org/10.1016/j.tsc.2021.100932

# The Main Models and Approaches in Creating of Innovations in the Post-COVID World in the Area of Information Technologies

Yulia Bolshakova[1]([✉]) [iD] and Victor Taratukhin[1,2] [iD]

[1] National Research University Higher School of Economics, Bolshaya Pecherskaya, 25/12, 603155 Nizhny Novgorod, Russia
yusbolshakova@edu.hse.ru, vtaratoukhine@hse.ru
[2] University of Muenster ERCIS, Leonardo Campus, 3, 48149 Muenster, Germany
victor.taratoukhine@ercis.uni-muenster.de

**Abstract.** The given paper elaborates on the main models and approaches for the innovation development applicable to IT companies in the context of Covid-19 pandemic impact for short and medium terms. Authors examine the experience of creating innovations in IT industry giving the context of epidemiological constraints with the aim to prepare a research base for the future implementation of the concept of a virtualization solution for design thinking as an approach to creating innovations. Along with an in-depth study of the latest papers on the impact of pandemic-induced virtualization on the workflow, an empirical study of the features of creating innovations in the IT industry during Covid-19 was also conducted. The paper conclusion offers the authors' proposal for adapting design thinking approach in accordance with the industry considered conditions previously obtained and processed by the research results, as well as the possibilities of their further practical implementation.

**Keywords:** Design thinking · Post-COVID · Virtualization · Innovation · Information technologies

## 1 Fundamentals of Creating Innovation in IT Companies

This paper is aimed at identifying the features of creating innovations in IT companies in the context of epidemiological constraints with the use of Covid-19 impact example. Table 1 shows the conceptual apparatus of the work. It includes the given by authors definition of concepts necessary to provide a research context. Theoretical foundations are based on studies of the impact of pandemic on the activities of organization, the creating innovations in general, as well as approaches and methods of creating innovations in IT sphere. Methodological foundations are subordinated to principles of business architecture approach.

V. Taratukhin et al. (Eds.): ICID 2021, CCIS 1539, pp. 52–65, 2022.
https://doi.org/10.1007/978-3-030-95494-9_5

**Table 1.** Conceptual apparatus of the research.

| Concept | Definition |
| --- | --- |
| Innovation | The result of intellectual activity characterized by the presence of novelty or improvement (in a form of a product, a process, or a business model) that creates additional value (in efficiency, quality superiority, leadership, profit) relative to existing solutions in a demanded market for an organization |
| Model | A coherent set of principles that subordinate certain methods of creative organization of intellectual activity to achieve a certain result; the innovation creation model provides added value for the organization's operations |
| Approach | A coherent set of techniques for organizing intellectual activity, the implementation of which is to provide the result of this activity; The result of the implementation of an approach to creating innovations is considered a solution that has additional value for the organization's activities |
| Post-pandemic | Designation of the period of economic recovery after prolonged exposure to epidemiological restrictions, which are official and universal |
| Onlineization | Purposeful and forced, due to external epidemiological restrictions, the digitalization of industrial communication and related processes because of the introduction of a remote mode of work as the main way of organizing work |
| Remote mode | The organization of the employee's work activity is fundamentally outside the workplace on the territory of the enterprise, carried out due to the widespread use of ICT (also: remote work) |

## 1.1 Rationale for the Relevance of Research

Companies with the priority to create innovations can quickly adapt to changing conditions, but the pandemic impact on such organizations was no exception [5, 10, 16, 23]. An impressive share of innovations created during the Covid-19 pandemic is the result of IT companies' activities [9]. According to the Coronavirus Innovation Map, the number of initiatives made by IT companies is 45% of the total number [7]. The question is more relevant than ever: how to most effectively create innovations demanded by society in the VUCA-environment of world and local markets, which will continue at least in the medium term, according to forecasts of industry experts [16]. It can be assumed that the ability to create innovations given the nature of the new reality will be a significant factor in obtaining the economic efficiency for IT companies, including the post-covid world perspective.

Due to the epidemiological situation, the need to ensure health safety has become a fundamental factor for remote work mode providing this mode is applicable giving the nature of job function. While IT industry is traditionally considered as one of the most predisposed to online organization of labor activity. Among the main factors are the widely shared experience of remote work mode before it had become a preventive measure, as well as the nature of the most job activities, and the current level of communication technologies used for work, cloud solutions, etc. [5]. Many studies note the pandemic impact on increased number of IT companies that are ready or have already implemented a remote mode (at least optionally) on a permanent basis [5, 23]. The largest share of remote work in the post-pandemic period as the main model of labor organization is expected belong to IT sphere [4, 16].

Presently, the most popular subject of scientific interest in creating innovation is its adaptation to the conditions of epidemiological restrictions (including expectations regarding the post-pandemic period), where the main principle is the absence of interaction in a non-virtual environment [3, 12]. Multiple papers show the different impact of epidemiological constraints across industries. The main factor of the unevenness of this impact is a predisposition to onlineization of key business processes inherent in a particular industry [16, 17, 25]. This impact is also tending to be considered either in general terms or as an illustration of extra cases. IT industry is not in the focus of research considered; however, few relevant studies consider its main challenges (an increasing demand to produce new solutions or to add innovative functionality to existing ones) and types of innovative solutions (product and process innovations) during the pandemic. Consequently, a high predisposition to digitalization of key business processes relative to other industries is assumed as the main reason for the non-proliferation of the IT industry as an object of research regarding the consequences of epidemiological restrictions. It may also indicate an insufficient degree of knowledge of the process of adaptation of IT companies to the conditions under consideration, therefore, to stimulate further research in this area.

## 1.2 Creating Innovation and the Pandemic Constrains: A Theoretical Overview

The process of creating innovations was influenced by the pandemic despite the general requirement for physical collaboration as a guarantee of effective results [25]. The emphasis is on the general change in goal setting: if before the pandemic, the main interest was the creation of solutions and mastery of technologies, while during – the adaptation of processes through their digitalization, as well as cost reduction [12]. A direct relationship between organizational maturity in digitalization (creating innovations is included) and economic indicators over the pandemic restrictions period [24] is an important finding as well. Effectiveness of interaction in geographically remote teams also remains a topical research issue. Insufficient ability of the current technological communication tools in realizing the experience of face-to-face interaction was found as the main limitation of remote team effectiveness [3, 12, 24]. An increase in the use of the open innovation model during the containment of the pandemic was revealed. Presumably, the reason is the previously demonstrated cost reduction [1, 21] (not necessarily on the own experience [18]) due to the principles of its organization, as well as the possibility of its easy onlineization.

Regarding the post-pandemic creating innovations, the main assumption is to preservation of the processes adapted during the constraints at least in the short term since the end of the pandemic [16, 17, 25]. The vector of goal setting for this process will probably remain, however it is expected to coincide with the pre-pandemic vector in the medium term [17, 25]. There is also a point to potential increase in organizational motivation to adhere to open policy in the field of innovation. Presumably, this may become a factor in an increase in the number of distributed teams, hence increased requirements for the efficiency of such interaction and its technological support. A few studies also indicate the possibility of using AR/VR technologies to create a new generation of communication tools for use in conditions of limited interaction [2, 25, 26].

A sufficient degree of knowledge of the models and approaches to creating innovations in the pre-pandemic period makes it possible to form a basis for conducting an empirical study. The focus of this work is the most used methods of organizing innovation in this industry. Models and approaches, mainly used for operational purposes, were excluded from the study (for example, Agile as a model and Scrum as a work organization approach) According to available research, the most key models for creating innovations in the IT field are Open Innovation (OI) and Customer Development (CD) [1, 12, 25]. OI model can be applied for a geographically distributed team, which is confirmed by some pre-pandemic experiences [18, 21]. However, there is not enough research on how IT companies have adapted the use of this model in the context of epidemiological constraints. Referring to the general trend, it can be assumed that the impact of the pandemic has not become a limiting factor for the use of OI model. Moreover, due to the need to reduce the time-to-market in relation to innovative solutions, the number of IT companies applied to this model has probably increased. The structure of CD model can be implemented in a geographically distributed team, which is also confirmed by the presence of relevant cases [12, 22]. As for OI, current research results are not enough to form an objective assessment of the features of CD implementation in the context of epidemiological limitations. Analogically, due to the IT companies' growing need to create the most satisfying solutions for end users to reach economic efficiency, such companies at least have continued to use this model, or even integrated into their innovation activities. Summing up the considered models, it can be assumed that the main possible obstacle to their use in the context of epidemiological constraints was technological support. The premise is to maintain employee performance in a virtual environment. Since such interaction is feasible solely due to technological support, it is logical to assume in the formation of the need to imitate a non-virtual environment. Also, relying on the research results, to date, these solutions are not able to fully replace real communication, which negatively affects the efficiency of business processes [23, 25].

Among the approaches to creating innovations in IT, it is worth noting Design Thinking (DT) among companies due to its wide distribution in this industry [1, 11, 19]. This approach can be consistently applied in the implementation of OI and CD models. The positive impact on the rate of creation of innovations with a specific focus on solving the problems of the end user ensured the widespread use of this approach, including in the IT environment [1, 19]. This approach can be used to create innovations of any type; however, it is mainly used in relation to product innovations [6, 11, 20]. This approach has many variations of its implementation, a comparative analysis of some of them was made (see Fig. 1). The approach developed at Hasso Plattner Institute of Design at Stanford was used as a reference model. Consequently, presumably DT has a sufficiently high degree of adaptability for certain conditions of use. However, this approach is usually used in the context of non-virtual interaction of participants [2, 8, 14, 15]. The main challenge for DT trainees in a virtual environment was the observance of all stages of the approach, as well as their full implementation, which is not inferior in quality to a relatively non-virtual environment [14, 15]. Despite this problem, to date there is an insufficient number of relevant research papers, which in turn does not contribute to its resolution. A similar situation is also for research on the use of DT in post-pandemic conditions, which implies a high degree of use of virtual communication tools. Regarding changes in the context of epidemiological restrictions, presumably there is a notable probability of rejection of its use. The value of the probability of such a failure may depend on the level of criticality of team interaction in a real environment in the activities of a particular IT company. Another possible factor in the abandonment of DT in a virtual environment is the limited means of virtual communication used in transferring non-virtual interaction that is identical to the experience. For example, the results of the available research have shown a qualitative deterioration in the implementation of the approach at the stages of empathy and testing (see d.school model in Fig. 1) in conditions of virtual interaction [14].

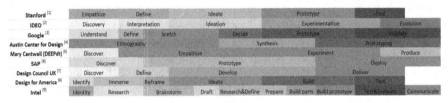

1. https://web.stanford.edu/~mshanks/MichaelShanks/files/509554.pdf
2. https://designthinking.ideo.com/
3. https://designsprintkit.withgoogle.com/
4. https://www.ac4d.com/worksheets
5. https://www.deepdesignthinking.com/
6. https://tegrous.com/sap-design-thinking-workshop/#:~:text=SAP%20Design%20Thinking%20is%20a,customer%2C%20resulting%20in%20innovative%20solutions.
7. https://innovationenglish.sites.ku.dk/model/double-diamond-2/
8. https://designforamerica.com/resources/
9. https://d1io3yog0oux5.cloudfront.net/_409081c3ac0d5430096de3c789c2cada/intel/db/887/8510/annual_report/2021+Form+10K_BMK.pdf

**Fig. 1.** A comparison of design thinking variants.

Thus, as a result papers review on the experience of creating innovations in the context of epidemiological constraints, as well as in the post-pandemic period, in relation to the IT field, the urgency of the need for an empirical study of such experience was confirmed. To implement the most localized solution to the problem under investigation, the empirical research focused on the smallest unit of organization of the innovation creation process adopted in this work. Such a unit is the approach to creating innovations; regarding the IT sphere, the DT approach is of the greatest interest. An additional motivation for focusing on this approach lies in the presence of difficulties in its full-fledged implementation in a virtual environment due to the lack of an initial predisposition to onlineization.

## 2 Research of Creating Innovations Experience in IT Companies in the Conditions of Epidemiological Restrictions

The purpose of the empirical study is to record the change in the image of creating innovations in IT in the context of the limitations of the pandemic, as well as assumptions about such changes in the post-pandemic period. For the most localized study of the present issue, a deepening in the study of the DT approach as widely used by IT companies in the pre-pandemic period has been implemented. Thus, the task of the survey is to find out the features of the process of creating innovations in IT companies in the conditions of remote work in relation to the conditions of physical presence in the office, with a focus on finding out the experience of using the DT approach.

The sample of this study included IT companies from the Volga Innovation Cluster (among these companies are Intel, Orion Innovation (ex. MERA), EPAM, Neuro.net, Harman, etc.) located in Volga Federal District (Russian Federation) that meet the following conditions: a) these companies were engaged in the creation of innovations as a main or additional activity; b) due to epidemic, employees of these companies were transferred to a remote mode of work (from March–April 2020); c) these companies continued to engage in innovative activities in remote mode; d) these companies have used models and\or approaches to create innovations before and during the pandemic; e) by the time of research conducting, its participants have been working in the given companies for at least one year (at least since March 2020); f) the company data must have been registered at least for a year at the time of research conducting (at least from March 2020). The empirical study involved 22 employees of companies that meet the above conditions. The survey was carried out from March 25th to April 11th, 2021. It included 26 questions divided into 3 parts: general info (4 questions about employer and employment details), innovation experience study (11 questions about the company's experience in creating innovations before and during the pandemic, as well as expectations for the post-pandemic), and DT experience study (11 questions about the company's DT experience before and during the pandemic, as well as expectations for the post-pandemic). Most participants (45%) have been working in these companies for 1 to 3 years, another third for more than 6 years, and the rest from 3 to 6 years. The main characteristics of these companies based on the survey results obtained are presented in Table 2.

**Table 2.** Main characteristics of the sample of companies.

| Characteristic | Value | Share of companies, % |
|---|---|---|
| Company age | 1–3 years | 14 |
| | 3–10 years | 18 |
| | More than 10 years | 68 |
| In total: 100 | | |
| Company geography | International | 68 |
| | Non-international | 18 |
| | In the process of entering the international market | 14 |
| In total: 100 | | |
| Number of employees | Less than 15 | 5 |
| | From 15 to 100 | 18 |
| | From 101 to 250 | 14 |
| | More than 250 | 64 |
| In total: 100 | | |

The largest aggregate share of the sample was made up of international IT companies that have been on the market for >10 years and have a staff of >250 people, which makes it possible to classify it as a large business. Presumably, the results obtained are most relevant to the dominant profile of the company, this remark should be considered regarding further research results.

## 2.1  Analysis of Research Results on Creating Innovations in IT Companies

Consider the results obtained regarding the experience of creating innovations in IT companies in the context of epidemiological constraints. For 73% of the respondents, it is the main economic activity, and for the rest of the sample it does not exceed 50% in the total share of the organization's activities. Regarding the main type of innovations created, more than 70% noted product innovations, about 20% is for innovative business models, and the rest is for innovations in processes. Consequently, the dominant aggregate profile is a company with the main activity is the creation of product innovations. In view of the assumed differences in innovation creation, depending on the share of activity and the prevailing type of innovation, this circumstance must be considered when deriving a general assessment of the creation of innovations in IT within epidemiological constraints.

The overwhelming majority of 86% noted that the use one any model or approach of creating innovations before the pandemic. Among such cases, only 10% noted the addition of new practices, half of them due to the ineffectiveness of previously used ones. Meanwhile, respondents who noted the absence of such practices before the pandemic indicated that they were initialized during pandemic; as an assumption, because of the need to organize the process due to the imposed limitations.

Table 3 presents a rating of adaptability of models and approaches in creating innovations to the usage in conditions of epidemiological constraints, as well as the fact of its adaptation among those who noted it before the pandemic. More than 40% find difficulties within facilitating adaptation to new reality. Only a third signals a successful experience of such adaptation. About 1/6 noted the lack of adaptation of ones used when convinced of difficulties in this process. The minimum share with a positive rating noted the absence of these actions. Perhaps it happened due to the presence of experience of such adaptation, or due to justification at the level of a key organizational unit in the absence of the need for such actions while maintaining the status quo. Evaluating the level of effectiveness, more than half noted a decrease relative to the pre-pandemic level. This answer was preceded by an attempt at such adaptation among the opinion about the process' complexity in 80% answers, and only 10% noted an increase of efficiency. A third noted that the efficiency of innovation creation processes remains at the pre-pandemic level. It is noteworthy that at least the preservation of the same level of efficiency of this process is typical for companies where the creation of innovations is a major activity. It can be assumed that the status of the company's core business process increases probably of optimal organization of the innovation creation process under the given constraints.

Regarding the post-pandemic forecast, depending on the quality of experience gained in their implementation under conditions of restrictions, a common decision was made to continue the practice. The more effective this experience turned out to be, the greater the probability of prolonging the use of the same models and\or approaches. Meanwhile, in the context of efficiency decrease, about 20% generally inclined towards a positive rate of its continuation. Perhaps it potentially made due to the lack of alternatives, or in anticipation of the pre-pandemic working conditions return. The proportions of positive and negative rates of its continuation in the whole sample are approximately equal. Probably, this fact signals the presence of certain problems during the adaptation to epidemiological restrictions, which in turn is the reason for rejection of ones previously satisfy the corresponding organizational need.

In view of the preliminary identification of the most common IT models and approaches, the structure of this study is consistently aimed at reflecting the experience of their application. From the calculation of the results, the opinion of respondents who noted the absence of certain models' usage before the pandemic were excluded. According to results, 50% and 60% of respondents indicated the use of OI and CD during the pandemic, respectively. A third noted the simultaneous use of these models. More than 40% generally noted the use of OI or CD in approximately equal proportions for each. The joint use of these models in the context of epidemiological constraints have the greatest positive impact in relation to the singe use results. As an assumption, the combination makes it possible to distant certain difficulties of their effective realization in conditions of limited physical interaction, which in turn allows maintaining the previously achieved level of productivity of the innovation creation process.

**Table 3.** Adaptation of the creation of innovations to epidemiological constraints.

| Adaptability rating | Fact of adaptation | Share of companies, % |
|---|---|---|
| It is possible to adapt | No | 10 |
| It is possible to adapt | Yes | 32 |
| It is impossible to adapt | Yes | 42 |
| It is impossible to adapt | No | 16 |
| In total: 100 | | |

Summarizing the results obtained, it is possible to report the confirmation of the theoretical assumptions proposed in paragraph 1.2 with regards to IT field related experience of creating innovations in the context of epidemiological constraints, as well as in the post-pandemic period. However, it is necessary to consider the prevailing aggerated company profile for which the present results are most relevant. Despite the participation of representatives of medium and small businesses, their sample share may be insufficient to form an objective picture of the experience regarding the possible organizational specifics of such companies.

## 2.2 Analysis of Research Results on the Application of Design Thinking in a Virtual Environment

The emphasis of research is made on the study of DT in the context of distributed teams. About 70% reported being used in a virtual environment, and 20% of them turned to

DT for the first time during a pandemic. It is noteworthy that in one third of cases this approach was used outside the previously considered models. In another third of cases with the combined use of OI and CD, the least DT was used within the separate application of a particular model.

Regarding the type of innovations created in relation to this approach, all respondents noted product innovations in the pre-pandemic period. In a virtual environment, the type of innovation tends to remain. It is worth noting the interest in creating innovative processes within the virtual application of DT. The number of such cases in sample is 20%, which does not allow to formulate objective conclusions about the features of DT usage in distributed teams. Meanwhile, regarding the experience of transition to a virtual environment, there is assumption about the insignificant influence of the approach implementation environment being made.

Considering the adaptation experience, more than half indicated facing the impossibility of onlineization of this approach. A quarter did not apply any actions on adaptation with the belief of its impossibility to obtain an interaction comparable to the intended experience. Less than 20% considers DT onlineization to be comparable, having directly gone through adaptation to a new reality. Regarding the change in implementation efficiency, 75% with prior DT experience in the pre-pandemic period noted a decrease in the quality in a virtual environment. In only one case the performance level had been improved, possibly due to several DT tools were used; while in the rest either one tool or non-DT communication tools were applied. Therefore, it can be assumed that the quality of virtual DT implementation depends not only on the use of specialized tools, but also on the functional distribution of these tools.

According to study results, the most popular DT implementation tools were selected (see Table 4). According to the table data, each tool is effective within the specific stages, not covering the entire approach structure. The criterion for determining the tool suitability for performing a particular stage of DT was the availability of functionality that is identical in nature to the means used in these stages in a non-virtual mode. The main functionality relates primarily to Define and Ideate phases. Meanwhile, there is a functional insufficiency of the full implementation of the Test, Prototype, and Emphasize phases within the single tool. It can be assumed that the use of a limited number of tools is the main risk factor for reducing the effectiveness of the virtual implementation of this approach.

According to the empirical results, in more than 70% of DT use cases in a virtual environment, any special DT tools were not used. Moreover, among all the considered cases of their use, there is a decrease in the effectiveness relative to the pre-pandemic experience. Therefore, it could be assumed that the use of communication tools for a virtual DT conducting is deficient in terms of experience itself affecting the results obtained, and there is a demand for a solution that would effectively implement each phase of DT virtually.

The following parameters were identified for assessing the satisfaction with DT tools: UI, UX, and API. The priority in choosing a particular tool was the quality of UX (93%), the next were for UI and API accordingly. Considering the most problematic areas of applied tools, which could be the reason for effectiveness decrease, most respondents gave the first place to UX (80%), the second was for both UI and API, 60% each relative

**Table 4.** Main tools for design thinking in a virtual environment.

| Tool | Characteristics | DT phase |
|---|---|---|
| Sprintbase | A virtual innovation platform that enables distributed teams to take DT approach in a virtual environment | Define Ideate Prototype |
| Stormboard | A team collaboration platform to make innovation techniques available for use in a virtual environment | Define Ideate |
| Mural | A virtual collaboration space that enables DT to be applied in a distributed team environment | Emphasize Define Ideate Prototype |
| Miro | A platform for interoperability in a virtual environment allowing DT implementation in a distributed team environment | Emphasize Define Ideate |
| Google Jamboard | Whiteboard for team interaction, integrated with Google services | Define Ideate |
| Conceptboard | A team collaboration platform to make innovation techniques available for use in a virtual environment | Emphasize Define Ideate Prototype |

to the priority. It is noteworthy that the ranking of the choice aspects coincides with the rating of the most problematic aspects of these tools. It can be assumed also that their use in terms of user interaction does not contribute to the effective virtual implementation of the approach.

The respondents were asked to rate the satisfaction degree with respect to the experience of using the considered tools remotely, on a 5-point scale. The average total score is 2.53 points, which indicates the unsatisfactory experience. Despite the prevailing negative assessment of the virtual DT experience, about 60% noted the will to continue using DT in the post-pandemic period even remotely. Moreover, almost 75% noted the feasibility of creating a DT tool that covers all phases. Thus, there is about confirming the hypothesis of this study about the need to create an alternative solution to virtualization of the innovation creation process using the example of DT approach.

### 2.3 Analysis of Possibilities to Adapt Design Thinking for Use in a Virtual Environment

As a result of empirical research on the experience of creating innovations in IT companies in the distributed interaction context, conclusions were formulated and substantiated about the need for an alternative solution, functionally covering all phases of DT approach. At this point in time, three ways of adapting DT to a virtual environment can be formulated, the consideration of each of them in more detail had been done.

The first way is to create a new version of this approach, the structure of which will be maximally adapted to the current specifics of DT implementation in the online environment. The obstacle is, there are already many variations of the most common DT model developed at the Institute of Design at Stanford (d.school); Fig. 1 presents a comparison of the most popular interpretations of the DT concept. Consequently, there is a potential variant of choosing one or another interpretation as the most relevant with respect to the practice of implementing DT in a virtual environment. Meanwhile, most often users turn to the d.school model [2, 13]. As it was found during empirical research, the existing DT virtualization tools do not allow a full-fledged implementation of this approach due to their focus on separate phases, mainly Define and Ideate. As a consistent assumption, due to the use of these versions, distributed teams had to modify the original structure of DT approach used, focusing on the previously mentioned phases, which could be the reason for the decrease in the effectiveness of this method relative to the experience in the pre-pandemic period.

The second way is to analyze existing virtualization solutions and develop a strategy for the effective use of these tools. In part, this analysis was implemented in this study (see Table 4), and simultaneously it implies a direct experiment to fix the rationale for the choice of certain tools for the implementation of every DT phases. Moreover, the given analysis may not provide scientific interest, being rather an integral part of a larger study of the experience of implementing DT in distributed teams.

The third way is to create an alternative virtualization solution for the passage of DT phases, using technologies to replace physical communication, such as VR. According to available studies, the use of these technologies could potentially contribute to a more efficient implementation of relatively other DT virtualization tools by simulating physical communication to a greater extent [2, 26]. The main prerequisite for an appropriate assessment of the prospects for developing an alternative solution is the result of an empirical study that demonstrates the probable inability of existing DT virtualization tools to transfer interaction experience corresponding to the implementation of this approach in a physical environment. It is hypothetically assumed that the experience of using such an alternative virtualization solution can match the level of efficiency of implementing DT in a physical environment. The fundamental condition for achieving this level of efficiency is the coverage of the functionality of the solution of activities inherent in all stages of the implementation of DT as a guarantee of the most optimized organization of the innovation creation process. It is also necessary to consider the specifics of the scope of this alternative solution, namely the implementation of DT in software development teams in IT companies. Thus, this solution must provide conditions for the sequential passage of all DT phases as well as comply with the specifics of IT industry. Meanwhile, an open question remains the applicability of technologies for replacing physical communication with respect to tasks typical for software development teams; at which DT phases the use of these technologies contributes to efficiency, as well as hinders.

Consequently, the key research question of the upcoming study is to determine the most effective design of such an alternative solution, considering the features of the experience indicated in this paper. The necessary conditions for successfully solving this problem are conducting an experiment of using the proposed DT virtualization solutions on a group of subjects, conceptual modeling of relations within the framework of the process under consideration and prototyping the proposed solution.

## 3  Conclusion

The pandemic impact on the organization of work and the prospects for this impact for IT industry was examined. The features of creating innovations, considering the impact of the pandemic and the prospects of this impact in the post-pandemic period, focusing on the practice of IT companies were also studied. Finally, the models and approaches for creating innovations common in the IT industry were considered, and assumptions were made about changing the corresponding practices in the context of epidemiological constraints.

An empirical study of the pandemic impact on the process of creating innovations in IT companies was carried out to record these changes and analyze them for this process in the post-pandemic period. To localize the solution to this problem, a deepening study of the experience of using DT approach, widely used by IT companies in the pre-pandemic period, has been implemented. Using this approach as an example, the possibilities of adapting the process of creating innovations to the epidemiological constrains were studied, including considering the most probable mode of organizing labor activity in the post-pandemic period. The formulated assumptions within the framework of the theoretical chapter of the study were generally confirmed by the results of empirical research.

Thus, the goal of this research, which consisted in studying the experience of creating innovations in IT companies in the context of epidemiological constraints and preparing a research base for the implementation of the concept of a virtualization solution for an approach to creating innovations using the example of DT, was achieved. The assumption about the need to create an alternative solution for virtualizing the process of creating innovations using DT approach as an example was confirmed, therefore the research hypothesis is also confirmed.

The emphasis of the future continuation of this study will be on creation of a concept of the most promising from the point of view of scientific novelty solution to virtualization of DT approach, considering the practical need for such a solution, demonstrated during the implementation of empirical research. To date, the necessary conditions for compliance have also been formulated as part of the creation of such a solution, ensuring that the needs of end users are met. Finally, considering the data obtained, the stages of implementation of the future research were outlined, the implementation of which requires additional research.

## References

1. Asikainen, A.-L., Mangiarotti, G.: Open innovation and growth in IT sector. Serv. Bus. **11**(1), 45–68 (2016). https://doi.org/10.1007/s11628-015-0301-2
2. Bader, L., Kruse, A., Dreßler, N., Müller, W., Henninger, M.: Virtual design thinking - experiences from the transformation of design thinking to the virtual domain. In: ICERI2020 Proceedings, pp. 9091–9099 (2020)

3. Belzunegui-Eraso, A., Erro-Garcés A.: Teleworking in the context of the Covid-19 crisis. Sustainability **12**(9), 3662 (2020)
4. Boston Consulting Group: Decoding Global Talent, Onsite and Virtual, 04 March 2021. https://www.bcg.com/ru-ru/publications/2021/virtual-mobility-in-the-global-workforce. Accessed 02 Apr 2021
5. Brem, A., Viardot, E., Nylund, P.A.: Implications of the coronavirus (COVID-19) outbreak for innovation: which technologies will improve our lives? Technol. Forecast. Soc. Change **163**, 120451 (2021)
6. Carlgren, L., Elmquist, M., Rauth, I.: Design thinking: exploring values and effects from an innovation capability perspective. Des. J. **17**(3), 403–423 (2014)
7. Coronavirus Innovation Map (2021). https://coronavirus.startupblink.com. Accessed 23 Mar 2021
8. Cousins, B.: Design thinking: organizational learning in VUCA environments. Acad. Strateg. Manag. J. **17**(2) (2018)
9. COVID-19 Innovation Hub (2021). https://covid19innovationhub.org/. Accessed 23 Mar 2021
10. Donthu, N., Gustafsson, A.: Effects of COVID-19 on business and research. J. Bus. Res. **117**, 284–289 (2020)
11. Fornasier, C., Demarchi, A.P., Martins, R.: Design Thinking and its visual codes enhanced by the SiDMe model as strategy for design driven innovation. Syst. Des. Beyond Processes Think., 686–699 (2016)
12. Hanna, N.K.: Assessing the digital economy: aims, frameworks, pilots, results, and lessons. J. Innov. Entrep. **9**(1), 1–16 (2020). https://doi.org/10.1186/s13731-020-00129-1
13. Hillner, M., Lim, S.: Design thinking — towards a new perspective. In: Proceedings of Academic Design Management Conference 1–2 August 2018, Design Management Institute of London, pp. 1–19 (2018)
14. Jolak, R., Wortmann, A., Liebel, G., Umuhoza, E., Chaudron, M.R.: The design thinking of co-located vs. distributed software developers: distance strikes again! In: Proceedings of the 15th International Conference on Global Software Engineering (ICGSE 2020), pp. 106–116 (2020)
15. Lattemann, C., Siemon, D., Dorawa, D., Redlich, B.: Digitization of the design thinking process solving problems with geographically dispersed teams. In: Marcus, A., Wang, W. (eds.) DUXU 2017. LNCS, vol. 10288, pp. 71–88. Springer, Cham (2017). https://doi.org/10.1007/978-3-319-58634-2_6
16. McKinsey & Company: The future of work after COVID-19, 18 February 2021. https://www.mckinsey.com/featured-insights/future-of-work/the-future-of-work-after-covid-19. Accessed 23 Mar 2021
17. McKinsey & Company: Next Normal: When will the COVID-19 pandemic end? 26 March 2021. https://www.mckinsey.com/industries/healthcare-systems-and-services/our-insights/when-will-the-covid-19-pandemic-end. Accessed 28 Mar 2021
18. Michelino, F., Caputo, M., Cammarano, A., Lamberti, E.: Inbound and outbound open innovation: organization and performances. J. Technol. Manag. Innov. **9**(3), 65–77 (2014)
19. Mitcheltree, C., Holtskog, H., Ringen, G.: Studying design thinking as a forthcoming source to innovation speed. In: Proceedings of the Design Society: International Conference on Engineering Design, vol. 1, pp. 2357–2366 (2019)
20. Nakata, C., Hwang, J.: Design thinking for innovation: composition, consequence, and contingency. J. Bus. Res. **118**, 117–128 (2020)
21. Parida, V., Westerberg, M., Frishammar, J.: Inbound open innovation activities in high-tech SMEs: the impact on innovation performance. J. Small Bus. Manag. **50**(2), 283–309 (2012)
22. Ramadi, K.B., Nguyen, F.T.: Rapid crowdsourced innovation for COVID-19 response and economic growth. NPJ Digit. Med. **4**(18) (2020)

23. Ramalingam, B., Prabhu, J.: Innovation, development, and COVID-19: challenges, opportunities, and ways forward. In: Innovation, Development, and Covid-19© OECD 2020, pp. 2–13 (2020)
24. Savić, D.: COVID-19 and work from home: digital transformation of the workforce. Grey J. **16**, 101–104 (2020)
25. Serbulova, N., Morgunova, T., Persiyanova, G.: Innovations during COVID-19 pandemic: trends, technologies, prospects. In: Proceedings of Innovative Technologies in Science and Education (ITSE-2020), vol. 210, pp. 1–10 (2020)
26. Stanford Engineering: Human Innovation Design. Current projects: Virtual Reality Simulation of High-growth Innovation Ecosystems (№2) (2021). https://hid.stanford.edu/research/research-example. Accessed 5 May 2021

# Using SAP Predictive Analytics to Analyze Individual Student Profiles in LMS Moodle

Anton N. Ambrajei⬦, Nikita M. Golovin⬦, Anna V. Valyukhova(✉)⬦, and Natalia A. Rybakova⬦

Peter the Great St. Petersburg Polytechnic University,
29 Polytechnicheskaya Street, St. Petersburg 195251, Russia
acc@acc-sap.ru

**Abstract.** The article is devoted to an important today's issue - the analysis of a student's digital footprint. The authors present their experience in research of a large amount of data collected within the Moodle learning platform during their hybrid course of continuing vocational education. The difficulties encountered in processing such data are considered, and a tool for overcoming them is proposed - the SAP Predictive Analytics Desktop software, available to universities for free within the framework of the SAP University Alliance global academic initiative. The result of using this tool to build models in order to find answers to such questions as the probability of successful completion of the course by a student, the influence of various factors on the level of motivation during training, the search for criteria for selecting students for the course, etc. is shown. The mechanism of clustering of trainees using SAP Predictive Analytics is also considered, which can be used to select leaders in educational process with the purpose of their further promotion (admission to internships, participation in projects, etc.). The article may be useful for university lecturers developing and conducting hybrid and online courses, as well as for any researchers interested in the analysis of the educational process.

**Keywords:** Student's digital footprint · Learning data analytics · Educational analytics · E-learning · Hybrid course · SAP S/4HANA

## 1 Introduction

Digital footprint analysis is now becoming one of the most important elements of the educational process [1, 12, 15]. This is especially true for hybrid and online courses, when the immediate reaction of the trainees is not available to the teacher, and every student has the opportunity to choose a strategy for taking the course [14]. Using LMS (Learning Management Systems) or other means of collecting a digital footprint allows the lecturer, in addition to monitoring the success of students, also to understand exactly how certain elements of the course are used, how students allocate their time, how their communication is arranged, how students' behavior affects the effectiveness of the course mastering [2, 3, 17].

In this article, the authors investigated and analyzed the results of the course and individual trajectories of the course by students.

© Springer Nature Switzerland AG 2022
V. Taratukhin et al. (Eds.): ICID 2021, CCIS 1539, pp. 66–77, 2022.
https://doi.org/10.1007/978-3-030-95494-9_6

## 2 Description of the Research Object

The object of research of this article is the digital footprint formed during the hybrid course "SAP S/4HANA Academy 2020", implemented with the direct participation of the authors of this article at Peter the Great St. Petersburg Polytechnic University (SPbPU).

This course is unique in its kind and deserves to be told about it in more detail. The initiator and customer of the course was SAP CIS with the support of the SAP University Alliances. The main goal of the Academy was to train a talent pool of specialists for the SAP partner's and client's ecosystem in Russia and the CIS countries. The creator of the main course, the performer and the coordinator of the program was the International Academic Competence Centre "Polytechnic-SAP" (ACC "Polytechnic-SAP"), National Technology Initiative Centre of Competence "New Manufacturing Technologies" of SPbPU.

The course program was formed in such a way as to make it relevant for young professionals studying in a variety of specialties: information technology, economics, logistics, engineering specialties [10], software development, mathematics and enterprise management. In total, more than 1000 students enrolled for the course, more than half of them (550, to be exact) have successfully completed it. Upon completion of the course, about 80 of its graduates found jobs in the SAP ecosystem, which can be considered a confirmation of the effectiveness of the course [9].

It is important to highlight several strengths of the Academy, which led to its success. First of all, this is, of course, the demand for training personnel with competencies in the field of technological innovations and current digital management technologies. Secondly, it is the long-term experience of ACC "Polytechnic-SAP" SPbPU in the field of successful implementation of complex practice-oriented courses in an online format [6, 7]. And thirdly, an important plus was the participation of partner companies from the SAP ecosystem in the preparation of additional materials for the Academy, which helped to acquaint students with real practical experience from the life of consultants and draw attention to such aspects as communication skills, development of soft skills, building teamwork, etc.

The duration of mastering the course was 108 h, the total duration was 3,5 months. The course combined training materials on the ERP system, webinars and presentations prepared by SAP partners, practice on the SAP S/4HANA system and a set of surveys. All content was included in a single learning environment on the Moodle platform, and all interaction between participants was also carried out through this environment. Therefore, a complete digital footprint of participants for all activities was available.

From the point of view of pedagogical design, the course consisted of two contours – academic training material for the implementation of basic business processes in the SAP S/4HANA ERP system and partner content delivered in the form of webinars and partner assignments. The practice was carried out on a real ERP system using data of a model enterprise.

The partner circuit was divided into hard-skills and soft-skills series related to SAP technologies or consultant skills, respectively.

Students could get grades in several categories: tests, evaluation of business process cases, evaluation for participation in webinars and good questions to speakers, as well as for completing partner tasks. To get a credit or a credit with honors, it was necessary to score a different number of points for each of the categories. At the same time, it was not mandatory to complete all cases or tests.

The course included a large block of onboarding, which helped students to get involved in the learning process and understand the rather complex technical infrastructure of the course. Also an important element was the knowledge base on the frequent mistakes that students make when doing practice; it was interesting to check how students used this element.

## 3   Justification of the Choice of SAP Predictive Analytics

Despite the fact that the need for express-analysis of course logs is almost obvious, there are a very few tools for processing large logs. Traditional office applications are not suitable for these purposes, at least because of the volume of information processed and the lack of predictive analytics tools. [20] Of course, specialized languages such as R [5] or Python can be used for these purposes, but this requires certain programming skills from the lecturer.

The main requirements for the analytical product were as follows:

- Easy to download data. Support of .csv files;
- Ability to work with a large amount of data;
- Advanced visualization tools;
- The possibility of content computing;
- Built-in data processing models (without programming).

The choice fell on the SAP Predictive Analytics package, which is available to universities for free under the SAP University Alliances program.

The basic functions (the Automated Analytics block), in addition to a built-in variety of visualizations, include automatic construction of classification models, regression, clustering, time series, as well as association rules. There is a Data Manager that helps with data preparation. It is also possible to create recommendations based on the designed models.

In the Expert Analytics module, deeper data analysis and integration with the SAP HANA platform and its libraries are possible, as well as the link with the R-language.

## 4   Description of Data Sources

Several datasets were used. The first one contained data about students, such as university, course, level of education, direction of training, presence in the target list from the home university, etc. Universities were grouped into enlarged groups based on such characteristic as the QS ranking. The second dataset contained data on students' academic performance, grades for theory and practice, as well as information on attendance on additional webinars, contacting the course support service, etc. Course logs (Moodle platform records) were also used, which show when and which elements of the online course were attended by students. There were about 900 thousand entries in the log.

The course log has a fairly simple format and includes the user, the date and time of the event, the type of the course element, the element name and a number of service information.

## 5    Visualization of Course Dynamics

First, the dynamics of user actions was analyzed. The figure shows by day the actions of users and the number of unique participants working on the course.

There is a sharp spike in attendance at the beginning of the course, which is due to introductory classes, then the students work fairly evenly and the next rise is observed around the first deadline. The number of unique users varies slightly, but the activity of work increases almost twice, reaching a peak of 80 actions on average per user on the day of the end of the course (Fig. 1).

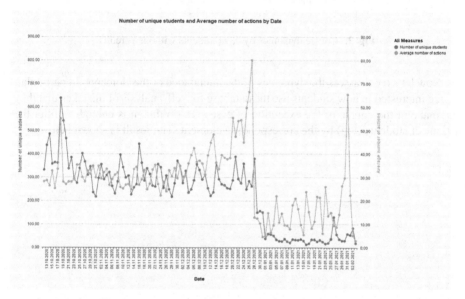

**Fig. 1.** Course dynamics by day (successful completion)

The course was extended for 1 month after its initial deadline, and the graphic shows that there were just a few users during this period, but their activity was very high – they were those who used the holidays to catch up. It can be concluded that the extension of the course helped only about 2% of students, and it is worth canceling such a practice.

We get an even more representative picture if we add a set with performance indicators. Below is the dynamics of actions on the course for a group of students who did not receive a credit (Fig. 2).

It can be seen that the number of active students decreased as they completed the course, i.e. they stopped studying, but individual students showed abnormally high activity during the course completion period.

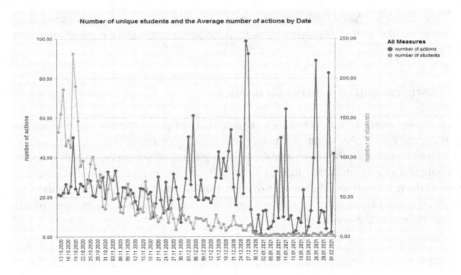

**Fig. 2.** Course dynamics by day (students without a credit)

Now let's try to assess the dynamics of the attendance of the elements. In particular, we are interested in how students use the database for self-analysis of mistakes that they can make in the course of the execution of cases. To do this, it is enough to filter the actions of students (log) by the corresponding element. The result is shown below.

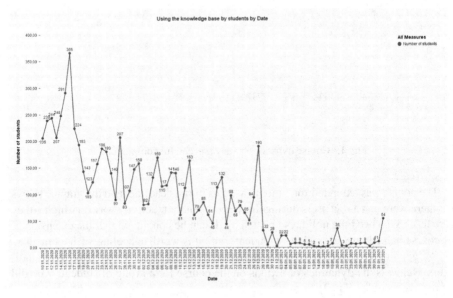

**Fig. 3.** "Knowledge base" usage by student by day

The curve in Fig. 3 contains two distinct peaks, one at the very beginning of the course, when students have just begun to master the course orientation and work with the training system, and the second - just before the deadline, when the lagging students began to complete tasks en masse.

In conclusion of the course dynamic review, let's check how the user actions are distributed over the days of the week. It can be assumed that mostly students study on weekends. This is easy to check, since SAP Predictive Analytics supports calculated variables, including converting a date to a day of the week. As you can see from the graph (Fig. 4), the maximum activity is observed on Sunday, and the minimum on Friday. It should be noted that the activity of students in the ERP system is not taken into account here.

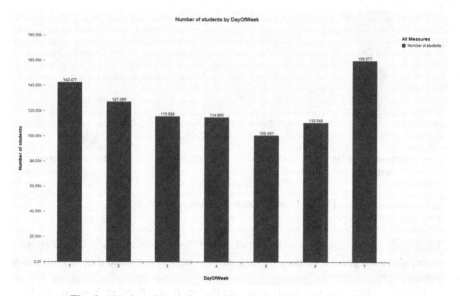

**Fig. 4.** Number of users by day of week (1 = Monday, 7 = Sunday)

This is quite simple, but nevertheless valuable information, since the assignments completed by students are checked manually, and understanding of this dynamic helps teachers to calculate their workload correctly.

## 6  Building a Regression Model

To determine the factors that influenced students' performance, the authors used an automatic regression model available in SAP Predictive Analytics. The whole process is automated, and the analyst's job is to select the target variable and dependent variables. The system itself can discard insignificant variables and show the degree of their influence. A report on the quality of the model is also issued.

For the initial design of the model the following variables we chosen: university, country, level of education (studying for a master's degree or bachelor's degree), direction of study, university group in the QS ranking, the presence of a student in the list controlled by home university curators and additional points for viewing webinars (they were not taken into account for scoring). The target variable – the fact of receiving the credit for the course.

As shown on Fig. 5, the fact of watching webinars significantly affects success, the university and its classification by QS ranking are also important.

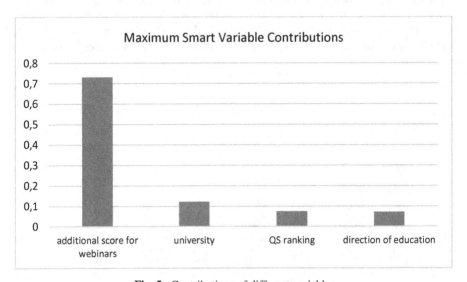

**Fig. 5.** Contributions of different variables

Let's look on the results in more detail. Since the influence of watching webinars rather indicates the general activity of students, the authors excluded this variable from the model.

In this case, the home university acquires the greatest influence. You can see how each of the parameters affects getting a credit. Figure 6 shows that students from a number of universities have a greater chance of completing the course. In the model, the target variable was "no credit" and anything below zero has a positive effect on getting a credit.

It is necessary to highlight that in these home universities there was either a strong curator, or SAP technologies were traditionally well taught.

If we do not take into account a specific home university, it turns out that the direction of study and the ranking of the university becomes significant. So, this fact was put into the model of pre-selection of the next course.

For example, as can be seen in the Fig. 7, students from higher-ranked universities were more likely to complete the course.

But the country and the level of education do not affect the success of students in any way.

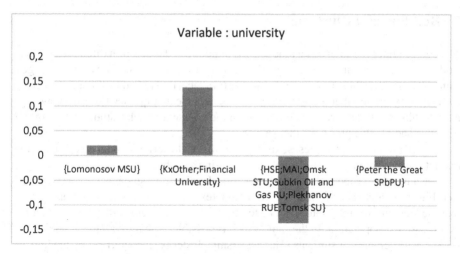

**Fig. 6.** Contributions of specific universities

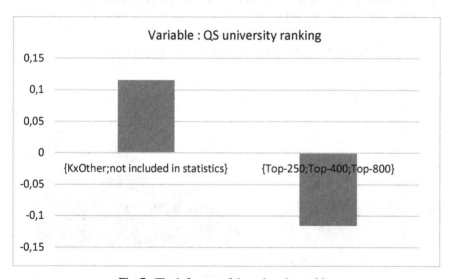

**Fig. 7.** The influence of the university ranking

Such an express analysis may allow the organizers to identify factors that can be used in selecting students for the next course, as well as show which students' motivation requires additional attention [4, 8, 16].

Despite the rather logical conclusions, for this dataset the model turned out to be of average quality and low predictive power, which allows it to be used only for qualitative evaluation.

# 7   Building of a Clustering

The course under research has a practical orientation, so the question often arises which students can be recommended to partner companies of the course. The authors suggest using clustering by three types of assessments (practice, theory and webinar attendance), as well as by the number of user actions on the course and queries to "the knowledge base" – this is a special reference section on the course that helps students to look for mistakes and resolve problems themselves.

In SAP Predictive Analytics, setting up automatic clustering is reduced to specifying restrictions on the number of clusters and selecting variables.

The authors selected 3 clusters that describe the groups of students very well. Below (see Fig. 8) is a cluster of "leaders" in the context of grades in theory, it can be seen that the students who got into this cluster passed the tests significantly better than the average for the population. In the same way it is possible to analyze all the variables and see the achievements of students. The leadership cluster is characterized by the highest academic achievements combined with very high activity on the course and participation in webinars. Such students can be recommended to employers. The remaining clusters were named "norm" and "not-studying".

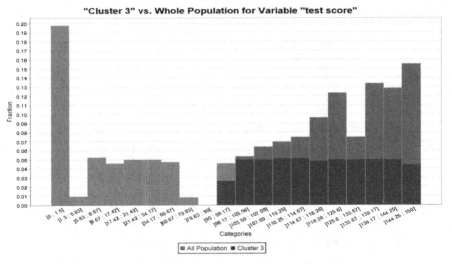

**Fig. 8.** The cluster of leaders compared to the whole population

The cluster of "leaders" is characterized, in addition to good grades, by a general high activity in mastering additional content (viewing webinars) and the ability to deal with errors independently, as evidenced by a large number of queries to the "knowledge base" (see Table 1).

**Table 1.** Characteristics of clusters. Average values by category.

| Cluster | Leaders | Norm | Not-studying |
|---|---|---|---|
| Share[a] | 25,86% | 23,27% | 46,06% |
| Grade for the theory (passing score is 90) | 125 | 110 | 13 |
| Grade for the practice (passing score is 70) | 114 | 104 | 5,26 |
| Extra grades for webinars | 16 | 1 | 1 |
| Actions during the course | 1602 | 1171 | 213 |
| Self-usage of the knowledge base | 16 | 0,76 | 0,9 |

[a]*the model did not assign some of the records.*

## 8  Conclusions and Key Outcomes

Using SAP Predictive Analytics, it is possible to perform an express analysis of an online course. Easy visualization allows to view the dynamics of the course both overall and by individual elements.

The analysis of the dynamics of the course allowed the authors to identify peak areas when students are especially active and need additional support. This applies both to the total number of students and the intensity of their work. The expected maximum activity is at the beginning of the course and before the deadline. This leads us to the conclusion that, despite the efforts of the organizers, students do not know how to apportion their time and capacity properly.

The analysis also showed that the prolongation of the course was used mainly not by those who wanted to finish, but by a small number of students who tried to complete the main volume of the course. The effectiveness of such training is questionable and extending this course after the main deadline for no particular reason is not effective.

The rapid and programming-free creation of regression models allowed to determine the key factors influencing the success of students. However, the available dataset was not enough to build a good predictive model.

In particular, the factors influencing success were identified. This is a specific university and its overall ranking. The assumption that the success is influenced by the country, the level of education (master's or bachelor's degree), as well as the direction of training was refuted because these variables were discarded by the model as not significant.

Clustering of students made it possible to identify clearly defined groups of students according to their characteristics on the course [18].

The course participants were divided into 3 clusters, which are well interpreted in terms of behavioral characteristics. The "Leaders" cluster contains only students who have successfully completed the course and most of them have received the certificate with honors. They are characterized by a high level of activity in non-mandatory educational activities and grades are on average 23% better than the minimum level required for the test. Thus we can say that students consciously wanted to study the material, and were not limited to passing scores. This group also used the "knowledge database" more than others for self-analysis of errors, which can be interpreted as a high level of ability to solve problems independently [11]. This is about 25% of the participants.

The second cluster «Norm» also demonstrates sufficient proactivity and the average score is 18% higher, but their participation in events outside the basic program (additional webinars) is almost zero, as well as the appeal to independent problem-solving tools. These are fairly typical students who have successfully and even with a margin completed the course, but were not ready to invest their time in additional activities or preferred to watch webinars on the record rather than participate in them (this parameter was not recorded in the study).

And the last, largest cluster of «Not-studying» includes those who have not started studying, although they signed up for the course, or dropped out at the very beginning. This is indicated by very low scores and a general lack of activity. Although the dynamics of their actions on the course says that there were those who were very actively trying to fix the situation [21].

Thus, the method of clustering can be recommended in the future for the selection of candidates for potential employers.

In addition, the implementation of an intermediate unloading of learning outcomes and preliminary clustering of students, without waiting for the completion of the course, allows to work with more "precision" with each category of students, identify their difficulties and support them in the process of mastering the course, which in general can have a positive effect on the overall course result [13, 14].

As a general conclusion, SAP Predictive Analytics can be recommended as an effective and relatively simple tool for analyzing a student's digital footprint.

# References

1. Conijn, R., Snijders, C., Kleingeld, A., et al.: Predicting student performance from LMS data: a comparison of 17 blended courses using Moodle LMS. IEEE Trans. Learn. Technol. **10**(1), 17–29 (2017)
2. Terbusheva, E., Piotrovskaya, K., Kalmykova, S.: Educational analytics based on Moodle SDO data. In: New Educational Strategies in the Modern Information Space: A Collection of Scientific Articles Based on the Conferences, St. Petersburg, 09–25 March 2020, pp. 98–104. Russian State Pedagogical University named after A. I. Herzen, St. Petersburg (2020)
3. Sheka, A., Larionova, V., Vasiliev, S., Pevnaya, M.: Early risers fail: behavioral patterns of students of online courses. In: Kulik, E. (ed.) Materials of the International Conference, Moscow, 05–06 December 2018, pp. 195–210. National Research University Higher School of Economics, Moscow (2018)
4. Azcona, D., Hsiao, I.-H., Smeaton, A.F.: Detecting students-at-risk in computer programming classes with learning analytics from students' digital footprints. User Model. User Adapt. Interact. **29**(4), 759–788 (2019)
5. Ambrajei, A., Tereshchenko, V.: Determination of factors affecting the success of the online course. In: Advanced Production Technologies: Computer (Supercomputer) Technologies and the Organization of High-Tech Industries: Collection of Abstracts of the National Scientific and Practical Conference with International Participation. SPb, Publishing house Polytech-Press (2021). 88 p.
6. https://www.vedomosti.ru/press_releases/2021/10/06/besplatnaya-obrazovatelnaya-progra mma-sap-s4hana-academy-dlya-studentov-proidet-vo-vtoroi-raz. Accessed 15 Oct 2021

7. Ambrajei, A., Golovin, N., Valyukhova, A., Rybakova, N., Zorin, V.: Experience in conducting a multi-format course on SAP technologies. In: XXI Proceedings of the International Conference on Digital Education, pp. 41–48 (2020)
8. Scherzinger, F., Singla, A., Wolf, V., Backenkohler, M.: Data-driven approach towards a personalized curriculum. In: Proceedings of the 11th International Conference on Educational Data Mining (EDM2018) (2018)
9. https://news.sap.com/cis/2021/03/итоги-sap-s-4hana-academy/ Results of the "SAP S/4HANA Academy": more than 50 graduates found a job in SAP ecosystem, SAP CIS Media (2021). Accessed 15 Oct 2021
10. Zotova, M., Likhouzova, T., Shegai, L., et al.: The use of MOOCS in online engineering education. Int. J. Eng. Pedagogy (iJEP) **11**(3), 157–173 (2021)
11. Klimenskikh, M.V., Maltsev, A.V., Lebedeva, Ju.V., Kaur, H.: Cognitive and emotional predictors of learning success on an online course. In: Cognitive Neuroscience - 2020: Materials of the International Forum, December 11–12, 2020. Yekaterinburg: Ural Federal University named after the first President of Russia B.N. Yeltsin, pp. 272–276. (2021)
12. Belonozhko, P.P., Karpenko, A.P., Khramov, D.A.: Analysis of educational data: directions and prospects of application. "Naukovedenie" Internet J. **9**(4), 57 (2017)
13. Hart, S.A.: Precision education initiative: moving towards personalized education. Mind Brain Educ. **10**(4), 209–211 (2016). https://doi.org/10.1111/mbe.12109
14. Yang, C., Chen, I., Hiroaki, O.: Toward precision education: educational data mining and learning analytics for identifying students' learning patterns with ebook systems. J. Educ. Technol. Soc. **24**(1), 152–163 (2021). 12p.
15. Marcu, D., Danubianu, M.: Learning analytics or educational data mining? This is the question... BRAIN Broad Res. Artif. Intell. Neurosci. **10**(Spec. Issue), 1–14 (2019). 14p.
16. Hu, Q., Rangwala, H.: Towards fair educational data mining: a case study on detecting at-risk students. In: International Educational Data Mining Society, Paper presented at the International Conference on Educational Data Mining (EDM), 13th Online, July 10–13, 2020 (2020). 7 p.
17. Namoun, A., Alshanqiti, A.: Predicting student performance using data mining and learning analytics techniques: a systematic literature review. Appl. Sci. (2076–3417) **11**(1), 237–237 (2021). 1p.
18. Martinez Navarro, A., Moreno-Ger, P.: Comparison of clustering algorithms for learning analytics with educational datasets. Int. J. Interact. Multimedia Artif. Intell. **5**(2), 9–16 (2018). ISSN: 1989-1660
19. Viberg, O., Hatakka, M., Balter, O., Mavroudi, A.: The current landscape of learning analytics in higher education. Comput. Hum. Behav. **89**, pp. 98–110 (2018)
20. Lerche, T., Kiel, E.: Predicting student achievement in learning management systems by log data analysis. Comput. Hum. Behav. **89**, pp. 367–372 (2018)
21. Ademi, N., Loshkovska, S.: Early detection of drop outs in e-learning systems. In: Conference Proceedings of the International Symposium on Innovative Technologies in Engineering & Science, pp. 1008–1015 (2019). 8p.

# Coordination of Models for Describing and Making Agro-Technological Decisions

Felix Ereshko$^{(\boxtimes)}$ and Vladimir Budzko

Federal Research Center "Computer Science and Control" of the Russian Academy of Sciences, Vavilova 44-2, 119333 Moscow, Russian Federation

**Abstract.** The analysis of transformations of ideas and management practices in modern conditions of the formation of digital economy models and information support in the agro-industrial complex is carried out. Since the main generator of data affecting the final result of production is external natural conditions, the description of the initial processes is carried out using an agrophysical model of plant growth. The formulation of strategic management tasks in the specified conditions is formulated, a scheme for combining agrophysical and economic descriptions is substantiated, mathematical decision-making models and fundamental simulation computational procedures are considered.

**Keywords:** Agro-industrial complex · Strategic management and control · Agrophysical description · Economic conditions · Data · Optimization · Models · Computational procedures

## 1 Introduction

The agro-industrial complex of the country is one of the most important industries, therefore, the digitalization of this industry is of great importance. This process is in line with global trends in precision farming.

Digital technologies based on hardware, software and networks are not an innovation, but every year moving further and further from the third industrial revolution, they are becoming more sophisticated and integrated, causing the transformation of society and the global economy [1].

According to this opinion, decision-making in economics will be based on computing platforms reflecting individual functional industries, which is very close to the ideas of the cybernetic school in the theory of management and management, and in particular, to the works of A. I. Kitova and V. M. Glushkov [2].

The main thesis of the Russian school was that the models of interaction of active elements are the initial framework of the platforms, and it is necessary to involve the theoretical constructions of control mechanisms in solving the problems of making coordinated decisions [3, 4].

## 2   Generation of Data in the Agro-Industrial Complex

In the agro-industrial complex (AIC), natural conditions are the main generator of information, and the main resource is land.

Examples of entities for which it is necessary to collect data to solve management problems in agricultural projects can be: different types of land, labor and water resources, technological equipment, construction materials, construction equipment, fuel, energy, chemical fertilizers, agricultural machinery, capital investments, operating costs, etc. The output factors of projects are products produced in agriculture: cereals, vegetables, meat, milk, etc.

The main research topics in AIC: gene pool, plant protection, variety testing, crop programming, agrochemical maintenance, the state of land resources, breeding, scientific and technical information, etc.

The regional factor is of great importance in the digital economy of AIC.

The development of a general program for the development and functioning of large agricultural regions is a very complex and multifaceted problem, in its analysis one has to take into account a large number of closely related natural, economic, social, organizational, environmental and other factors [5].

At the same time, a systematic approach presupposes a comprehensive analysis of the complex problem under study, including both informal and formalized assessment methods and combining the advantages of rigorous research based on mathematical models, simulation based on computers and the knowledge of specialists in specific areas.

The characteristics of these factors can act both in the form of resource constraints and in the form of indicators, which should be improved.

Based on the results of studies of plant growth processes, we present a generalized record of dynamic relationships and describe the agreement between agrophysical and economic models.

Let us give the PCP-Model [5, 6], as an example of an agrophysical description, which describes the growth of plants with a time step of ten days and the transformation of soils with a time step of one year. It is enough to consider the dynamics of the process with a step of one year to formulate management tasks, since it is with this step that the main economic decisions are made. Let us introduce phase variables, control variables and parameters of the PCP-Model into the description. The symbol t stands for the number of the year and refers to its end. Consider a certain area of the region with homogeneous characteristics, the meaning of which is described below.

### 2.1   Phase Variables

The vectors that make up the vectors of phase variables for this section:

- $Ph^t$ – vector of variable physical characteristics, the components of which are: thickness, porosity, density of soil horizons;
- $Ch^t$ – vector of variable chemical characteristics, the components of which are for each of the three soil horizons: organic matter content in the soil, nitrogen content,

acidity, concentration of available inorganic phosphorus, concentration of available potassium, soil quality (ratio of carbon to nitrogen);
- $Oh^t$ – vector of variable characteristics of soil organic matter;
- $Ph^t$ – vector of variables characterizing soil moisture for soil horizons.
- The vector of phase variables is represented as $z^t = (Ph^t, Ch^t, Or^t, Ws^t)$.

## 2.2 Control Variables

The control vectors of the model include:

- $(N^t, P^t, K^t)$ – vector of volumes of nitrogen, phosphorus, potash fertilizers applied per year;
- $O^t$ – vector characterizing the use of organic fertilizers, the components of which are: amount of applied organic fertilizers, decade of fertilization, structure of fertilizers;
- $W^t$ – vector of variables characterizing water consumption for irrigation systems: total volume of water available for consumption, maximum throughput of the irrigation system;
- $C^t$ – the number of the crop cultivated in this area;
- $A^t$ – vector of agrotechnical measures, the components of which are determined by the number of the culture Ct, the type of land cultivation by its characteristics.

Thus, the control vector has the form: $u^t = (N^t, P^t, K^t, O^t, W^t, C^t, A^t)$.

## 2.3 Uncontrollable Factors

The vector of uncontrollable factors $\xi^t$ is determined by weather conditions and consists of a series of mean-decade values throughout the year:

- air temperatures,
- relative air humidity,
- wind speeds,
- the number of hours of sunshine,
- the amount of precipitation.

## 2.4 Model Parameters

The parameters of the model are the physical characteristics of the soil.

The dynamics of phase states in the PCP-Model for a given section of the region's territory is determined in the form of a controlled dynamic system with undefined parameters $z^t = F(z^{t-1}, u^t, \xi^t, p)$. In this case, for the section under consideration, the output vector $y^t$ is calculated, the components of which are the output of the main and by-products, $y^t = \Psi(z^{t-1}, N^t, P^t, K^t, O^t, W^t, C^t, \xi^t, p)$.

The model also calculates water erosion et of lands, the level of which is determined by the phase state, crop number, agrotechnical measures, so that in general it is possible to write down a general simulation system for the region.

Output, phase state, and soil erosion obtained in the agrophysical PCP model are inputs to the economic and decision-making blocks.

We divide the territory of the entire region into a set L of homogeneous areas, numbering them with the index l, so that $\sum_l s_l = S$, where $s_l$ is the area of the site number l, S is the total area of the region. The homogeneity of a site means that the physical, chemical, and other characteristics that appear in the PCP model and the economic model are the same within it.

In addition, we will assume that only one technology h for crop c is used in this area. Under this condition, the set of homogeneous sections L remains unchanged at all times t, while without this condition the number of homogeneous sections can vary from t to t + 1. The set of numbers of areas occupied by the technology h of culture c at time t is denoted, then the set of numbers of areas occupied by culture c, $L_{ch}^t = \bigcup_h L_{ch}^t$, and

$$L = \bigcup_c L_c^t.$$

The number of elements in the set L is determined, on the one hand, by the homogeneity of soils in terms of physical and chemical characteristics and, on the other hand, by economic considerations, since sufficient representativeness of technologies and crops is required to ensure, for example, a given production output. Therefore, the dimension of the set L can be quite significant.

In this case, the general model for the entire region will look like

$$z_l^t = F\left(z_l^{t-1}, u_l^t, \xi_l^t, p_l\right) \tag{1}$$

$$v_l^t = \Psi\left(z_l^{t-1}, N_l^t, P_l^t, K_l^t, O_l^t, W_l^t, C_l^t, \xi_l^t, p_l\right) \tag{2}$$

where $y_l^t$ – release of products per unit area of a site with a number l, $e_l^t = \Phi\left(z_l^{t-1}, u_l^t, \xi_l^t, p_l\right)$.

## 2.5 Economic Model

$\sum_l s_l y_l^t = y^t$ – production vector; $y^{t,j}, j \in J_1$ – production of main products; $y^{t,j}, j \in J_2$ – release of by-products.

Let's designate through $d^{t,j}$ the part of the main product that goes to feed. By-products are completely used for feed. Then the release of feed with number v is determined

$$\sum_{j \in J_1} \beta_j^v d^{t,j} + \sum_{j \in J_2} \gamma_j^v y^{t,j} = d^{t,v}, \tag{3}$$

where the coefficients $\beta_j^v$, $\gamma_j^v$ determine the output of the product j into the feed with the number v. $\sum_{ch} \sum_{l \in L_{ch}^t} r_{ch}^k s_l = r^{t,k}$ – request in the k-th resource, $k \in K$, $r_{ch}^k$ is the consumption of the k-th resource by the technology h of the crop c during cultivation per unit area, K is the set of indices for the resources, the specific indicators for which have been prepared expertly: electricity, fuel, pesticides, vehicles.

$\sum\limits_i k_i^{\nu} g_i^t$ – demand in feeds of the $\nu$-th type, $k_i^{\nu}$ is the specific consumption of $\nu$-th feed by the $i$-th type of animals, $g_i^t$ – the number of structural units of animals of the $i$-th type (cattle, pigs, sheep, poultry) in the region.

The relationships between the economic model and the PCP model together constitute a description of the general simulation model.

Simulation experiments.

The controls in this model are sets $L_{ch}^t$ and vectors $N_l^t, P_l^t, K_l^t, O_l^t, W_l^t, A_l^t, g_l^t$.

By setting various scenarios for choosing these controls, and calculating step by step (integrating) the general system, we will receive series of values for the quantities $y^t, r^{t,k}, b^t, a^t, e^t$ on which we can calculate the values of indicators of interest to experts. We will focus here on the following quantitative indicators:

- output of consumer products in a given proportion (or gross output of consumer products),
- soil erosion,
- imbalance in the demand and release of feed.

As a rule, the possible amounts of fertilizer and water are limited and therefore experts need to choose controls provided that

$$\sum_l N_l^t \le F^{1,t}, \sum_l P_l^t \le F^{2,t}, \sum_l K_l^t \le F^{3,t}, \sum_l Q_l^t \le F^{4,t}, \sum_l W_l^t \le W^t. \quad (4)$$

where the values $F^{1,t}, F^{2,t}, F^{3,t}, F^{4,t}, W^t$, as well as $L$, are set at the beginning of the experiments.

It is very difficult to pose problems of an optimization or multicriteria nature under these conditions, due to the large dimension of the model and the discrete nature of the controls, since one of the control components is a set of sets $L_{ch}^t$.

### 2.6   Crop Version of the Model

To facilitate scripting in GM, we introduce the following simplifying condition. Let us assume that for each site of the territory numbered $l$ there is such an initial state $z_l^0$ of the soil, such a time interval $T$, such a sequence of controls $u_l^t$ that under certain stationary weather conditions $\xi_l^*$ (for all $t \in T$) the final state of the soil coincides with the initial one, i.e.

$$z_l^T = \tilde{F}\left(z_l^{T-1}, z_l^{T-2}, ..., z_l^0, u_l^T, ..., u_l^1, \xi_l^*, ..., \xi_l^*, p_l\right) = z_l^0 \quad (5)$$

We use this property of periodicity as follows. We split the given segment $l$ into $T$ identical parts and implement the given sequence of controls on each of them. Let $C = \left(c^1, ..., c^T\right)$ be the corresponding sequence of cultures. We take the initial state for each part so that at the moment of time $t = 1$ the initial phase state of the part of the number $i$, $i = 1, 2, ...T$ is $z_l^{i-1}$ and on it the culture $c_i \in C$ is located. Then the phase state of the site $l$ at the moment of time $t = 1$ will be $\left(z_l^0, z_l^1, ..., z_l^{T-1}\right)$.

With such an organization of production, at each moment of time $t$, all cultures from the set $(c^1, c^2, ..., c^T)$ will be present in the section of the room. This organization of production will be called crop rotation. Crop rotations are widespread in practice, and for computational experiments, the corresponding sets can be extracted from the relevant literature.

In this case, the ratios describing the transformation of soils on the room number $l$ remain the same for all its parts. Therefore, we will formulate the general relations of agrophysical dynamics for the part for which the initial culture is $c^1$ and the initial state is $z_l^0$.

$$z_n^t = F(z_n^{t-1}, u_n^t, \xi^t, p) \tag{6}$$

$$y_n^t = \Psi(z_n^{t-1}, N_n^t, P_n^t, K_n^t, O_n^t, W_n^t, c_n^t, \xi^t, p) \tag{7}$$

$$e_n^t = \Phi(z_n^{t-1}, u_n^t, \xi^t, p) \tag{8}$$

where the index $n$ means the number of the crop rotation.

Let the territory of the region be divided into $L$ plots for which the sets of crop rotations are adopted: $N_l^1$ – for irrigation and $N_l^2$ – for rainfed lands. Let us denote $x_n^{1,l}$ – the area occupied by the crop rotation of the number $n$ on the plot of the number $l$ during irrigation, and $x_n^{2,l}$ – the area occupied by the crop rotation $n$ on the plot of the number $l$ in rainfed conditions. Let us assume that only one production technology is associated with each crop rotation.

Let $y_n^{1,l}$ be the vector of production on irrigated lands, $y_n^{2,l}$ – the vector of production on rainfed lands. Then vector of production in the region

$$\sum_l \left( \sum_{n \in N_l^1} y_n^{1,l} x_n^{1,l} + \sum_{n \in N_l^2} y_n^{2,l} x_n^{2,l} \right) = y. \tag{9}$$

Resource request

$$\sum_l \left( \sum_{n \in N_l^1} r_n^{1,l} x_n^{1,l} + \sum_{n \in N_l^2} r_n^{2,l} x_n^{2,l} \right) = r^k, k \in K, \tag{10}$$

limitation on irrigation of the plot number $l$

$$\sum_{n \in N_l^1} x_n^{1,l} \leq S_l^1, \tag{11}$$

general restriction of the plot number $l$

$$\sum_{n \in N_l^1} x_n^{1,l} + \sum_{n \in N_l^2} x_n^{2,l} \leq S_l, \tag{12}$$

request for feed in the region

$$\sum_i k_i^v g_i = b^v, \tag{13}$$

production of livestock products under the number $m$

$$\sum_i a_{im} g_i = a^m. \tag{14}$$

## 3   The Task of Making a Decision

The formulation of the auxiliary crop rotation problem is closed if the rules for the choice of control variables are formulated. Note that due to the introduction of crop rotations, we have passed from a discrete control problem to a continuous one. Nevertheless, the possibilities of an exact solution of optimization or multicriteria problems are very limited in this case; therefore, we will use the approach of scenario modeling and heuristic decomposition.

Let us describe a way to use optimization models to filter out non-rational options and build scenarios for a general simulation model.

We will consider the problem of increasing the output of consumer products in a given proportion (or increasing gross output) at a given level of soil erosion and imbalance in feed.

Let's build a discrete set of technologies on the PCP model and move on to the economic block with these technologies. We will obtain an estimate of the given criteria by solving an auxiliary linear programming problem. Thus, we will make an approximate decomposition of the crop rotation problem.

A set of technologies will be obtained by solving the general system with a discrete set of possible volumes of fertilizers $N, P, K, O$ and water $W$ for different crop rotations $c_n$ for given sequences $\xi^1, ..., \xi^{T_n}$ reflecting the experts' views on the uncertainty of weather conditions. The following table gives an idea of the procedure for generating technologies for model decomposition.

From here we get: the values for the consumption of fertilizers and water

$$f_n^1 = \frac{1}{T_n} \sum_t N^t, f_n^2 = \frac{1}{T_n} \sum_t P^t, f_n^3 = \frac{1}{T_n} \sum_t K^t, f_n^1 = \frac{1}{T_n} \sum_t O^t, v_n = \frac{1}{T_n} \sum_t W^t, \tag{15}$$

soil erosion

$$e_n = \frac{1}{T_n} \sum_t e^t \tag{16}$$

and yield

$$y_n = \frac{1}{T_n} \sum_t y^t, \tag{17}$$

**Table 1.** Data in computational experiments.

| Phase state | $z$ | $z^0$ | $z^1$ | ... | $z^{t-1}$ | ... | $z^{T_n-1}$ |
|---|---|---|---|---|---|---|---|
| product yield | $y$ | $y^1$ | $y^2$ | ... | $y^i$ | ... | $y^{T_n}$ |
| culture number | $c$ | $c^1$ | $c^2$ | ... | $c^i$ | ... $c^{T_n}$ | |
| weather | $\xi$ | $\xi^1$ | $\xi^2$ | ... | $\xi^i$ | ... | $\xi^{T_n}$ |
| amount of applied nitrogen fertilizers | $N$ | $N^1$ | $N^2$ | ... | $N^i$ | ... | $N^{T_n}$ |
| phosphoric | $P$ | $P^1$ | $P^2$ | ... | $P^i$ | ... | $P^{T_n}$ |
| potash | $K$ | $K^1$ | $K^2$ | ... | $K^i$ | ... | $K^{T_n}$ |
| organic fertilizers | $O$ | $O^1$ | $O^2$ | ... | $O^i$ | ... | $O^{T_n}$ |
| water volume | $W$ | $W^1$ | $W^2$ | ... | $W^i$ | ... | $W^{T_n}$ |
| erosion | $e$ | $e^1$ | $e^2$ | ... | $e^i$ | ... | $e^{T_n}$ |

which will be included as specific indicators in the technological processes of the economic block.

Comment. The volumes of water cannot be specified, but obtained from the calculations as the volumes of the request of plants in water.

Let us now supplement the ratio of the economic model with fertilizer requests

$$\sum_l \left( \sum_{n\in N_l^1} f_n^{1,k} x_n^{1,l} + \sum_{n\in N_l^2} f_n^{2,k} x_n^{2,l} \right) = f^k, k = 1, 2, 3, 4 \qquad (18)$$

Where $f_n^{1,k}, f_n^{2,k}$ – specific indicators of consumption of nitrogen $k = 1$, phosphorus $k = 2$, potassium $k = 3$, organic fertilizers $k = 4$ for irrigated and non-irrigated crop rotations, obtained as described above, and for water

$$\sum_{n\in N_l^1} v_n x_n^{1,l} \le W^l \qquad (19)$$

Let's define the total soil erosion in the region

$$l = \sum_{l,n} \left( e_n^1 x_n^{1,l} + e_n^2 x_n^{2,l} \right). \qquad (20)$$

Now let us formulate an auxiliary optimization problem

$$\max \min \left[ \min_{\substack{j\in J}} \frac{y^j - d^j}{Y^j}, \min_{m\in M} \frac{a_m}{A_m}, \frac{E - e}{E} \right] \qquad (21)$$

where $Y^j$ is a given level of crop production, $A_m$ is a given level of livestock production, $E$ is a given level of erosion with restrictions on land and restrictions on resources higher:

This task is reduced to a linear programming task:

$$\max p$$
$$y^j - d^j \geq pY^j, j \in J; a_m \geq pA_m, m \in M,$$
$$E - e \geq pE$$

(22)

under the described resource constraints.

Experimental calculations were carried out on data from work [5].

# 4  Linear Case [8]

To illustrate the general provisions, we present the corresponding constructions for the case that is most adequate for agricultural situations, taking into account the following [5]:

- the fundamental presence of uncertainty caused primarily by natural factors,
- the effect of decentralization, which is caused by the prevailing market relations,
- the use of specific indicators (standards), which corresponds to common economic practice.

Let there be active agents $n$. We will denote them by numbers from 1 to $n$.

Each agent can produce $m$ types of products, while spending some resources. The number of resources will be denoted by a letter $k$. For the production of a unit of a product of a type $j$, an agent $i$ spends a resource of a type $l$ in quantity $p^i_{lj}$. The agent $i$ has its own stock of a resource of the type $l$ in the amount $b^i_l$ ($l = 1, 2, ..., k$).

In addition, there are general stocks of resources in quantity $r_l$ ($l = 1, 2, ..., k$).

Products of the type $j$ can be sold on the market at a price $c_j$.

Thus, if an agent $i$ produces products of a kind $j$ in quantity $x^i_j$ ($j = 1, 2, ..., m$), then he will spend a resource of a kind $l$ in quantity $p^i_{l1}x^i_1 + p^i_{l2}x^i_2 + ... + p^i_{lm}x^i_m$, and he will be able to sell these products for an amount $c_1x^i_1 + c_2x^i_2 + ... + c_mx^i_m$.

Prices $c_j$ are positive in their meaning. We assume that the cost coefficients are non-negative $p^i_{lj}$, and for any $i$ and every $j$ at least one coefficient $p^i_{1j}, p^i_{2j}, ..., p^i_{kj}$ is positive. Stocks $b^i_l$ and $r_l$ are assumed to be non-negative.

Further, it will be convenient to use the following matrix notation. Will denote $x^i$ a column vector $(x^i_1, x^i_2, ..., x^i_m)^T$ through $x^i$ (superscript $T$, as usual, denotes transposition). Let $c = (c_1, c_2, ..., c_m)$, and

$$P^i = \begin{pmatrix} p^i_{11} & p^i_{12} & \cdots & p^i_{1m} \\ p^i_{21} & p^i_{22} & \cdots & p^i_{2m} \\ \cdots & \cdots & \cdots & \cdots \\ p^i_{k1} & p^i_{k2} & \cdots & p^i_{km} \end{pmatrix}.$$

(23)

Let denote $b^i = (b^i_1, b^i_2, ..., b^i_k)^T$, $r = (r_1, r_2, ..., r_k)^T$.

In this notation, the previous formulas will look like this. If the agent $i$ releases products, in quantity $x^i$, then he will be able to bail out the amount $cx^i$ for it, and at the same time resources in the quantity $y^i = P^i x^i$ will be spent, where $y^i$ is the vector column $(y_1^i, y_2^i, ..., y_k^i)^T$.

Let $Y$ be the set of all sets of vectors $y^1, y^2, ..., y^n$ satisfying the conditions

$$y^1 + y^2 + ... + y^n \leq r, y^i \geq 0, \ i = 1, 2, ..., n \tag{24}$$

(hereinafter, vector inequalities must be satisfied componentwise).

In this setting, we assume that the Center does not know exactly the technological matrices $P^i$ and the agents' own reserves of resources $b^i$. It is only known that they belong to parametric families $P^i(\alpha)$ and $b^i(\alpha)$, where $\alpha$ they belong to some set $A$. For simplicity, in this model, we will assume that the stocks $r_1, r_2, ..., r_k$ are strictly positive.

Agents, on the other hand, know for sure their own technologies and capabilities.

We will assume that the Center disposes of the division of "general" resources, but in this case it allocates resources for specific production programs. Agents have the right to choose these programs on their own. Of course, the center is not obliged to ensure the feasibility of any program proposed by the agent. On the contrary, the agent is forced to choose his program based on the resources allocated to him. In addition, agents can transmit information to the Center about the realized value of the undefined factor. These reports are not necessarily reliable and the Center is aware of this.

These considerations are formalized as follows.

The center chooses a collection $y_* = (y_*^1, y_*^2, ..., y_*^n)$ of functions $y_*^i : R_+^m \times A \to R_+^k$ such that

$$y_*^1(x^1, \beta^1) + y_*^2(x^2, \beta^2) + ... + y_*^n(x^n, \beta^n) \leq r \tag{25}$$

for any plans $x^1, x^2, ..., x^n$ from $R_+^m$ and any messages $\beta^1, \beta^2, ..., \beta^n$ from $A$. The set of all such sets will be denoted by $Y_*$.

After that, the $i$-th agent ($i = 1, 2, ..., n$) selects a vector $x^i$ and a message $\beta^i$ from the set

$$X^i(y_*^i, \alpha) = \left\{ \left( z^i, \beta^i \right) \in R_+^m \times A : P^i(\alpha) z^i \leq b^i(\alpha) + y_*^i(z^i, P^i(\alpha) z^i) \right\}. \tag{26}$$

After these choices have been made, the Center receives a payoff

$$g(x^1, x^2, ..., x^n) = cx^1 + cx^2 + ... + cx^n, \tag{27}$$

and the $i$-th agent receives a payoff $h^i(x^i) = c^i x^i$.

Thus, we get a game with forbidden situations $\langle N, Y_*, X^1, X^2, ..., X^n, g, h^1, h^2, ..., h^n, A \rangle$, where $N = \{C, 1, 2, ..., n\}$ is the set of players (the symbol $C$ is reserved for the Center, and the agents, as before, are numbered from 1 to $n$). Let's use the notation $\Gamma_*$ for this game.

To close the model, it is necessary to describe the relationship of the Center to uncertainty. As before, we will assume that the Center has the right of the first move; it considers all agents to be rational and is careful with respect to the remaining uncertainty.

We formalize what has been said as follows.

Let $H^i(y_*^i, \alpha)$ be the least upper bound for the values of the function $h^i(x^i) = c^i x^i$ on the set $(x^i, \beta^i)$ of pairs satisfying the conditions

$$P^i(\alpha) x^i \leq b^i(\alpha) + y_*^i \left( x^i, \beta^i \right), x^i \geq 0. \tag{28}$$

If this upper bound is attained, then we define the set $BR^i(y^i_*, \alpha)$ as the set of pairs $(x^i, \beta^i)$ satisfying the conditions

$$c^i x^i = H^i(y^i_*, \alpha) P^i(\alpha) x^i \le b^i(\alpha) + y^i_*(x^i, \beta^i), x^i \ge 0. \tag{29}$$

Otherwise, we define the set $BR^i(y^i_*, \alpha)$ by the conditions

$$c^i x^i \ge H^i(y^i_*, \alpha) - \kappa P^i(\alpha) x^i \le b^i(\alpha) + y^i_*(x^i, \beta^i), x^i \ge 0, \tag{30}$$

where $\kappa$ is a given positive number.

Then the maximal guaranteed result of the Center $R(\Gamma_*)$ is equal to

$$\sup_{(y^1_*, y^2_*, \dots, y^n_*) \in Y_*} \min_{\alpha \in A} \min_{((x^1, \beta^1), (x^2, \beta^2), \dots, (x^n, \beta^n)) \in BR^1(y^1_*, \alpha) \times BR^2(y^2_*, \alpha) \times \dots \times BR^n(y^n_*, \alpha)} \left( cx^1 + cx^2 + \dots + cx^n \right). \tag{31}$$

The resulting expression reflects the fact that the Center, choosing its program of action, focuses on some guaranteed payoff, the uncertainty of which in the initial expression is determined by the multiple choice of many participants and the uncertain external situation.

Direct calculation of the quantity $R(\Gamma_*)$ is quite difficult, but by now a set of effective tools for solving problems of this class has been developed, either by reducing them to optimization problems or by using the scenario approach and simulation experiments.

## 5 Conclusion

This article reflects the first phase of research for agro-production when ontologically oriented models of a conceptual nature are created. This will be followed by a practical stage of research, which is described at the end of this conclusion. The article demonstrates an approach to the formation of digitized decision-making technologies in one of the sections of the agro-industrial complex associated with the production of crop products. A systematic approach to the problems of decision-making provides for the development of conceptual models of the controlled object, the assessment of the data obtained, the formalization of the control problem, the development of algorithms for solving the assigned problems. All these stages are reflected in the proposed text. The essential point of the presentation is that the technological data are generated on the basis of models of agrophysical description, and the normative data for the economic model on the use of production resources are prepared by experts. Further, the investigation work will be followed by a research stage, which includes a series of computational experiments on a selected object, an analysis of the results obtained with decision-makers, and a stage of using the proposed solutions.

**Acknowledgement.** This work was supported by a grant from the Ministry of Science and Higher Education of the Russian Federation, internal number 00600/2020/51896, agreement No. 075-15-2020-914.

# References

1. Schwab, K.: The Fourth Industrial Revolution. Preface Gref G. O. Eksmo, M. (2016). 138 p.
2. Glushkov, V.M.: Macroeconomical models and principles of construction of the OGAS. Statistics, M. (1975). 160 p.
3. Moiseev, N.N.: Mathematical Problems of System Analysis. Nauka, M. (1981). 488 p.
4. Germeyer, Yu.B.: Introduction to the Theory of Operations Research. Nauka, M. (1971). 384 p.
5. Ereshko, F.I.: System analysis in the Stavropol project of agricultural management. Vestn. s. - kh. nauki (1), 40–49 (1984)
6. Frans, J., Thornley, J.H.: Mathematical Models in Agriculture. Agropromizdat, M. (1987). 400 p.
7. Ereshko, F.I., Lebedev, V., Parikh, K.S.: Decision-making and simulation strategies for the system of models for agricultural planning of the Stavropol region: mathematical description. In: IIASA, Laxenburg, Austria, WP-83-93, pp. 1–21 (1983)
8. Gorelov, M., Ereshko, F.: On models of centralization and decentralization of control in a digital society. In: Ivanov, V.V., Malinetsky, G.G., Sirenko, S.N. (eds.) The contours of digital reality: The humanitarian and technological revolution and the choice of the future, pp. 187–202. Lenand, Moscow (2018)
9. Ereshko, F., Gorelov, M.: Information and hierarchy. In: Recent Advances of the Russian Operations Research Society, pp. 2–28. Cambridge Scholars Publishing, Cambridge (2020)
10. Budzko, V., Ereshko, F., Gorelov, M.: Mathematical models of control in digital economy platforms. In: 2020 Annual International Conference on Brain-Inspired Cognitive Architectures for Artificial Intelligence (BICA*AI 2020), the Eleventh Annual Meeting of the BICA Society, Natal, Brazil, held on October 10–11 and November 10–15 as a virtual-only event (2020)
11. Gorelov, M.A.: On a quantity of information required for efficient control. Large Scale Syst. Control **88**, 41–68 (2020)
12. Budzko, V.I., Gorelov, M.A., Ereshko, F.I.: Models of decision making with limited volume of processed information. In: Proceedings of XIII «Data analytics and management in data intensive domains» Conference (DAMDID), MISIS, Moscow (2021)
13. Pospelov, I.G.: Variational principle in the description of economic behavior. In: Mathematical Modeling: Processes in Complex Economic and Ecological Systems, pp. 148–163. Nauka, Moscow (1986)
14. Budzko V.I., Ognivcev S.B., Ereshko F.I., Shevchenko V.V.: On the system of models for supporting decision-making in the agro-industrial complex. Plenary report. In: Sigal (ancestry), A.V., et al. (eds.) Analysis, Modeling, Management, Development of Socio-Economic Systems: A Collection of Scientific Works of the XV All-Russian with the International Participation of the School-Symposium AMUR-2021, 14–27 September 2021/ed. Council, pp. 77–81 (2021). ISBN 978-5-6046168-1-9

# Virtual Reality-Built Prototype as a Next-Gen Environment for Advanced Procurement Reporting and Contract Negotiation

Artem Levchenko[1]([✉]) [iD], Vyacheslav Ivanov[2] [iD], and Victor Taratukhin[3] [iD]

[1] Voronezh State University, Voronezh, Russia
artem.levchenko@sap.com
[2] Higher School of Economics, Moscow, Russia
[3] University of Muenster, Münster, Germany

**Abstract.** Lockdowns and remote working styles create demand for specialized software for business communication. This article describes the design and results of user testing of a Virtual Reality (VR) collaborative environment for advanced procurement reporting and contract negotiation between a supplier and a buyer. We have designed and implemented a VR application that allows analyzing operational, tactical, and strategic procurement reports in a virtual room, meeting with a supplier in the virtual café, discussing the contract data, demonstrating the 3D model of the goods, and signing the contract. The prototype included integrated procurement software SAP Ariba what contains all necessary transactional and master data, and SAP BTP as an integration tool. We developed an introduction and testing scenario for users and presented the prototype at business conferences in EMEA and NA regions. Structured feedback about the desirability, feasibility, and usability of the prototype was collected. We used the Multi-Criteria Decision Making (MCDM), including Analytic Hierarchy Process (AHP) methods, to summarize the feedback scores. The feedback was positive, indicating potential for further research; we share insights, discuss improvement areas and directions for future design efforts.

**Keywords:** Virtual Reality · User interface design · User studies · Enterprise information systems · Procurement

## 1 Introduction

### 1.1 Problem Statement

Business communication and collaboration has undergone unexpected change due to novel coronavirus pandemic. Remote workstyle became part of HR policies transformation strategies for major industries, such as Software & IT Services, Media & Communications, Education, Finance, Retail, etc. Disrupted face-to-face communication between buyers and suppliers crucial for some cultures and industries impacts the core business processes, such as supplier contract negotiation. Replaced mainly by remote video calls, interaction still has challenges to be resolved, such as lack of "hands" during the negotiations, when a supplier can demonstrate a 3D product model in real-time.

Virtual reality (VR) is naturally well-suited to address these communication challenges. VR allows business users to interact with peers in virtually generated surroundings that replace traditional office meeting rooms and other areas. Examples of implemented VR cases are remote Design-Thinking workshops with unlimited 3D visualization capabilities [1], employee training at metal and mining plants to protect workers from industrial accidents [2], social networking platforms such as the project Sansar [3].

## 1.2 Our Contribution

This paper explores possibilities of using VR applications in procurement, with focus on reporting and contract negotiation. We describe design and development of a proof-of-concept VR application prototype and share findings of the user test with procurement managers. Based on these findings and our insights, we attempt to formulate recommendations and directions for future work on VR products in procurement.

## 1.3 Related Work

This part covers a scientific literature overview in virtual reality, procurement field, user interfaces, prototypes evaluation areas.

Last year, the academic community described a significant VR technology appliance in various industries. Ball et al. claim that the pandemic's perceived impacts influenced the likelihood of acquiring VR for education, tourism, and work [4]. Several case studies that have been developed for flooding, wildfire, transportation, and public safety were described by Sermet [5]. The author says that "using the solution to enhance existing web-based cyberinfrastructure systems with the integration of immersive geospatial capabilities to assist the development of next-generation information and decision support systems powered by virtual reality". Seers applies the technology for outcrop geology [6]. The author mentioned that DT "allows users to fuse powerful 3D visualizations of photo-realistic outcrop models with geological interpretation and data collection, fulfilling the early promise of 'virtual outcrops' as an analytical medium that can emulate traditional fieldwork" [6]. Kim evaluated the VR appliance for the retail industries and considered VR as a promotion tool for small independent stores, resulting in the conclusion that compared with the classical website experience, VR "enhanced their (shops) flow state, which, in turn, increased interest and visit intention toward the store" [7]. De Regt also highlights the advantages of VR applicating for the sales processes as it offers immersive and interactive encounters tools that excite and engage their customers in novel ways [8]. That proves VR gains not only for the buyer from the procurement processes point of view but also as a beneficial tool from the supplier's sales processes perspectives.

Relatively less knowledge is obtained in the intersection of the procurement field and VR. H. Wang designs a government procurement system under cross-border E-commerce with virtual image technology as the basis to study the key technology of updating massive terrain data in real-time [9]. R. Muntean explores if and how new immersive visual technologies might better communicate and transmit values and the importance of sustainability efforts, including circular procurement cases [10]. H. Hassan describes the benefits of VR appliances for complex procurement processes in the construction industry to support the construction planning in the virtual environment and is expected to improve efficiency at the bidding stage [11]. Virtual purchasing assistance based on AI platform using natural language processing technologies was designed by teamed up Korea's and Japan's companies to build a new user purchasing experience [12]. This example inspires to include AI-block for the VR-based procurement processes. A "child" of VR technology is augmented reality (AR) [13]. AR add value in logistics areas, such as sales, outdoor logistics, human resource management, warehousing, and manufacturing [14], the procurement is out of this scope. J. Du considers VR as a powerful tool for the project communication of remote users [15]. That's why it makes technology popular nowadays.

Overall, the applications we have reviewed seem to be at a proof-of-concept stage, with only a few examples of established successful solutions. Reported outcomes of early user tests are favorable, indicating that VR environments can facilitate communication in a variety of business and industrial contexts. There is little evidence though directly related to procurement processes, and business negotiations, which are of interest to us.

## 1.4  Research Question and Approach

Our study examined whether a VR application can facilitate communication specifically in the context of procurement processes. We were interested in evaluating perceived usefulness, feasibility, and usability of a virtual environment where procurement managers can review reports and participate in negotiations. To our knowledge, there was no existing research or solutions with this particular focus, so our research was exploratory and open-ended. We did not specify formal hypotheses or use comparative measures, instead focusing on direct feedback from potential end users. Our primary questions were:

1. Does VR application add value to the user compared to existing products?
2. What are the limitations and downsides of a VR application compared to existing products?
3. What are the usability problems and implementation issues of this application?

The following chapter describes the solution proposal, including business processes scoping and description, prototype functionality and architecture, user interface design, testing environment design, and user onboarding features. The third chapter includes feedback collections with the following steps: feedback collection methods and context, recruitment and sample, the user test procedure, and data analysis. The fourth chapter is results and discussions, including limitations and future work.

## 2  Proposed Solution

### 2.1  Business Processes and VR Capabilities Definition

To define the scope of the prototype, we decided to focus on a small number of business processes. We interviewed nine procurement experts working at SAP. We chose experts representing different industries, with knowledge of the digital procurement processes and aware of the relevant technological and business trends. We asked them to identify most interesting procurement business processes. The main question was, "Where in the end-to-end procurement process can VR technology add the most value?". Two business processes got the most votes: viewing the advanced reporting as a CPO (Chief Product Officer), and contract negotiation between buyer and supplier.

Contract negotiation was chosen as an area with the greatest difference between online and in-person communication.

- In online communication, it is harder to direct one's attention to something, like pointing to a piece of data, a paragraph, and so on. VR naturally enables the use of pointing hand gestures to address this issue.
- During in person negotiations, parties demonstrate 3D models of objects or samples. In traditional online call settings, it is more difficult both to demonstrate and to view those models, while in VR, samples can be viewed as 3D models and any scale.
- Setting and mood matter for negotiations. Being in a dedicated meeting room creates immersion and working mood that are more difficult to achieve in a call. VR meeting room minimizes distractions and mimics familiar settings.

Advanced procurement reporting was selected because it requires the user to inspect and compare multiple types of data from multiple sources (for instance, geographical locations, positions in supply chains, volumes of supply over time, risk levels, etc.). We assumed that in a 3D environment the user will be able to simultaneously view more layers of data, plus use spatial cues to navigate and to group reports and objects. VR has already been actively used to improve the user experience of viewing item catalogs [16] and reports [17]. It means that the technology is mature enough to cover the selected business process.

The essence of the VR application is to place a user in the limited and fenced virtual environment to run the devoted activity. With using in combination with a headset and a chair, virtual reality isolates the person and captures sight, hearing, and touching experience completely. It allows being entirely concentrated on the AR application scenario. UX designer gains unlimited capabilities to create any space with 3D objects. Compared with 2D screens, VR does not require enormous intellectual work to construct a 3D mental model [18]. We considered all these VR features to build the prototype.

### 2.2  Prototype Architecture and User Stories Modeling

The technology stack for the VR prototype includes SAP BTP (Business Technology Platform) to build the integration, SAP Ariba to store the business transactional and master data, and Unity as a tool to design the VR environment and work with virtual reality glasses [19]. Unity has a visual development environment, cross-platform support, and modular component system with 3D models libraries. SAP solutions were selected because they were available to the article authors. The research aims to examine the VR capabilities, focusing more on VR experience and less on integrations and transactional system selection. The connection of the technology stack with physical VR glasses, headset and controllers is presented in Fig. 1.

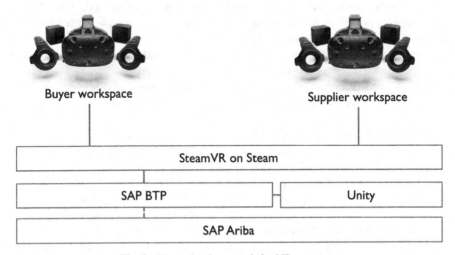

**Fig. 1.** The technology stack for VR prototype

We defined a user story for each VR scene for the selected business processes. The first user story starts with placing a user (acting as a Chief Procurement Officer) sitting in the CPO's office chair. The CPO sees a table with three buttons to start the dashboards: Operational, Tactical, and Strategic procurement. The user can physically move controllers and press the virtual button. Once the button is pressed, the VR application shows moving graphs and diagrams for each procurement dashboard and prerecorded AI-style voice comments on the analytics.

Operational procurement dashboard is a map with online delivery tracking, geographical areas with risks of disruptions, and aggregated analytics of several placed purchase orders in progress and open invoices.

By pressing the "Tactical Procurement" button, CPO sees the numbers of open-sourcing activities, such as open RFIs, RFPs, and Live Auctions.

Strategic dashboard presented by visualized 3D graphs with spend analytics by months and categories and highlights the unexpected deviations compared with historical data. For this scene in the first release of the VR prototype, we decided to minimize the list of actions that users can do and limit it by pressing the three buttons. We considered the first scene a "warming up" exercise to allow the user to adapt to the VR environment by perceiving what is happening rather than influencing it.

The Fig. 2 represents functional areas of the CPO dashboards' scene (#1).

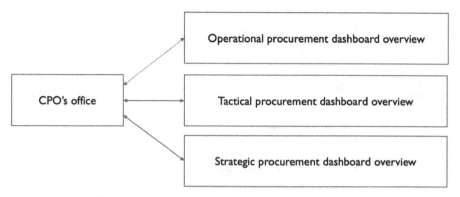

**Fig. 2.** Functional areas for CPO dashboards' scene (#1)

The timer was set to 5 min for the first scene before switching to the second scene. We selected a limited and fixed timing only for the prototype demonstration purpose.

The second scene starts with placing a user (acting as a buyer or supplier based on pre-configuration) on the outdoors terrace in the city café. The user sees a negotiation party in the front, a table between, and a screen assistant on the right-hand side. There are four actions for the buyer role and four actions for the supplier role that can be chosen by virtual hand.

The buyer has following options to initiate by the screen assistant's menu: supplier check-in media, market price analysis, AI recommendation, and contract signature. By clicking the "supplier check" option, recent news about the supplier is popped up, including the risk factor built by press and data sources analyses. Selection of the second AI recommendation option recommended negotiation mechanics with a supplier based on historical transactional data. The third option is the market price analysis shows the graphs about the commodity price's fluctuation. And the last ones – the contract signature options offer three prices for selection and the Sign button.

The options available for the supplier are analysis of production costs and buyer check using media, contract signature, and 3D product's model demonstration. The first three options are relatively similar to the buyer's option. The product demonstration function triggers a 3D object on the table, which can be picked up and turned in the hands of any negotiation participants. As a 3D object, we selected an industrial plant model because it's complicated to show on the 2D images, and the VR environment benefits the visualization of multi-element constructions.

The expected difficulties were entering text without a physical keyboard for the contract price entering and text editing. That's why we choose the price selection option. We recorded audio tracks that assisted each selected option at the screen menu to avoid reading long texts.

The Fig. 3 represents functional areas of the Contract negotiation' scene (#2).

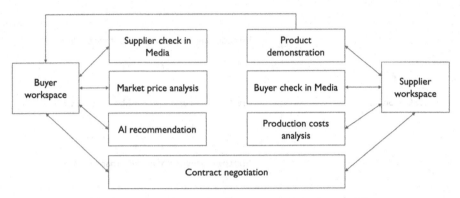

**Fig. 3.** Functional areas for Contract negotiation' scene (#2)

After finishing the prototype architecture, user stories, and functional modeling, we started to develop user interfaces.

### 2.3 User Interface Development

We defined general principles guiding the UI design for the prototype:

- Audio background with guiding voice during all scenes
- Both parties can observe counterpart's gestures and movement during the negotiation
- Both parties can see a 3D model of the product, can grab and move it
- Both parties can see a Spending Analysis report during negotiation

User interface (UI) was developed according to VR-technology specifications and included Russian and English languages. The language model can be selected before the prototype execution in a configuration file. The UI design covers two main parts – visualization and voice acting. For the visualization part, we faced the challenge of moving the 2D graphs to a VR environment because of the limited resolution of screens in virtual glasses. Also, in VR environment it is easier to inspect a 3D object from different points of view, and we wanted to take advantage of that. The two phases of the UI development for scene #1 – a simplified graph mockup and a final 3D version of the graph – are presented in Fig. 4.

About fifty audio tracks were written and recorded by professional announcers in Russian and English. Because all graphs were simplified, we used voice guidance to explain the content of the graphs (presented as an "AI" analytics system). Timing for audio tracks was aligned with the sequence of the graphs and diagrams demonstration to make it aligned with the user stories.

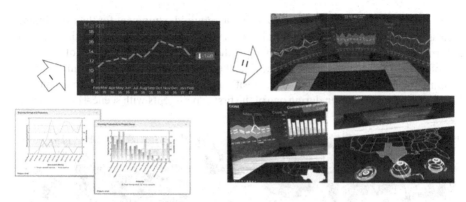

**Fig. 4.** Two phases of the UI development (example of scene #1)

## 2.4  Environment and Setup for Prototype Testing

In order to establish the requirements to the user testing environment, we developed a 3D model of the area where user testing would have taken place. We wanted the user test to go smoothly and without external disruption from outside of VR space. Via modeling, we developed an understanding of the user's physical movement trajectory and special physical requirements of how much space are required for each user in each scene. In both scenarios, users sit in a chair during the whole performance. It makes the unique modeling easier and the experiment safer for the participants. Figure 5 shows two persons in the chair who will come into VR, and the one sitting near the 60″ plasma is a trainer.

**Fig. 5.** Spatial modeling of the user test facility

Spatial modeling also helped to identify the required equipment for the experiment, such as HTC Vive helmet, grabs controllers, servers, chairs, screens for tracking how the experiment is going and what is happening in the VR environment, a couple of chargers, and other devices. Several pieces of advice about the spatial setup, for example, Base Stations for HTC Vive location, chair position, and additional hints, are presented in Fig. 6. The photo was taken during one of the first experiments with scene #1 in Moscow.

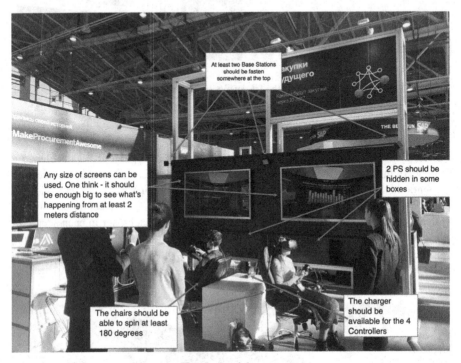

**Fig. 6.** Spatial set-up

## 2.5   User Training Procedure

Audio instructions were prerecorded and embedded in the VR application and guided the user to take necessary actions, such as looking on the left, clicking the virtual button, etc. The moderator spent two minutes before the sequence started to explain the essentials of using controllers, which physical buttons to click, and what to do if something happens unexpectedly, for example, if the person feels dizzy or voice instructions were not clear enough. Figure 7 presents the picture from the experiment in Austin (USA), with one VR user and one trainer who explains possibilities to move in the chair to adapt to the VR environment quickly. Designed auditory instructions in the application made it possible to simplify the learning process and minimize user questions significantly.

**Fig. 7.** User training procedure

## 3 Feedback Collection

### 3.1 Methods and Context

The prototype was presented at five business conferences organized by SAP in the following sequence: Moscow (Russia), Kyiv (Ukraine), Minsk (Belarus), Austin (USA), Barcelona (Spain). The target audience were industry experts in the procurement field working with procurement analytics and purchasing contract negotiations daily. More than 100 industry experts successfully finished onboarding and experienced the VR application. Twenty-four experts agreed to participate in the interview and provide written feedback on the prototype. The set of users includes procurement managers and CPOs, who are decision makers about the VR technology and use cases relevant to their business.

### 3.2 Procedure

The virtual reality-based prototype was advertised as a next-gen environment for advanced procurement reporting and contract negotiation. It was tested in two modes depending on the number of testers.

Figure 8 presents scene #1, procurement reporting, experiencing a user at Moscow's "SAP Forum" conference.

Scene #1 can be tested by one user, and scene #2, contract negotiation, requires two testers. In most cases, we asked two persons to join, have two-minute instructions, experience scene #1 in a five-minute timeframe simultaneously, and start scene #2 in

**Fig. 8.** Experiencing the scene #1 at the "SAP Forum" conference in Moscow

different roles of buyer and supplier that takes about 10 min. Figure 9 presents scene #2 execution by two participants at SAP Forum in Moscow.

**Fig. 9.** Prototype examination during SAP Forum in Moscow (scene #2).

After the test the users were asked to fill the questionnaire consisting of 13 questions in three areas: Desirability, Usability, and Feasibility. The questions relate to two scenes, so an interviewee should provide answers for Desirability and Usability areas. Feasibility area was evaluated by a developer. The answers type was a score from 1 to 5 where one means "not at all," two – "more no than yes," three – "find it difficult to answer," four – "more yes than no," five – "definitely yes." The complete list of questions is presented in Table 1. In addition to qualitative questions, we asked verbal questions about the general impressions of the prototype and thoughts on its applicability.

**Table 1.** The questions for user's feedback collection.

| Area | Factor | Question | Scene 1 | Scene 2 |
|------|--------|----------|---------|---------|
| Desirability | D1 | Would you like to use the application in your enterprise now? | [1–5] | [1–5] |
| | D2 | Would you like to use the application in your enterprise in the future? | [1–5] | [1–5] |
| | D3 | Would you like to apply VR technology in the proposed processes? | [1–5] | [1–5] |
| Usability | U1 | Were the principles of use/play clear? | [1–5] | [1–5] |
| | U2 | Was the scenario clear? | [1–5] | [1–5] |
| | U3 | Was the control comfortable? | [1–5] | [1–5] |
| | U4 | Was the voice acting clear? | [1–5] | [1–5] |
| | U5 | Were the visuals clear? | [1–5] | [1–5] |

## 4 Results and Discussion

### 4.1 Results

The structured feedback was used for the Analytic Hierarchy Process (AHP) as an exanimating Multi-Criteria Decision Making (MCDM) method [20]. Calculated average scores for each factor are presented Table 2.

For the negotiation scene, the score is generally neutral, but users agreed with the statement about appliance VR technology in the proposed procurement processes. It means that the prototype did not meet the users' expectations, but the research area and direction have potential. The reporting scene is more desirable. Some comments from the users include statements that the dashboard features are easier to implement. Users said that implementing negotiations VR application required supplier involvement. Suppliers are generally wary to accept even simple IT tools, such as supplier web-portal registration, and it's too early to discuss the new technology appliance for all supply chain parties.

The principles of use, represented by the user in-app instructions and training procedure, were not clear for both scenes and had room to improve. Some users commented that it was their first VR experience. They were lost in a new space, forgot trainer introduction, and did not listen to audio guidance.

The sequence of actions the user should take during the negotiation scene was less clear than during the reporting dashboards scene. As most users said, the first scene is

**Table 2.** Average scores for each question (Desirability and Usability)

| Area | Factor | Scene 1 | Scene 2 |
|---|---|---|---|
| Desirability | D1 | 3.042 | 2.167 |
| | D2 | 3.708 | 2.958 |
| | D3 | 3.792 | 3.792 |
| Usability | U1 | **3.958** | **3.292** |
| | U2 | 4.083 | **3.208** |
| | U3 | **3.667** | 4.083 |
| | U4 | **3.708** | 4.708 |
| | U5 | 4.042 | **3.542** |

simple to operate by just clicking three buttons, while the second scene requires special activities to check the news, inspect the 3D model and sign the contract.

In the reporting dashboards scene, the control was evaluated as less comfortable. The user said that it was not easy to press the button for the first time, because physically you should "pull the trigger of the gun, "not to press the button from the top. We think the problem is that this operation caused a problem with the mental model. Interestingly, after the 5 min adoption, this problem disappeared. No one has complained about the virtual operations with the physical controller differences. We think learning the controller took a few minutes of learning in the actual use.

Voice acting scored neutral for the dashboard's scene and highly favorable for the negotiation's scene. Users did not provide any feedback or explanation of this difference.

The visuals (graphic elements) for the negotiation's scene are comparatively less clear. In our prototype, the buyer and supplier have presented the same avatar, and several users dislike it. Mostly it's due to the limitations of the first version of the prototype and can be technically resolved.

### 4.2  Feasibility Evaluation and Next Steps Definition

We ran a different interview when we finished the prototype testing. We asked developers who coded the VR prototype application to fill the questionnaire. It was necessary for understanding for simplifying a decision making about the next steps for the prototype improvement and the planning of further VR application's releases. Table 3 consists of questions in the Feasibility area.

**Table 3.** The questions for developer's feedback collection.

| Area | Factor | Question | Scene 1 | Scene 2 |
|------|--------|----------|---------|---------|
| Feasibility | F1 | Is it easy to customize the user instructions? | [1–5] | [1–5] |
| | F2 | Is it easy to customize the user's script? | [1–5] | [1–5] |
| | F3 | Are the user controls easy to customize? | [1–5] | [1–5] |
| | F4 | Is it easy to modify the sound elements? | [1–5] | [1–5] |
| | F5 | Are the graphic elements easy to modify? | [1–5] | [1–5] |

Table 4 presents the developer's interview results.

**Table 4.** Average scores for each question (Feasibility)

| Area | Factor | Scene 1 | Scene 2 |
|------|--------|---------|---------|
| Feasibility | F1 | 5.000 | 5.000 |
| | F2 | 4.000 | 3.000 |
| | F3 | 2.000 | 1.000 |
| | F4 | 4.000 | 4.000 |
| | F5 | 2.000 | 3.000 |

To improve usability elements in the prototype and define the next step for prototype enchantments, we determined the alternatives (Table 5).

**Table 5.** Alternative's definition

| Code | Alternatives for the application enhancements |
|------|----------------------------------------------|
| S1 | Refinement of Scene 1 instructions |
| S2 | Refinement of Scene 1 Controls |
| S3 | Improvement of the voice acting for Scene 1 |
| S4 | Refinement of instructions for Scene 2 |
| S5 | Finalization of the user script Scene 2 |
| S6 | Renovation of visual elements Scene 2 |

The values of normalized vectors per alternative after applying the AHP method are presented in Table 6. Consistency relation is less than 10% for all factors shows good consistency of ratings.

The calculated consistency relation is less than 10% for all factors shows good consistency of ratings. Based on the number's the most weight has the strategies S1,

**Table 6.** Analytic Hierarchy Process method numbers

| Name | D1 | D2 | D3 | U | F | Result |
|---|---|---|---|---|---|---|
| Normalized factors vector | 0.122 | 0.060 | 0.105 | 0.296 | 0.417 | Not required |
| Normalized vector S1 | 0.173 | 0.172 | 0.167 | 0.185 | 0.227 | 0.199 |
| Normalized vector S2 | 0.194 | 0.185 | 0.167 | 0.172 | 0.091 | 0.141 |
| Normalized vector S3 | 0.217 | 0.200 | 0.167 | 0.173 | 0.182 | 0.183 |
| Normalized vector S4 | 0.123 | 0.137 | 0.167 | 0.154 | 0.227 | 0.181 |
| Normalized vector S5 | 0.138 | 0.148 | 0.167 | 0.150 | 0.136 | 0.144 |
| Normalized vector S5 | 0.155 | 0.159 | 0.167 | 0.166 | 0.136 | 0.152 |

S3 and S4. It defines the next steps for the prototype improvements – Refinement of Scene 1 instructions, Improvement of the voice acting for Scene 1, and Refinement of instructions for Scene 2.

### 4.3  Limitations

The main research limitations are.

1. Target audience is loyal to company brand, so the perception may be skewed
2. The questionnaire we used was not standardized
3. We used dummy data as opposed to users' own data, so they could only partially evaluate if the data was useful and presented in a usable manner

### 4.4  Conclusions, Recommendations, and Future Work

Based on collected feedback to the VR prototype, we conclude that VR can be applied to add value to the procurement departments compared to existing products, but not on the current release of our prototype. The reporting tools are more desirable rather the contract negotiation process.

The VR product limitations are the resolutions of VR glasses screens that can't display a relatively immense amount of data. The construction of the controllers considers only having a couple of buttons that make possible data selection, operating with existing in VR environment objects, but not suitable for entering the data to VR environment and working with texts. Adding the audio guidance during the negotiation can disturb the verbal communication with other users in VR. These limitations force us to investigate alternative user training and data presentation methods for the collaboration scenarios. For the reports, the visualization scenario VR prototype has potential for the coming future and satisfies users in terms of usage clarity and visualization instruments.

For the development of the VR prototypes, it's beneficial to model a special physical environment only after designing the VR environment. The controls set up and visual elements creation requires the most effort for the development. The first process got constructive feedback on improving the user instructions, controls, and visual elements. The second process is more complicated to realize in the VR environment from a technical perspective. Still, users believe in VR technology's future as an appliance to the proposed procurement processes.

The generally positive perception indicates a possibility for the follow-up research. Industry experts' feedback defines the next steps for the prototype improvements – a refinement of the reporting scene's instructions and voice acting and refinement of instructions for Contract negotiation's scene. It's defined as a scope for future work.

**Acknowledgments.** The authors of the study express their sincere gratitude to Alexander Alferov, a senior SAP developer. Alexander coded, tested, implemented a complex technical project in a short timeframe, and offered valued ideas to improve the prototype.

# References

1. Vogel, J., Schuir, J., Koßmann, C., Thomas, O., Teuteberg, F., Hamborg, K.-C.: Let's do design thinking virtually: design and evaluation of a virtual reality application for collaborative prototyping. In: Proceedings of the Twenty-Ninth European Conference on Information Systems. A Virtual AIS Conference, pp. 1–19 (2021)
2. Tripathy, D.: Virtual reality and its applications in mining industry. J. Mines Met. Fuels **62**, 184–195 (2014)
3. Kwon, H., Hudson-Smith, A.: Redesigning experience consumption in social VR worlds: decentralised value creation, mobilisation, and exchanges. In: 21st Academic Design Management Conference, pp. 1–17 (2018)
4. Ball, C., Huang, K.-T., Francis, J.: Virtual reality adoption during the COVID-19 pandemic: a uses and gratifications perspective. Telemat. Inform. **65**, 101728 (2021)
5. Sermet, Y., Demir, I.: GeospatialVR: a web-based virtual reality framework for collaborative environmental simulations. Comput. Geosci. **159**, 105010 (2021)
6. Seers, T.D., Sheharyar, A., Tavani, S., Corradetti, A.: Virtual outcrop geology comes of age: the application of consumer-grade virtual reality hardware and software to digital outcrop data analysis. Comput. Geosci. **159**, 105006 (2022)
7. Kim, G., Jin, B., Shin, D.C.: Virtual reality as a promotion tool for small independent stores. J. Retail. Consum. Serv. **64**, 102822 (2022)
8. De Regt, A., Plangger, K., Barnes, S.J.: Virtual reality marketing and customer advocacy: transforming experiences from storytelling to story-doing. J. Bus. Res. **136**, 513–522 (2021)
9. Wang, H., Fang, F.: Research on E-commerce supply chain design based on MVC model and virtual image technology. IEEE Access **8**, 98295–98304 (2020)
10. Muntean, R., Park, M.-L., Rubleva, Y., Hennessy, K.: Sustainable production and consumption in 360. In: 2019 IEEE Conference on Virtual Reality and 3D User Interfaces (VR), p. 1400 (2019)
11. Hassan, H., Taib, N., Rahman, Z.A.: Virtual design and construction: a new communication in construction industry. In: Proceedings of the 2nd International Conference on Digital Signal Processing, pp. 110–113 (2018)
12. Kang, I.: Clova: services and devices powered by AI. In: The 41st International ACM SIGIR Conference on Research & Development in Information Retrieval, p. 1359 (2018)
13. Faisal, A.: Computer science: visionary of virtual reality. Nature **551**, 298–299 (2017)

14. Rejeb, A., Keogh, J.G., Wamba, S.F., Treiblmaier, H.: The potentials of augmented reality in supply chain management: a state-of-the-art review. Manag. Rev. Q. **71**(4), 819–856 (2020). https://doi.org/10.1007/s11301-020-00201-w

15. Du, J., Shi, Y.M., Zou, Z.B., Zhao, D.: CoVR: Cloud-based multiuser virtual reality headset system for project communication of remote users. J. Constr. Eng. Manag. **144**(2), 04017109 (2018)

16. Allal-Chérif, O., Simón-Moya, V., Ballester, A.C.C.: Intelligent purchasing: how artificial intelligence can redefine the purchasing function. J. Bus. Res. **124**, 69–76 (2021)

17. Schleußinger, M.: Information retrieval interfaces in virtual reality—a scoping review focused on current generation technology. PLoS ONE **16**(2), e0246398 (2021)

18. Matthews, D.: Virtual-reality applications give science a new dimension. Nature **557**, 127–128 (2018)

19. Yen, B.P., Ng, K.Y.M.: Web-based virtual reality catalog in electronic commerce. In: Proceedings of the 33rd Annual Hawaii International Conference on System Sciences, vol. 2, p. 10 (2000)

20. Emrouznejad, A., Marra, M.: The state of the art development of AHP (1979–2017): a literature review with a social network analysis. Int. J. Prod. Res. **55**(22), 6653–6675 (2017)

# Integration of Research on Resilience of Energy and Socio-Ecological Systems Using Artificial Intelligence Methods

Liudmila V. Massel$^{(\boxtimes)}$ ⓘ, Aleksei G. Massel ⓘ, and Dmitrii V. Pesterev

Melentiev Energy Systems Institute SB RAS, 130, Lermontova Street, Irkutsk 664033, Russia
amassel@isem.irk.ru

**Abstract.** Resilience research of the of energy and socio-ecological systems are becoming more and more relevant, since in the first place are not only issues of the stability of these systems (sustainability), but questions resilience as possibility and speed of the return of these systems to a stable state after disturbances, which these systems are exposed.

Resilience is the ability of a system to return to an equilibrium state after a temporary disturbance; the faster it returns to balance and the less it loses, the more stable it is. The level of resilience is proportional to the speed of returning back (recovery). According to the ecological approach, resilience is a measure of the constancy of ecosystems and their ability to adapt to changes and disturbances and still maintain the same relationships between a population or a state.

The novelty of the proposed approach is determined, firstly, by the integration of resilience research of energy and socio-ecological systems using the concept of "quality of life"; secondly, the use of the concept of situational management in resilience research in the formation of control actions to return systems to a stable state after disturbances; third, the integration of qualitative methods (based on semantic and predictive modeling) and quantitative methods (based on mathematical modeling) in resilience research.

**Keywords:** Resilience · Quality of life · Ecological and energy security · Cognitive modeling · Artificial intelligence

## 1 Introduction

Recently, scientific teams and individual scientists abroad have shown great interest in the scientific area defined by the term "resilience". The Russian language is usually translated closer to sustainability, but more precisely translated as elasticity (in physics).

In Russia, this area is mostly based on research in the field of technical sustainability, but in Western Europe, this area is considered more broadly and also includes environmental, psychological, social, and economic resilience. At the same time, the factors that determine social stability in foreign studies have something in common with the factors used in assessing the quality of life in Russian studies.

© Springer Nature Switzerland AG 2022
V. Taratukhin et al. (Eds.): ICID 2021, CCIS 1539, pp. 107–119, 2022.
https://doi.org/10.1007/978-3-030-95494-9_9

At the same time, when considering the resilience of technical systems, it is necessary to assess the risks of both natural and man-made threats. In the works of the Melentiev Energy Systems Institute SB RAS, until recently, these threats were considered as threats to energy security (ES). The implementation of these threats can cause emergencies, aggravated by the likelihood of multiple accidents, including cascading accidents, in the energy sector. This industry is one of the critical infrastructures that directly affect the quality of life of the population.

The need for an interdisciplinary study of these issues determines the relevance of the work. The basis of this study is a systematic analysis of natural and man-made factors affecting the resilience of both energy and social systems, as well as the ability of these systems to adapt to existing and possible new threats.

The fulfillment of the tasks of system analysis requires the development and integration of appropriate methods and the use of modern information technologies. These technologies include, among other things, intelligent information technologies (for example, cognitive modeling), developed and used by a team of researchers at MESI SB RAS (Department of Artificial Intelligence Systems in Energy Sector).

## 2  Approaches to Determining Resilience

Currently, there is no clear definition of resilience due to its widespread use in different fields with different meanings and consequences. Some of them are given in the report of the International Institute for Applied Systems Analysis (IIASA) prepared for the Virtual Competency and Training Center on the Protection of Critical Energy Networks from Natural and Man-Made Disasters, created based on the Organization for Security and Cooperation in Europe (OSCE). Our team has prepared a section for this report, which was named "Russian approach to resilience". Materials from this section were also used in joint work with colleagues from IIASA [1, 2].

Here are the definitions of resilience used in the report:

- Resilience is often defined as the ability of a system to return to equilibrium, or rather, the ability to return to equilibrium and develop despite further tremors and disruptions.
- Resilience may be related to the ability to withstand stress and "bounce back".
- Resilience can be the ability to achieve some new stages of dynamic equilibrium after shock, readiness for dynamic and inter-scale interactions of the paired system: man-environment.
- Resilience can be a person's ability to successfully cope with traumatic experiences and avoid negative developmental trajectories.

One of the successful attempts to define resilience was made in [3], we will use this definition:

«Resilience is the ability of a system to return to an equilibrium or steady-state after a disturbance, which could be either a natural disaster, such as flooding or earthquakes or a social upheaval, such as banking crises, wars or revolutions».

Here, resilience is defined as the ability of a system to return to a certain equilibrium state after suffering a temporary disturbance. In this case, the system is more stable, the faster it returns to equilibrium, and the less it loses. This definition is based on the root of the term "resilience" in the Latin word "Resilio", which literally means "to bounce back." In a broad sense, resilience refers to the ability to recover from some shock, insult, or anxiety. At the same time, it is believed that the level of resilience is proportional to the rate of recovery.

Environmental scientists in their works define that resilience as a measure of the constancy of ecosystems and their ability to adapt to changes and disruptions and still maintain the same relations between the population or the state [4]. And ecosystem resilience refers to the ability to absorb disturbing factors and reorganize while the system is changing.

In the study of social resilience, the following factors are most often encountered: moral values, realistic optimism, a stable role model, receiving social support, mental and emotional flexibility, the meaning of life and goals, spiritual practices, physical activity, the ability to resist fear. Disaster risk reduction is also considered to be an important factor that has a significant impact on resilience.

The Multidisciplinary and Multi-hazard Earthquake Engineering Research Center (MCEER) has identified four aspects that increase resilience [5–7]:

- Robustness: the strength or ability of elements, systems, and other measures of analysis to determine the ability to withstand a given level of stress or need, without suffering degradation, or loss of their function.
- Redundancy: the ability to meet functional requirements in the event of destruction, degradation or loss of functionality.
- Rapidity: the ability to meet priorities and goals promptly to contain waste, restore functionality, and avoid future failure.
- Resourcefulness: the ability to identify problems, prioritize and mobilize external resource alternatives when conditions exist that threaten to disrupt an element or system.

## 3  Energy and Ecological Safety

Recently, the need to study the totality of resilience of ecological, social, and economic factors has increased. This fact is supported by one of the global principles of responsible investment - checking a company for compliance with ESG criteria (environmental - ecology, social - social development, governance - corporate governance). ESG is a triad of groups of parameters, based on which the company provides sustainable development management aimed at ensuring a decent quality of life (Fig. 1) [8].

The need to study the resilience of energy systems together with socio-ecological systems is due to the fact that the power industry is a critical infrastructure that directly affects the sustainable and safe development of society. At the same time, the influence of energy on the quality of life is obvious. However, a comprehensive study of this issue is required, since this influence can obviously be positive if the population is provided with energy resources of adequate quality and in the required volumes. But it can also be negative, for example, the negative impact of emissions from energy facilities on the environment.

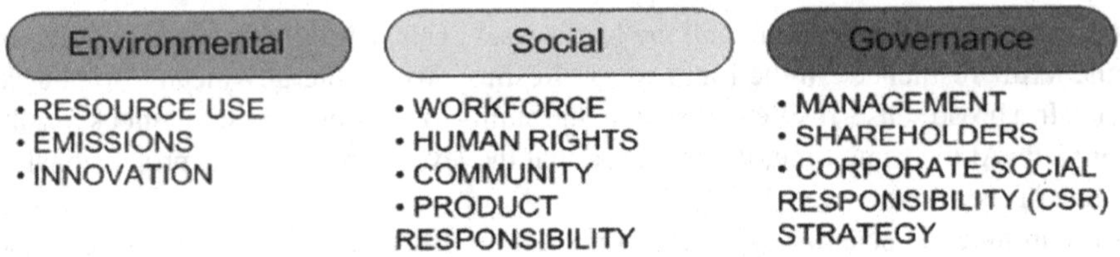

| Environmental | Social | Governance |
|---|---|---|
| • RESOURCE USE<br>• EMISSIONS<br>• INNOVATION | • WORKFORCE<br>• HUMAN RIGHTS<br>• COMMUNITY<br>• PRODUCT<br>RESPONSIBILITY | • MANAGEMENT<br>• SHAREHOLDERS<br>• CORPORATE SOCIAL<br>RESPONSIBILITY (CSR)<br>STRATEGY |

**Fig. 1.** ESG sustainability assessment criteria

The resilience of energy systems is based on the study of threats to energy security. In Russia, ES threats are defined in the Energy Security Doctrine of the Russian Federation (approved by Decree of the President of the Russian Federation No. 216 of 05/13/2019). In accordance with it, ES threats include adverse and dangerous natural disasters, environmental changes leading to disruption of normal functioning, and destruction of infrastructure and facilities of the fuel and energy complex. The doctrine defines the risks in the field of energy security, including insufficient level of protection of infrastructure and facilities of the fuel and energy complex from acts of unlawful interference and dangerous natural phenomena. In addition to this, the Doctrine states that the consequences of the implementation of threats to energy security are, including causing harm to the life and health of citizens. When studying the resilience of technical, ecological, and social systems, based on ES studies, it is necessary to take into account the described threats, the risks of their occurrence, and their consequences.

Melentiev Energy Systems Institute SB RAS is one of the leading Russian centers in the field of energy security problems research. In the studies of the Institute's researchers, the strategic threats to electronic security are identified [9]. At the same time, it is noted that one of the most important is natural threats. Previously, foreign policy, economic and managerial threats were identified as more significant threats. Recently, an important aspect of ensuring the country's energy security is the study of the negative impact of natural disasters on the electric power system (EPS). Research in this direction is carried out in order to reduce the risks of major systemic accidents, which can be cascade in nature. The occurrence of such accidents has a significant impact on the quality of life. The following natural risks are named as the main causes of cascade accidents in power systems: earthquakes, storms, floods, periods of extreme heat.

## 4  Adaptation of the Concept of Situational Management

The authors propose the idea of joint use of cognitive and mathematical modeling to assess the quality of life. To implement this approach, it is required to adopt the concept of situational management used in the study of ES [2].

The concept of "situational management" appeared at the end of the 60s of the last century. The founder of this direction is the Russian scientist D.A. Pospelov [10]. Situational management was proposed for large (complex) systems in which it is impossible and/or impractical to formalize the control process in the form of mathematical models, while there is only its description in the form of a sequence of sentences in natural language using logical-linguistic models. Situational management is based on the concepts of the situation, the classification of situations, and their transformation.

The current situation C is a set of the current state of the object (X) and its external environment (F). Then $C = <X, F>$. The complete situation is denoted as $S = <C, G>$, where C is the current situation, G is the management goal. The management goal is represented in the form of a target situation $G_g$, to which the existing current situation must be reduced. Then $S = <C, G_g>$.

Assuming that the current situation C belongs to some class Q', and the target situation $G_g$ belongs to the class Q", a control (vector of control actions U) is sought that belongs to the set of admissible controls $\Omega_u$ and provides the required transformation of situations of one class in the situation of another.

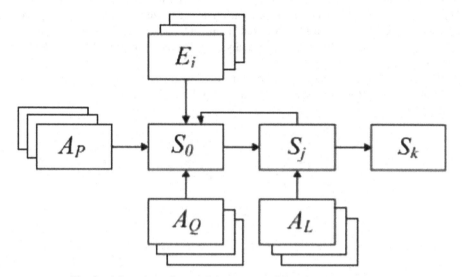

**Fig. 2.** Adaptation scheme of the concept of situational management.

Figure 2 shows a scheme for the study of resilience based on the concept of situational management, where

- $S_0$ – initial state of the system (current situation C);
- $E_i$ – i-th emergency scenario (set of scenarios - vector of disturbances F);
- $A = A_p, A_Q, A_l$ – a set of preventive, operational, and response measures to neutralize or mitigate the consequences of an emergency situation (EmS) (vector of control actions U);
- $S_j$ – the state of the system after an emergency $E_i$ taking into account the implementation of a set of activities $A_p$ and/or $A_Q$ (possibly steady state);
- $S_k$ – the state of the system after carrying out liquidation measures (steady-state);
- $S_j$ и $S_k$ can be considered as analogs of the corresponding target situations $G_g$.

Adaptation of the situational approach to studies of the resilience of energy and socio-ecological systems will allow the use of approaches, methods, and tools (including cognitive modeling) that were previously developed for research in ES.

## 5  Cognitive Modeling Is One of the Tools for Studying the Sustainability of Energy and Socioecological Systems

Cognitive modeling - building cognitive models in the form of directed graphs, in which the vertices correspond to factors (concepts). The arcs of the digraph denote connections between factors, in the simplest case with a "+" or "−" sign, depending on the nature of the cause-and-effect relationship. A graphical representation of a cognitive model is called a cognitive map.

The term "Cognitive Map" (CM) was proposed by E. Tolman. in his work "Cognitive maps in rats and in humans" when studying the behavior of rats in an impromptu maze [11]. The researcher believed that rats, to navigate the maze and memorize the places they have already visited, build a so-called "map" in their heads. R. Axelrod suggested using cognitive models in relation to the analysis and decision-making in poorly defined situations [12].

One of the first cognitive maps connecting energy, environmental and social factors is the now-classic Roberts-Axelrod cognitive map (Fig. 3) [12,13].

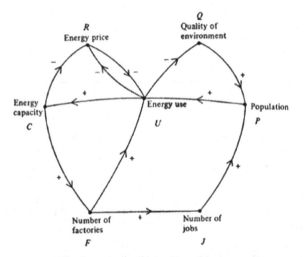

**Fig. 3.**  Roberts-Axelrod cognitive map.

It was proposed by Roberts as a signed digraph and was later cited in Axelrod's book as a cognitive map; therefore, in several sources, it is cited as Axelrod's cognitive map.

In the study of ES, it is proposed to use cognitive modeling for situational analysis of the ES problem and modeling of ES threats, which are understood as unfavorable events for the energy sector. To develop a cognitive map, you can use the methodology for constructing a CM proposed in [14].

CM construction technique.

1. Identification of the main factors - concepts influencing the development of the fuel and energy complex or the energy system of the country or its region.

2.  Establishing causal relationships between factors, placing the weights of these rela-
    tionships, and building a cognitive model of the fuel and energy complex or energy
    system (country or region). In the simplest case, the weights can be "+1" or "–1".
3.  Identification of strategic threats - factors that negatively affect the development of
    the fuel and energy complex/energy system or the development of an emergency
    situation in the fuel and energy complex/energy system.
4.  Identification of factors - preventive, operational, and liquidation measures affecting
    the scenarios for the development of the fuel and energy complex/energy system,
    directly for each threat.
5.  Changing the weights (or signs) of cause-and-effect relationships, depending on the
    influence of threats and measures on the factors of development of the fuel and
    energy complex/energy system.

In Fig. 4 shows a previously developed cognitive map in the notation of the cognitive
modeling tool CogMap [15], which in studies of resilience illustrates the relationship of
ecological factors (cold snap), energy factors affecting the resilience of energy systems
(an increase in energy consumption and the occurrence of their deficit) and factors
(activities), reducing the likelihood of a critical situation (increasing the production of
heat and electricity).

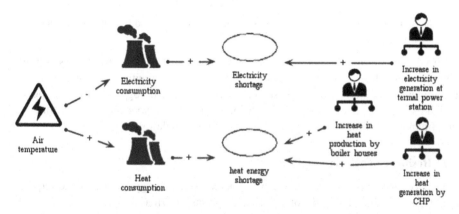

**Fig. 4.** Cognitive map linking ecological and energy factors

# 6  Quality of Life as a Factor in the Integration of Studies on the Resilience of Energy and Ecological Systems

The world scientific community understands the quality of life as a set of objective and
subjective parameters that characterize the maximum number of aspects of a person's
life, his position in society, and his satisfaction. The integral indicator of the quality of
life summarizes the indicators of health, social well-being, subjective social well-being,
and well-being [16].

The quality of life is differentiated from the widely used concept of "standard of
living", which means exclusively the material component. J. Forrester suggested that

the level and quality of life are inversely related to one another: the higher the standard of living associated with the growth rate of industrial production, the faster mineral resources are depleted, the faster the natural environment is polluted, the higher the population density is, worse health of people, more stressful situations, that is, the quality of life worsens.

The quality of life is determined not only by financial well-being, but also takes into account the state of security, health, a person's position in society and, most importantly, his own assessment of all these factors (Fig. 5).

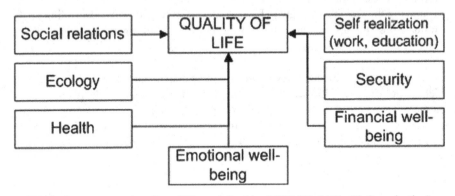

**Fig. 5.** Components of quality of life as defined by WHO (World Health Organization).

To determine the integral indicator of the quality of life, it is also proposed to use cognitive modeling. Earlier, the team of authors built a cognitive map of the integral assessment of the quality of life, based on the SF-36 methodology used by sociologists, which includes the external factor "Provision with energy resources"; to determine it, a questionnaire was formed for a social survey of the population (Fig. 6) [17, 18].

### 6.1 Resilience Criteria

For energy systems, such criteria can be indicators of energy supply of the required quality and in the required volume (i.e. quantitative criteria - indicators of energy security).

For ecological systems: MPC values (maximum permissible concentrations) can act as quantitative resilience criteria. Qualitative criteria can be, for example, total and specific (per person and GDP) $CO_2$ emissions; the trend of greenhouse gas emissions; energy production and consumption efficiency; the share of renewable energy sources (RES) and nuclear power plants, etc. For social systems, resilience criteria can be the main indicators of the quality of life: the state of the environment, health, social relations, self-realization (work, education), safety, emotional and financial well-being; provision of energy resources.

When forming sustainability criteria for socio-ecological and socio-economic systems, it is advisable to take into account also the ESG parameters [8].

E-factors include: an assessment of the company's environmental policy; the impact of the company's activities on the atmosphere; impact on the aquatic environment; impact

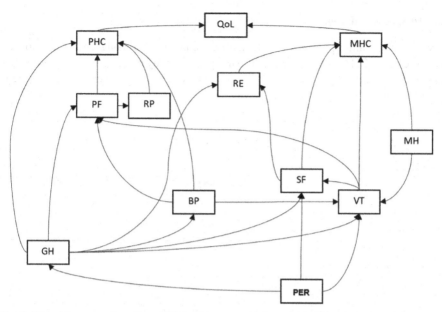

**Fig. 6.** Cognitive map of quality of life indicators according to the SF-36 methodology (all connections are positive "+").Comments to Fig. 6: PF - physical functioning; RP (Role-Physical
Functioning) - role-based functioning due to the physical condition; BP (Bodily pain) - pain intensity; GH - general health; VT - vital activity; SF - social functioning; RE (Role-Emotional) -
role functioning due to the emotional state; MH - mental health; PHC (Physical health) - general
component of physical health; MHC (Mental health) - general component of mental health; QoL
(QualityofLife) - an integral indicator of the quality of life; PER - Provision with energy resources.

on the soil; waste disposal; the use by the company in its management of indicators to
assess the impact of the company's activities on the environment; the company has a
plan to reduce the negative impact on the environment; the presence of "green projects"
in the loan portfolio (rating of banks' activity).

S-factors include: an assessment of the company's corporate social responsibility
policy; remuneration of employees; social security and professional development of
employees; staff turnover; labor protection and industrial safety; work with clients;
availability of a plan to improve socially significant indicators.

G-factors include: assessment of business reputation, development strategy, efficiency of the Board of Directors, activities of executive bodies, availability of a risk management system, the degree of transparency (transparency) of information, protection of
the rights of owners [19].

For us, in terms of the considered formulation of the problem, E-factors and S-
factors, which are mainly qualitative, are important. Obviously, a new toolkit is needed
to describe the qualitative criteria, for which semantic modeling tools (cognitive, event
and probabilistic) can also be used [20].

## 6.2  Instrumental Tools

Earlier, in the team represented by the authors, a knowledge management technology was developed [21] to support decision-making in the study of energy security problems (Fig. 7). It integrates ontological and cognitive modeling tools, as well as expert system technologies.

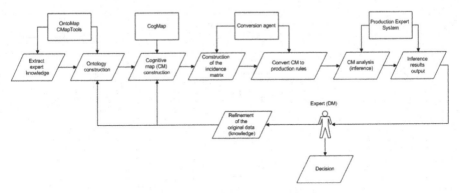

**Fig. 7.** Knowledge management technology

Within the framework of this technology, cognitive maps are converted into a set of production rules (rules of the "If-then" type) using the conversion module. Based on these rules, the shell of the expert system makes a logical conclusion about the mutual influence of the factors of the transformed cognitive map. The decision-maker can base on these results, draw a conclusion about the stability of the system under study. Moreover, if he considers the results to be unsatisfactory, he can make changes to the original cognitive map. Table 1 shows the stages of technology.

**Table 1.** Contents of the stages of knowledge management technology.

| Stage No. | Stage content | Instrumentation (performers) | Stage result |
|---|---|---|---|
| 1 | Domain analysis and knowledge extraction | Expert, researcher (cognitive engineer) | Basic concepts of the domain and the relationship between them |
| 2 | Building ontologies | OntoMap, CMapTools, Protégé | Knowledge presented as an ontology |
| 3 | Building cognitive maps (CM) based on ontologies | CogMap | Knowledge represented as a CM |

*(continued)*

**Table 1.** (*continued*)

| Stage No. | Stage content | Instrumentation (performers) | Stage result |
|---|---|---|---|
| 4 | Constructing an incidence matrix for a CM | CM conversion agent | Knowledge presented as an incidence matrix for CM |
| 5 | Converting CM into production rules of an expert system | CM conversion agent | Knowledge represented as production rules |
| 6 | CM analysis (logical inference in expert system) | Expert system | Delivery of results (interpretation of knowledge in the form of CM) |
| 7 | Analysis of output results | Expert | Solution recommendation or knowledge refinement (go to stages 2, 3) |

# 7 Conclusion

The article presents the formulation of the task of integrating studies of the resilience of energy, socio-ecological and socio-economic systems using the quality of life as an integration factor. The basic concepts are defined, the adaptation of the concept of situational management to these studies is illustrated. Examples of the application of cognitive modeling, criteria for the quality of life, and the relationship of energy, environmental and social factors are shown, the possibilities of adaptation and application of previously developed tools for resilience research are considered.

**Acknowledgements.** Research is carried out within the framework of a state assignment project MESI SB RAS no FWEU-2021-0007 (no AAAA-A21-121012090007-7) of the Fundamental Research Program of Russian Federation 2021-2030 using the resources of the HighTemperature Circuit Multi-Access Research Center (Ministry of Science and Higher Education of the Russian Federation, project no 13.CKP.21.0038). As well as with partial financial support from the Russian Foundation for Basic Research, grant no 19–07-00351, no 20-07-00195.

# References

1. Massel,' L.V., Komendantova, N.P.: Risk assessment of natural and man-made threats to the sustainability of energy, ecological and social systems based on intelligent information technologies: Informacionnye i matematicheskie tekhnologii v nauke i upravlenii - Information and Mathematical Technologies in Science and Management, **4**(16), 31–45 (2019). (in Russian)
2. Massel,' L.V., Massel,' A.G., Komendantova, N.P.: An Approach to Research on the Sustainability of Energy and Ecological Systems Based on Intelligent Information Technologies: Trudy Mezhdunarodnoj nauchnoj konferencii Ustojchivoe razvitie energetiki respubliki

Belarus': sostoyanie i perspektivy - Proc. of the International Scientific Conference Sustainable Energy Development of the Republic of Belarus: State and Prospects, Minsk, Publ. Belaruskaya navuka, pp. 33–43 (2020). ISBN 978–985–08–2654–1. (in Russian)

3. Davoudi, S.: Resilience: A Bridging Concept or a Dead End / Planning Theory and Practice. – vol. 13, no. 2, pp. 299–307 (2012)
4. Holling, C.: Engineering Resilience Versus Ecological Resilience/Engineering Within Ecological Constraints. Ed.: Peter Schultz, National Academy Press, Washington D.C., pp. 31–43 (1996)
5. Bruneau, M., Chang, S., Eguchi, R., Lee, G., O'Rourke, T., Reinhorn, A.M., et al.: A framework to quantitatively assess and enhance the seismic resilience of communities. Earthq. Spectra **19**(4), 733–752 (2003)
6. Cimellaro, G.P., Reinhorn, A.M., Bruneau, M.: Resilience of a health care facility. In: Proceedings of Annual Meeting of the Asian Pacific Network of Centers for Earthquake Engineering Research. ANCER (2005)
7. Cimellaro, G.P., Fumo, C., Reinhorn, A.M., Bruneau, M.: Quantification of seismic resilience of health care facilities. MCEER technical report-MCEER-09–0009. Buffalo (NY): Multidisciplinary center for earthquake engineering research (2009)
8. Chen, J.: Environmental, Social, and Governance (ESG) Criteria, Investopedia. https://www.investopedia.com/terms/e/environmental-social-and-governance-esg-criteria.asp. Accessed 14 Oct 2021
9. Pyatkova, N.I., Rabchuk, V.I., Senderov, S.M., Cheltsov, M.B.: Energy security of Russia: problems and solutions. Novosibirsk: Publ. SB RAS, p. 198 (2011). (in Russian)
10. Pospelov, D.A.: Situational Management. Theory and Practice, Moscow, Publ. Nauka, p. 284 (1986). (in Russian)
11. Tolman, E.C.: Cognitive maps in rats and men. Psychol. Rev. **55**(4), 189–208 (1948). https://doi.org/10.1037/h0061626
12. Axelrod, R.: Structure of Decision: The Cognitive Maps of Political Elites. Princeton University Press, Princeton, p. 422 (1976)
13. Roberts, F.S.: Discrete Mathematical Models with Applications to Social, Biological, and Environmental Problems, p. 559. Prentice-Hall, Englewood Cliffs, N.J. (1976)
14. Massel,' A.G.: Cognitive Modeling of Energy Security Threats: Gornyj informacionno-analiticheskij byulleten - Mining Informational and Analytical Bulletin, no. 17, Moscow, Publ. Gornaya kniga, pp. 194–199 (2010). (in Russian)
15. Massel, A.G., Pjatkova, E.V.: Intelligent Information Technologies for Research on Energy Security Problems: Trudy Vserossijskogo seminara s mezhdunarodnym uchastiem «Metodicheskie voprosy issledovaniya nadezhnosti bol'shih sistem energetiki» - Proc. of the All-Russian Scientific Workshop Methodological problems in reliability study of large energy systems, no.64, ISBN 978–5–93908–115–3, Irkutsk, Publ MESI SB RAS, pp. 472–483 (2014). (in Russian)
16. Finogenko, I.A., D'jakovich, M.P., Blohin, A.A.: Metodologija ocenivanija kachestva zhizni, svjazannogo so zdorov'em [Methodology for assessing the quality of life associated with health] // Vestnik Tambovskogo universiteta. Serija: Estestvennye i tehnicheskie nauki = Bulletin of the Tambov University. Series: Natural and Technical Sciences, vol. 21, no. 1, pp. 121–130 (2016). (in Russian)
17. Massel', L.V., Blohin, A.A.: Kognitivnoe modelirovanie indikatorov kachestva zhizni: predlagaemyj podhod i primer ispol'zovanija [Cognitive modeling of indicators of quality of life: the proposed approach and example of use] // Vestnik NGU. Serija: Informacionnye tehnologii = Bulletin of NSU. Series: Information Technology, vol. 14, no. 2, pp. 72–79 (2016). (in Russian)

18. Massel,' L.V., Blohin, A.A.: Metod kognitivnogo modelirovanija indikatorov kachestva zhizni s uchetom vneshnih faktorov [The method of cognitive modeling of quality of life indicators taking into account external factors] // Nauka i obrazovanie. Nauchnoe izdanie MGTU im. Baumana = Science and Education. Scientific publication of MSTU Bauman, no. 4, pp. 65–75, (2016). (in Russian)
19. Soboleva, O.V., Steshenko, A.S.: ESG-faktory kak novyj mekhanizm aktivizacii otvetstvennogo investirovaniya i dostizheniya celej ustojchivogo razvitiya [ESG Factors as a New Mechanism to Promote Responsible Investment and Achieve Sustainable Development Goals] / Ustojchivoe razvitie: vyzovy i vozmozhnosti = Sustainable Development: Challenges and Opportunities // Sb. nauchn. statej pod red. kand. ekon. nauk E.V. Viktorovoj. = Sat. scientific. articles ed. Cand. econom. Sciences E.V. Viktorova. SPb.: Publishing house of UNECON 2020, pp. 246–255. (in Russian)
20. Massel, L.V., Massel, A.G.: Tehnologii i instrumental'nye sredstva intellektual'noj podderzhki prinjatija reshenij v jekstremal'nyh situacijah v jenergetike [Technologies and tools of intelligent decision-making support of in emergencies in the energy sector] // Vychislitel'nye tehnologii = Computational technologies, vol. 18, pp. 37–44 (2013). (in Russian)
21. Massel,' L.V., Massel,' A.G., Pesterev, D.V.: Knowledge management technology using ontologies, cognitive models and production expert systems. In: Izvestiya YUFU. Tekhnicheskie nauki - Izvestiya SFedU. Engineering sciences, no. 4, pp. 140–152 (2019). (in Russian)

# Dynamic Reconfiguration of a Distributed Information-Computer Network of an aircraft

A. M. Solovyov[1]([✉]) [ID], M. E. Semenov[2] [ID], N. I. Selvesyuk[3] [ID], V. V. Kosyanchuk[3] [ID], E. Yu. Zybin[3] [ID], and V. V. Glasov[3] [ID]

[1] JSC "Concern "Sozvezdie", Plekhanovskaya Street 14, 394018 Voronezh, Russia
a.m.solovev@sozvezdie.su

[2] Zhukovsky-Gagarin Air Force Academy, Starykh Bolshevikov Street 54, Voronezh, Russia

[3] State Research Institute of Aviation Systems, Viktorenko Street 7, 125167 Moscow, Russia

**Abstract.** The paper presents the concept of configuration an automated information system, which is part of a complex of onboard equipment built on the basis of a distributed information-computer network. The presented system is intended for intelligent support of the crew in the event of emergency situations and allows in real time to fend off failures that arise during the execution of the task by the method of dynamic reconfiguration of the onboard equipment complex. At the same time, both problems arising in the operation of radio-electronic equipment (failures of critical components of the complex) and problems arising in the computer on-board network (failure of elements of application software) are considered as a failure. The application of the proposed concept in practice makes it possible to increase the reliability and fault tolerance indicators of the on-board information-computer network of the aircraft as a whole.

**Keywords:** Integrated modular electronics · Dynamic reconfiguration · Optical on-board network · Avionics · Distributed computing systems

## 1 Introduction

Modern avionics systems, built on the basis of the onboard distributed information-computer network (ODICN), implemented according to the principles of the second generation distributed modular electronics (DME) concept, are currently the subject of a large number of studies. The demand for such equipment (OBE) is due not only to the possibility of installation on various types of objects of application (air, land and marine equipment), but also to their ability to provide an increased level of reliability in the future while reducing costs for development, production and maintenance. Various manufacturers of OBE DME, offering their products on the market, have different approaches to the implementation of the main provisions of the design of integrated complexes [1–5].

The most famous representatives of the integrated systems launched into mass production are OBE for Airbus A380 and Lockheed/Boeing F-22 Raptor [6–8]. The complex

This work was supported by Russian Foundation for Basic Research (RFBR), projects 18-08-00453a, 20-08-01215a, 19-29-06091mk.

processes information from aircraft sensors (radar, pressure sensors, etc.), controls electronic countermeasures systems, communication and navigation equipment, as well as identification equipment. The key feature of this complex is to provide fault tolerance and recoverability by means of reconfiguration, which is achieved by reassigning and/or reprogramming modular resources. Restoration of full functionality or functionality of only the selected flight mode depends on the number of remaining modules that support reconfiguration. In this case, reconfiguration is divided into two types:

- secondary reconfiguration (minor reconfiguration) – carried out when one or more modules are lost as a result of failures;
- primary reconfiguration (major reconfiguration) – carried out in case of failure of the whole crate (for example, as a result of mechanical damage in combat).

One of the currently existing approaches to organizing problem-oriented reconfigurable multiprocessor computing systems is to represent the stream problem in the form of a set of input data vectors, which are processed according to a fixed algorithm. In this case, the solution to the problem is the transformation of a tuple of input data vectors into a tuple of output data vectors [9]. It is proposed to increase the efficiency of the problem being solved by organizing so-called "structural-procedural" calculations, in which the original information graph is divided into several independent subgraphs, each of which is implemented on separate computing cells that form a "multiconline". The structural nature of the computations is to implement the structure of each subgraph on a multipipeline, and the procedurality is to organize a certain sequence of displaying these subgraphs in a multipipeline. Reconfiguration in this case is carried out by switching the cells of the multicoreline when changing the problem being solved at the hardware level using FPGA arrays that simultaneously perform the functions of calculators. That is, the system in this case is a combination of computing units using various types of commutation (orthogonal and hierarchical).

Another approach to reconfiguring multiprocessor systems (MS) at the hardware level is proposed in the paper [10]. It consists in selecting from the set of processors of the system a certain subset of calculators, which will carry out partial execution of the functions of the failed element. The algorithm for distributing tasks in the event of a failure provides for the use of such system parameters as the connectivity of the components (topologies "ring", "daisy chain", "each with each", etc.), the number of failed and operational processors, as well as additional time spent to solve the problem of the failed element by one of the "reserve" ones.

There is another approach to the principles of reconfiguration, developed in papers. It is based on identifying three types of degradation in US systems (containing both specialized S-modules and universal U-modules): functional, structural, and structural-functional. Functional degradation occurs when a function fails without increasing the number of faulty modules. Structural degradation provides for the preservation of all functions in the event of module failures. Structural and functional degradation occurs when one or more modules fail with a corresponding decrease in the number of functions performed. Depending on the types of degradation, the so-called "strategies" of functional restructuring of the system are proposed. These strategies involve taking into account the following parameters:

- the set of states of the system requiring restructuring;
- a lot of working and failed modules;
- many functions related to healthy and failed modules;
- the possibility of excluding functions from the system or changing the algorithms of their work towards simplification in the event of a failure.

In this case, all possible combinations of reducing the number of functions performed and simplifying the algorithms for their implementation are considered, as well as possible distributions of these combinations by modules. The formation of the set of the sought options follows in the formal formulation of the following problem: to find a set of options for the distribution of functions that provide the minimum values of the error in their implementation for any state of the system, subject to restrictions on the time of their execution.

We especially note the works [11, 12] devoted to the dynamic reconfiguration of the OBE of an aircraft (AC). These works consider the process of reconfiguring the switching environment based on the agent-based approach. So, if a failure is detected, the agents determine the type of failure and use the corresponding previously prepared (before the system starts operation) scheme for reconfiguring the communication environment, reconfiguring the communication equipment. The scheme for reconfiguring the communication medium is prepared in advance, in the same way as the scheme for reconfiguring the computational nodes (CN) is prepared.

In this case, several options for the implementation of such actions are possible:

- reconfiguration of all switches is carried out by one agent (this option can be used if the agent somehow has access to all communication equipment that requires reconfiguration); the reconfigurator agent is assigned, is selected in advance, at the stage of creating the reconfiguration scheme, for example, it may be an agent with the maximum value of the agent's unique identifier;
- the reconfiguration of switches is carried out by a limited set of agents, each of which has direct access to the reconfigurable equipment (this option can be used if agents are not able to transfer configuration data through other communication equipment);
- switch reconfiguration is also carried out by a certain set of agents, while the number of agents involved in the process tends to the number of reconfigurable switches (this option can be used both with random access of agents to reconfigurable equipment, and with limited access).

Thus, one of the most promising methods of increasing the reliability of the designed OBE with the DME is the joint solution of the problems of optimal use of the resources of computing modules, the distribution of functional tasks and configuration management of the complex in the event of failures or changes in the tasks being solved. To create effective methods and algorithms for the design of OBE with DME, it is necessary to develop a special mathematical model of the complex, which will combine all its structural components and allow obtaining, analyzing and evaluating various architectural and structural options for constructing the complex.

## 2  The Concept of an Automated Dynamic Reconfiguration System

Based on the specifics of the OBE organization based on the DME, in order to solve the problem of dynamic reconfiguration in order to increase the reliability and fault tolerance of the OBE, as well as to implement the ability to fend off failures of on-board electronic equipment (avionics) directly during the flight, we will formulate the following requirements for ODICN equipment and software:

- there must be a subset of computational tasks $A$ from the set of all possible ones for a given ODICN $B$ (i.e. $A \subseteq B$), the implementation of which is supported in more than one computer in the network;
- at each moment of time, each task of the subset $A$ is implemented (launched) only in one of the computers supporting it, in the rest it is in the hot standby (i.e., the support of the hot standby in the ODICN computers is necessary);
- ODICN equipment should support the ability to automatically diagnose and monitor the health of all communication channels (both transmitting and receiving), as well as resources, nodes and interfaces that ensure the successful implementation of computational tasks of the set $B$;
- all ODICN computers must have support for the implementation of a specialized *supervisor* task that continuously collects up-to-date information on the state of the avionics of the entire ODICN, and is also capable of initiating the process of dynamic reconfiguration;
- the supervisor of each computer must be able to exchange with other network supervisors through any information channel of this computer, and also, there must be a possibility of switching between these channels;
- it is necessary to have a system of uniform network time available for all OBE computers.

Also, you can formulate requirements that are not necessary, but significantly simplify the work of the supervisor in terms of analysis and decision-making:

- the computing resource of each ODICN computer must be sufficient for the simultaneous implementation of all tasks supported by it (including tasks from the hot standby);
- ODICN software (expert system of each supervisor) should be able to quickly search for a suitable configuration in the event of an emergency situation (it is possible to use hardware support in the form of a neurocontroller or other solutions that speed up the search in the knowledge base containing failure precedents and ways to solve them).

Obviously, the implementation of the process of dynamic reconfiguration of the ODICN should be ensured by adding specialized functional tasks (FO) to the OBE and software and hardware for the administration and monitoring of avionics. Let's formulate a list of such tools and tasks:

*Knowledge base "Precedents"* (KB). It contains a set of independent configurations of the ODICN, each of which is a description of the distribution of the FO and avionics systems among the OBE computers, as well as a set of parameters and characteristics, both included in this avionics configuration, and the configuration itself as a whole. The KB is formed on the basis of the experience of qualified professional pilots and, in fact, is a database of precedents containing information on possible emergency situations (failures) and methods and solutions due to the dynamic reconfiguration of the ODICN with full or partial restoration of the OBE operability, leading to the successful completion of the flight.

*Supervisor.* It is a specialized FO implemented by each ODICN computer. ODICN supervisors organize a closed network and allow continuous analysis of the OBE operability, record the occurrence of an emergency situation and initiate the process of dynamic reconfiguration. The network of supervisors can be either decentralized – supervisors are independent, or centralized – at a time, only one supervisor of the network is the master, the rest are slaves. If the lead supervisor refuses, his role is assigned to one of the slave supervisors.

*Expert system* (ES). Is a part of every supervisor. Depending on the current contingency situation, it selects suitable configurations from the knowledge base and selects one of the most suitable ones. In the absence of suitable configurations in the knowledge base, the ES synthesizes a new configuration that allows to fully or partially restore the operation of the ODICN.

*Function of universal time* (UT). Specialized FO, implemented in each OBE computer, which synchronizes them in time. The UT functions organize a centralized network. One of these functions is assigned as the master, the others as slaves. The leading function is the source of UT, the slave functions receive and apply the obtained value of UT to the clock (timers) of the OBE calculators. In case of failure of the slave function UT, its role is assigned to one of the slaves.

*Switching functions of the memory area* (SMA). A specialized FO that allows the supervisor on each specific computer to be independent of possible failures of the communication channels of this computer, flexibly switching between the available working channels.

*System.* A set of interconnected FOs designed to implement a specific aviation functionality.

*Retranslator function.* An optional auxiliary FO, which retransmits the data stream between the specified two optical channels inside the selected OBE calculator. This function allows you to restore the connection lost as a result of failure between two functions located in different computers through its mediation.

Given the complexity of organizing data exchange between computers in a decentralized network, as well as the increased likelihood of various collisions, it is advisable to opt for the centralized structure of the supervisor network, as the simplest and most reliable. It should also be noted that a centralized network of supervisors can be combined with a network of UT functions (the lead supervisor is, among other things, the source of UT). This makes it possible to simplify the structure of the organization of the OBE administration and monitoring software.

Let us consider in more detail the composition and algorithmic support of the dynamic reconfiguration automated system being developed within the framework of this work, as well as the mathematical model underlying it.

As described earlier, the system being developed is based on the knowledge base of precedents formed by professional pilots on the basis of their experience in parrying avionics failures that occurred during the flight. At the same time, when filling out the knowledge base, the pilots are invited to solve hypothetical emergency situations by turning off some of the secondary systems or redistributing the computing resource of the ODICN, that is, by the method of dynamic reconfiguration. Thus, each such hypothetical contingency (precedent) generates a new ODICN configuration with its own set of parameters.

The developed model assumes the following classification of KB configurations:

***Basic configuration*** (BC). Independent configuration of ODICN, providing full OBE functionality with proper operation of all its systems.

***Decomposition of the basic configuration*** (DBC). Subsidiary configuration, which is a variant of a certain BC, which differs from it by a different distribution of functional tasks among computers;

***Emergency configuration*** (EC). A subsidiary configuration, which is a variant of the BC with a certain list of disabled secondary functional tasks (or systems) that do not affect the successful completion of the flight, but the disconnection of which entails a deterioration in some parameters compared to the BC.

Thus, the KB has the following composition and structure shown in Fig. 1.

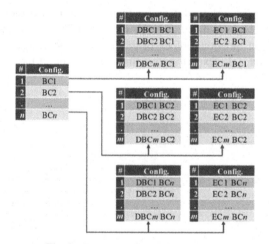

**Fig. 1.** Structure of KB of precedents.

In addition to the configuration, as a description of the distribution of computational tasks among the ODICN computers, each entry in the knowledge base has the following parameters, in aggregate, characterizing the reliability of this configuration:

***Accuracy.*** Configuration estimate (from 1 to 10), which characterizes the order of accuracy of all systems included in a given configuration.

***Ergonomics.*** Configuration assessment (from 1 to 10), summarized, characterizing the convenience of using various OBE interfaces from the composition of this configuration, actively interacting with the pilot, and, as a consequence, affecting the speed of decision-making response in the event of an emergency situation.

***Reliability.*** Assessment of the configuration (from 1 to 10), which generally characterizes the fault tolerance and reliability of the avionics of the given configuration.

Also, additional parameters of KB configurations are resource characteristics of functional tasks and ODICN computers involved in this configuration. These parameters will be used in the automated dynamic reconfiguration system to assess the uniformity of the distribution of computing resources (load) across the ODICN computers.

The developed model, at the current level of abstraction, assumes the following types of avionics failures:

- failure of the communication channel (transmitting or receiving);
- failure of a software application (i.e. FO);
- failure of equipment (sensors, interfaces, etc.) associated with a certain computational task and, as a consequence, the inability to implement this computational task in a particular computer (in other words, it is a failure of a certain binding between the computer and the task).

In the future, complicating and detailing the model, it is possible to expand the list of failures by adding faults at different levels of the ODICN organization.

Another important functional block of the model is the supervisor. As mentioned earlier, it also integrates the expert system and the function of the organization of the UT (Fig. 2).

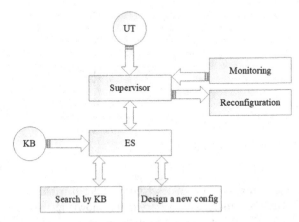

**Fig. 2.** Supervisor structure.

The supervisor continuously monitors the state of the ODICN, which consists in transmitting information about the state of his computer to other network supervisors, receiving similar information from other supervisors and aggregating this data in RAM for its further analysis. The analysis of monitoring information carried out in the supervisor's ES consists in assessing the operability of the avionics and, in the event of an abnormal situation, in finding a suitable configuration from the knowledge base, redistributing the computing resource and the list of active communication channels in such a way that the failed ODICN elements are in the area of idle (reserve). In the absence of suitable configurations, the ES proceeds to the synthesis of a new configuration, relying on the generalization of all the information available in the knowledge base and, using iterative search algorithms, tries to find ways to restore the lost connections between the functional tasks of the ODICN, taking into account their possible redistribution and, if necessary, introducing them into network auxiliary functions-repeaters. At the same time, from the wide range of such iterative algorithms for finding a solution, we will single out the ant colony optimization algorithm as the most appropriate in the context of the presented model. An alternative to the ant algorithm can be an approach using a recurrent neural network (similar to the solution of the classical traveling salesman problem).

## 3 Algorithmic Support

Let's formalize the ODICN model using a specialized functional graph (SFG). SFG ODICN is multilayer, weighted and oriented. For clarity, let us consider a simplified example of ODICN, implemented on the basis of an all-optical on-board environment (AOBE) [13–15], consisting of 3 computers (CPU1, CPU2, CPU3), i.e. $N = 3$, each of which has 2 transmission channels, i.e. $M = 2$. Then, based on the logic of constructing an OBE, each calculator will have $M \times N$ optical channels, of which $M$ are transmitting and $M \times (N - 1)$ receiving (Fig. 3).

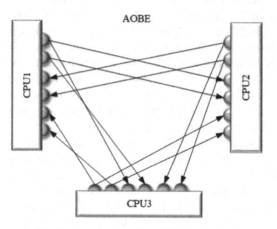

**Fig. 3.** An example of an AOBE structure.

We will consider a ODICN configuration consisting of 2 independent systems, each of which contains 2 computational problems. In addition to the main computational tasks, the developed model assumes the presence of a centralized network of supervisors. The SFG describing one of the possible configurations of such an ODICN is shown in Fig. 4.

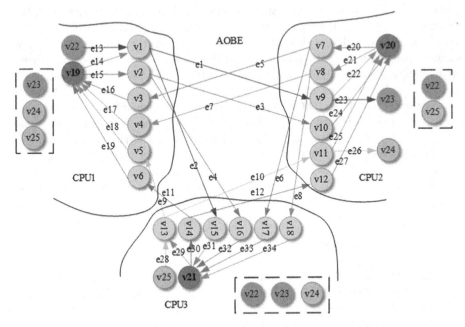

**Fig. 4.** Specialized functional graph.

SFG consists of 3 interconnected subgraphs – the AOBE subgraph (blue vertices), the ODICN subgraph (green and orange vertices) and the supervisor network subgraph (red vertices). The gray arrows outside the calculators show working, but inactive channels (redundancy), and inside the calculators – secondary exchange paths for supervisors through the SMA function and the mechanism for temporal division of the optical channel resource. The dashed blocks show the hot spare tasks that are in a particular calculator.

Within the framework of graph theory, SFG can be described using adjacency matrices and incidence modified for the specificity of ODICN on the basis of AOBE. Also, to describe the list of functional tasks supported by a certain calculator, both activated and in hot standby, an implementation table is needed.

The above description of the model makes it possible to identify and modify data routing in ODICN, as well as, in the event of emergency situations, to find a set of knowledge base architectures that allow to fend off avionics failures with varying efficiency. Also, according to the integrity of the SFG, after the completion of the process of dynamic reconfiguration, it is possible to assess the correctness of the operation of the entire reconfiguration model as a whole, which can serve as a tool for self-control of the ODICN.

Adjacency matrices, incidence matrices, and a table of realizations are necessary, but not sufficient, elements of the description of the ODICN based on the AOBE. For the correct choice of configurations from the list that are suitable for implementation in the current emergency situation, we also introduce resource parameters that allow us to evaluate the uniformity of the distribution of the computing resource over the ODICN computers (load), as well as the load on the optical channels of the AOBE. For this purpose, we introduce resource parameters for functional tasks: storage capacity of $RAM_f$, storage capacity of $ROM_f$, the number of $N_T$ system cycles, call period $T$, the number of free transmitting channels (without failure) $Nf_{tch}$, the number of free receiving channels (without failure) $Nf_{rch}$; as well as resource parameters for calculators: storage capacity of $RAM_{CPU}$, storage capacity of $ROM_{CPU}$, clock frequency of $F_{CPU}$, number of transmitting channels $Nc_{tch}$, the number of receiving channels $Nc_{rch}$.

For each calculator, the following rule must be fulfilled:

$$
\begin{cases}
sRAM = \sum_{i=1}^{L} (RAM_f)_i \leq RAM_{CPU}, \\
sROM = \sum_{i=1}^{L} (ROM_f)_i \leq ROM_{CPU}, \\
sV = \sum_{i=1}^{L} \frac{(N_T)_i}{(T)_i} \leq F_{CPU}, \\
sNf_{tch} = \sum_{i=1}^{L} (Nf_{tch})_i \leq Nc_{tch}, \\
sNf_{rch} = \sum_{i=1}^{L} (Nf_{rch})_i \leq Nc_{rch},
\end{cases}
\tag{1}
$$

$L$ is the number of tasks in this computer.

Let us introduce a special vector of parameters characterizing each configuration of the knowledge base:

**Table 1.** Vector of configuration parameters

| $Ka$ | $K_e$ | $K_p$ | $Flt$ | $\sigma RAM$ | $\sigma ROM$ | $\sigma V$ | $\sigma Nf_{tch}$ | $\sigma Nf_{rch}$ |
|------|-------|-------|-------|--------------|--------------|------------|-------------------|-------------------|

$K_a$ – calculation accuracy (1–10), $K_e$ – ergonomics (1–10), $K_p$ – priority of use (1–10), $Flt$ – status (0 - BC/DBC, 1 - EC),

$$
\begin{aligned}
\overline{RAM} &= \frac{1}{N} \sum_{i=1}^{N} (sRAM)_i, \\
\sigma RAM &= \sqrt{\frac{1}{N} \sum_{i=1}^{N} \left[ (sRAM)_i - \overline{RAM} \right]^2},
\end{aligned}
\tag{2}
$$

$$\overline{ROM} = \tfrac{1}{N} \sum_{i=1}^{N} (sROM)_i,$$

$$\sigma ROM = \sqrt{\tfrac{1}{N} \sum_{i=1}^{N} \left[(sROM)_i - \overline{ROM}\right]^2}, \qquad (3)$$

$$\overline{V} = \tfrac{1}{N} \sum_{i=1}^{N} (sV)_i,$$

$$\sigma V = \sqrt{\tfrac{1}{N} \sum_{i=1}^{N} \left[(sV)_i - \overline{V}\right]^2}, \qquad (4)$$

$$\overline{Nf_{tch}} = \tfrac{1}{N} \sum_{i=1}^{N} (sNf_{tch})_i,$$

$$\sigma Nf_{tch} = \sqrt{\tfrac{1}{N} \sum_{i=1}^{N} \left[(sNf_{tch})_i - \overline{Nf_{tch}}\right]^2}, \qquad (5)$$

$$\overline{Nf_{rch}} = \tfrac{1}{N} \sum_{i=1}^{N} (sNf_{rch})_i,$$

$$\sigma Nf_{rch} = \sqrt{\tfrac{1}{N} \sum_{i=1}^{N} \left[(sNf_{rch})_i - \overline{Nf_{rch}}\right]^2}, \qquad (6)$$

where $N$ is the number of calculators.

Thus, the KB organization and the structure of descriptions of each configuration (use case) have the form shown in Fig. 5.

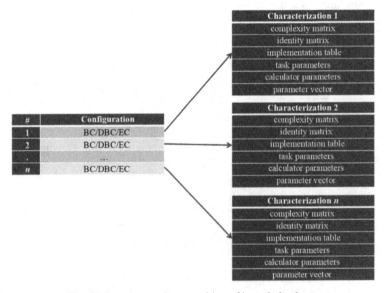

**Fig. 5.** Structure and composition of knowledge base.

We use the above description of the ODICN model based on AOBE to develop an algorithm for dynamic reconfiguration in the event of an emergency situation (failure of avionics elements). For the algorithm to work, it is necessary to form a KB of precedents with a set of descriptions (Fig. 5). The generated knowledge base must be stored in the ROM of each OBE calculator. Also, in ODICN, a centralized network of supervisors with a unique priority should be implemented, while the serviceable supervisor with the highest priority is always the leading one. The work of the ODICN elements responsible for monitoring and administration for the purpose of dynamic reconfiguration, as a reaction to an emerging emergency situation, can be divided into the following stages:

- Both during the initialization (start) of ODICN, and in the process of its operation, the supervisor on each computer monitors the operation of its avionics and transfers information about its state to the network, as well as receives similar information from other supervisors of the OCN and aggregates complete information about the ODICN;
- During initialization or upon the event of an emergency situation, the leading supervisor transmits monitoring information to the ES, which, based on the analysis of the situation, selects the most suitable configuration from the knowledge base or, in the absence of a current one, synthesizes a new one. As a result, the ES sends to the leading supervisor a description of the configuration designed to counter the avionics failures that have arisen, the supervisor initiates the reconfiguration process and transmits to all slave supervisors a description of the resulting architecture (each slave supervisors apply a new configuration on their own computer);
- In the event of such a failure of the avionics, which no longer allows the leading supervisor to transmit data to the network, his role begins to be performed by the next serviceable supervisor in priority (the process of transfer of authority is initiated after a certain timeout of the leader's failure to communicate, indicating his refusal). At the same time, during the timeout, on a computer with a failed avionics, the supervisor tries, using the SMA function, to transfer data to the network through secondary channels (channels that he shares with other ODICN tasks through the feature switching mechanism or time division multiplexing). If attempts to use secondary channels during the timeout period do not lead to success, the next good supervisor in priority becomes the lead supervisor;
- The ES searches in the knowledge base according to the following principle: the new configuration must have such an architecture that the failed network elements are in the area of the unused resource of the ODICN. From the set of suitable configurations found, the ES chooses the most suitable one according to the following principle: the most accurate one in terms of the parameter $K_a$ is selected, if architectures with the same accuracy parameter are found, the ES selects from them the configuration with the best ergonomics in terms of the parameter $K_e$, then a similar choice is made according to the resource parameters and if, as a result, again there are several suitable configurations, the ES chooses from them the one that has the highest priority $K_p$ (the priority is unique in this case, the existence of configurations with the same priority in the knowledge base is not allowed).

## 4 Conclusion

In this article, requirements were formed for the hardware and software implementation of the OBE system of the reconfigurable ODICN, as well as the software for administration and monitoring, a mathematical model and algorithmic support for the ODICN were developed, methods for constructing an automated dynamic reconfiguration system based on the principles of the precedent approach were proposed.

The results obtained are applicable in the modeling, testing and study of the properties of high-performance OBE structures and general aircraft systems based on the DME using promising information interfaces and can be used in the design of the OBE architecture of civil aviation aircraft, as well as in the development of specialized software. Potential consumers of the results obtained are aircraft design bureaus that develop civil aviation aircraft, aircraft construction teams when creating promising aircraft, as well as companies seeking to modernize aircraft avionics in order to improve its compliance with promising international requirements for reliability, efficiency and flight safety.

## References

1. Liang, M., Delahaye, D., Maréchal, P.: Integrated sequencing and merging aircraft to parallel runways with automated conflict resolution and advanced avionics capabilities. Transp. Res. Emerg. Technol. **85**, 268–291 (2017)
2. Prisaznuk, P.J.: ARINC 653 role in integrated modular avionics (IMA). In: 2008 IEEE/AIAA 27th Digital Avionics Systems Conference. IEEE (2008)
3. Sánchez-Puebla, M.A., Carretero, J.: A new approach for distributed computing in avionics systems. ISICT **3**, 579–584 (2003)
4. Zhou, Q., et al.: A two-phase multiobjective local search for the device allocation in the distributed integrated modular avionics. IEEE Access **8**, 1–10 (2019)
5. Neretin, E.S., et al.: Application of distributed integrated modular avionics concept for perspective aircraft equipment control systems. In: Journal of Physics: Conference Series vol. 1353, no. 1, p.012005. IOP Publishing (2019)
6. Airbus. Global Market Forecast 2016–2035 (2016). <http://www.airbus.com/company/mar ket/global-market-forecast-2016-2035.pdf>
7. Boeing, Current Market Outlook 2015–2034, August 2015
8. CAAC, China National Civil Aviation Efficiency Report 2015, July 2016
9. Levin, I.I., Dordopulo, A.I., Pisarenko, I.V., Melnikov, A.K.: Parallelization of Algorithms in Set@l Language of Architecture-Independent Programming. In: Sokolinsky, L., Zymbler, M. (eds.) PCT 2020. CCIS, vol. 1263, pp. 31–45. Springer, Cham (2020). https://doi.org/10. 1007/978-3-030-55326-5_3
10. Degtyarev, A.R., Kiselev, S.K.: Hardware reconfiguration algorithm in multiprocessor systems of integrated modular avionics. Russ. Aeronaut. **60**(1), 116–121 (2017). https://doi.org/ 10.3103/S1068799817010172
11. Bukov, V., et al.: Determination of functional efficiency of an airborne integrated navigation system for the purpose of reconfiguration of it in flight. In: Proceedings of the 4th European Conference for Aero-space Sciences, Saint Petersburg, Russia, Report 818–1245–1. (2011)
12. Ageev, A.M., Bronnikov, A.M., Bukov, V.N., Gamayunov, I.F.: Supervisory control method for redundant technical systems. J. Comput. Syst. Sci. Int. **56**(3), 410–419 (2017). https://doi. org/10.1134/S1064230717030029

13. Kosyanchuk, V.V., Semenov, M.E., Solovyov, A.M., Novikov, V.M., Karpov, E.A., Mishchenko, I.B.: Concept of construction of a decision support system when performing dynamic reconfiguration of airborne equipment complex. Theor. Technol. Radio Commun. **1**, 5–18 (2021)
14. Zhao, Y., et al.: On-board artificial intelligence based on edge computing in optical transport networks. In: 2019 Optical Fiber Communications Conference and Exhibition (OFC). IEEE (2019)
15. Zhao, Y., et al.: Coordination between control layer AI and on-board AI in optical transport networks. J. Optic. Commun. Netw. **12**(1), A49-A57 (2020)

# Cyber Threat Risk Assessment in Energy Based on Cyber Situational Awareness Techniques

Daria Gaskova$^{(\boxtimes)}$ ⓘ, Elena Galperova ⓘ, and Aleksei Massel ⓘ

Melentiev Energy Systems Institute SB RAS, Lermontov Street 130, 664033 Irkutsk, Russia
gaskovada@isem.irk.ru

**Abstract.** The high rate of digitalization into all spheres of human life including the energy sector, on the one hand, increases the efficiency of the physical infrastructure management at facilities and simplifies the interaction between system users, and on the one hand, it incites risks associated with interaction in the cyber environment. Both aspects referred to the relevance of this research. In this regard, it's imperative to utilize artificial intelligence methods and technologies to assess cyber threat risk in the energy sector. It's connected with the tendency of systems to increase the sophistication and the increase of the amount of information that is important in assessing risks in the cyber environment. The article proposes a method and model for assessing the risks of the cyber threat in the energy sector, developed by the authors. The proposed method differs by the integration of semantic modeling methods, expert systems technology, and cognitive graphics. Semantic methods serve for analyzing the cyber threats' impact on energy facilities account for energy security for current research. Such methods offer efficiencies in the absence or incompleteness of data in modeling the behavior of systems that defies formal description or sufficiently accurate prediction. The kind of semantic modeling type as Bayesian Belief Network serves to probabilistic risk assessment at an energy facility. The graph's nodes are interrelated events corresponding to the vulnerabilities of the assets of the local area network, cyber threats, technogenic threats to energy security, and the consequences of their implementation. The probabilistic component of risk is understood as the possibility of a meaningful effect (intentional or accidental) on the physical infrastructure of a digital energy facility from cyberspace in this context. That sort of effect is considered an extreme situation, including critical and emergency situations. The risk assessment is based on the posterior probabilities of the nodes corresponding to the consequences are used, as well as components of damages, taking into account the possibility of an extreme situation. The feasibility of methods and models is illustrated with an example at the end of the article.

**Keywords:** Cyber threats · Cyber risk · Extreme situations · Semantic modeling methods

## 1 Introduction

Cybersecurity and cyber situational awareness in energy research generally are related to critical infrastructures. Disruption of the functioning or destruction of energy facilities leads to extreme situations namely disastrous consequences, and energy security

© Springer Nature Switzerland AG 2022
V. Taratukhin et al. (Eds.): ICID 2021, CCIS 1539, pp. 134–145, 2022.
https://doi.org/10.1007/978-3-030-95494-9_11

violations. Meanwhile, cyber threats are identified as a group of strategic threats in the energy sector [1]. Ensuring cybersecurity in the energy sector is a considerable and multifaceted problem, complicated by different levels of the hierarchy in energy management. That way the problems are separated by different management levels from cybersecurity issues of particular facilities (for example, digital substations) to automated control systems for energy enterprises and also the industry generally.

It is seen from studies of attacks on the fuel and energy complex of Russia [2] for 2018–2019 that the main targets of cybercriminals are a destructive effect on infrastructure, as well as industrial espionage. That effect in energy security studies is considered an extreme situation.

Documents of federal significance reflect the results of these studies at the strategic management level. Specifically, in Russia, in the Energy Security Doctrine [3] computer attacks and the unlawful use of information and telecommunication technologies that can disrupt energy facilities' functioning among the threats to energy security are highlighted. At the tactical level of management, one of the traditional study methods is scenario analysis, which deals with predicting the operation of facilities, investigating accidents that have occurred, and so on. A large number of events are difficult to characterize by one scenario reflecting a large-scale accident, especially a cascade one. The scenario is divided into several stages of the development of an accident for this reason. The stages are considered sequentially and thus reproduce the entire accident [4].

The paper considers the stage of an extreme situation covering events in the facility's cyber environment including guest, corporate and technological domains of local area network (LAN), and also the possible consequences associated with technological threats to energy security.

## 2   Cyber Situational Awareness of Digital Energy Facilities

Although the term "Situational Awareness" appeared in the 80s of the last century, research related to it remains relevant today. The traditional definition of situational awareness proposed by Mica Endsley [5] is gaining momentum in the field of cybersecurity. According to M. Endsley [6], the growing interest in situational awareness is associated mainly with the problems of a new class of technologies. One of these problems is associated with the processing of large amounts of data generated by modern systems. "The problem with today's systems is not a lack of information, but finding what is needed when it is needed." [6].

### 2.1   Cyber Situational Awareness Basics

Cyber situational awareness is interpreted as part of situational awareness, limited by the cyber environment [7]. Systems and networks operating in cyberspace include vulnerabilities that drive significant risks to both particular organizations and national security [8]. Accordingly, cyber situational awareness models are considered at the national and facility levels. The paper deals with the facility level. The main activities related to the level of a facility are highlighted in the [9]. There are vulnerability management, patch management, event management, incident management, malware detection, asset management, configuration management, network management, license management, and information management.

There are two components of cyber situational awareness: i) technical one and ii) cognitive one. In the first case, cyber situational awareness is provided by the improvement of intelligent protection technologies and systems, for example, anomaly detection systems, intrusions, user activity monitoring, etc. One such technology is security monitoring. Security monitoring covers both levels of network security monitoring and endpoint security monitoring. Different flows of information about the states of local computing networks and assets are allocated for this purpose. This component also includes the exchange of information between various stakeholder groups in order to prepare for and manage possible incidents [10]. Data exchange formats under develop and distribute like the Incident Object Description Exchange Format (IODEF) [11], Common Event Expression (CEE) [12], Malware Metadata Exchange Format (MMDEF) [13], Intrusion Detection Message Exchange Format [14] for instance and another ones besides that. In the second case, one implies the ability of the analyst to understand information to make informed decisions. Visualization and visual analytics are proposed for a better human understanding of security aspects as follows in [15].

The article discusses methods of semantic modeling appearing, on the one hand, tools for the expert's work and visualizing their knowledge about the subject area, and, on the other hand, tool for working in conditions of uncertainty and incompleteness of data.

## 2.2  Proposal Method and Models

The work presents cyber situational awareness of energy facilities which is understood as an awareness of the state of the cyber environment of energy facilities, including information about i) critical vulnerabilities of energy facilities from the point of view of cybersecurity; ii) cyber threats that initiate these critical vulnerabilities, as well as iii) technogenic threats to energy security caused by cyber threats. From this perspective threats to energy security constitute a deficit of resource requirements with acceptable quality under normal conditions and extreme situations, the violation of stability and uninterrupted energy supply [16].

The scenario model for energy emergencies caused by cyber threats is proposed in [17]. The model includes both four types of concepts and types of relations between them. The main types of concepts are i) vulnerabilities ($V$) of assets of the facility's cyber environment; ii) cyber threats ($T$); iii) technogenic threats ($W$) to energy security; iv) consequences ($C$) from the implementation of cyber and technogenic threats. The relations between the concepts reflect the cause-and-effect relationships between them and demonstrate the vectors of cyber threats that can potentially be considered as causes of extreme situations. Arcs represent the relations.

Bayesian Belief Network (BBN) is used to model cyber threat vectors, in particular, attack vectors, within the cyber environment of an organization in the presented work. Bayesian Belief Network allows one to interpret and combine information fragments about cyber threats into plausible knowledge for predicting the risks of cyber threats at enterprises [18], including energy facilities. It is proposed to combine information fragments using an expert system and display the risk assessment using cognitive graphics.

Modeling probabilistic scenarios of extreme situations caused by cyber threats using the Bayesian Belief Network demands preliminary preparation and includes several stages combined into the technology of cyber threats analysis in the energy sector proposed in [19].

# 3 Cyber Threat Risk Assessment

The development and implementation of new technologies, business processes, and models contribute to the existing trend of system complexity. To meet the needs of analyzing and assessing risk new methods have been proposed instead of traditional ones in this regard. Examples of relatively newly introduced methods are Bayesian Belief Networks, Binary Digit Diagrams, multi-state reliability analysis, Petri Nets and advanced Monte Carlo simulation tools [20]. Cybersecurity is closely related to risk assessment and decision-making to protect the cyber environment from cyber threats. Probabilistic models are built based on Bayesian Networks is often in such studies. For instance, in [21] Zhang Q., Zhou C., and co-authors propose to carry out a dynamic assessment of cybersecurity risks using a multi-layered Bayesian Network containing models of attacks, functions and incidents. The studies [22, 23] of Dantu R. and Kolan P. are founded on the idea that the sequence of an attacker's actions in the network depends on his social behavior. To build their model based on BBN they considered several profiles of attackers respond to cost, computer and hacking skills, attitude, time, tenacity, perseverance, and motives. Liu Y. and Man H. in [24] presented a method for modeling potential attack paths on the present network using BBN. The authors refer to three advantages of using BBN as follows i) providing cause-effect relations to display the stages of the attack, i.e. the success of the next step depends on the success of the previous one; ii) compact representation of attack paths in comparison with traditional attack trees and, accordingly, better scalability; iii) formalism of reasoning in conditions of uncertainty.

## 3.1 Cyber Threat Risk and Cyber Risks

Analyzing the definitions provided by the NIST Glossary [25] advances the understanding that cyber risk covers both the risks of unauthorized access to assets of the cyber environment, cyber attacks, cyber negligence, and issues of reliability of assets and systems. Hence, the work uses the concept of cyber threat risks, limiting research to threats that can lead to extreme situations.

Carfora M. and Orlando A. in [26] observe cyber risks present some specific characteristics including i) a fluid cyber risk landscape seeing the rapid development of information systems and technologies; ii) the difficulty of assessing the frequency of cyberattacks; iii) the difficulty of assessing the economic and financial impact of a cyber attack (for the nature of information assets); iv) interdependence of assets (the security level of one system may affect the security level of another).

The attack and cyber threat vectors model is a probabilistic scenario of an extreme situation represented by the graph model of the Bayesian Belief Network. Table 1 shows the states of nodes corresponding to vulnerabilities, cyber threats, threats to energy security (technogenic threats) and consequences.

**Table 1.** Vulnerabilities, cyber threats, technogenic threats, and consequences nodes.

| Node type $X_i$ | Notation | States | Parents limitation $pa(X_i)$ | Probability |
|---|---|---|---|---|
| Vulnerability | $X_h^V$, $h = \overline{1,H}$ | $v_1$: exploit<br>$v_2$: omit | $X^V$, $X^T$ | $\mathcal{P}\left(X_h^V\right) = P\left(X_h^V \lvert pa\left(X_h^V\right)\right)$ |
| Cyber threat | $X_k^T$, $k = \overline{1,K}$ | $t_1$: exploit<br>$t_2$: omit | $X^V$, $X^T$ | $\mathcal{P}\left(X_k^T\right) = P\left(X_k^T \lvert pa\left(X_k^T\right)\right)$ |
| Technogenic threat | $X_m^W$, $m = \overline{1,M}$ | $\{w_{11},\ldots,w_{2u}\}$, where $w_{1i}$ is percentage of generating capacities failure | $X^V$, $X^T$, $X^W$ | $\mathcal{P}\left(X_m^W\right) = P\left(X_m^W \lvert pa\left(X_m^W\right)\right)$ |
| Consequence | $X_z^C$, $z = \overline{1,Z}$ | $\{c_{11},\ldots,c_{2q}\}$, where $c_{1i}$ is an event that characterizes the damage | $X^W$ | $\mathcal{P}(X_z^C) = P(X_z^C \lvert pa(X_z^C))$ |

The purpose of such modeling is to answer the question "How will the likelihood of consequences change if one assumes that several vulnerabilities and/or threats have been implemented?". Usage of the Bayesian Belief Network allows one to determine the probabilistic components of the risks of cyber threats, represented by the posterior probability of the consequences of the scenario. The risk is considered as the product of the probability of the consequence by damage. The structure of damages of an extreme situation is formed to the consequences. The damage can be expressed initially in kind, and then estimated in monetary terms.

## 3.2 Case Study

In the process of digital transformation of the energy sector, enterprises are divisible into three groups: i) that have not yet reached the level of application of advanced information technologies; ii) already started using digital technologies, but not aware of their dangers; iii) applying digital technology and taking security measures by standards and fundamental documents, as well as existing practices. The most vulnerable are enterprises of the second group. The following is a small example of an abstract combined heat and power station and a risk assessment for it.

**Analysis of Vulnerabilities and Cyber Threats.** For each asset in the cyber environment of the facility in question, a list of vulnerabilities that have not been eliminated is determined. Table 2 shows an example of such a list.

**Table 2.** List of assets and their vulnerabilities.

| Symbol | Remark | LAN domain | Vulnerabilities |
|--------|--------|------------|-----------------|
| R1 | WiFi router. Performs segmentation into guest and corporate domains | Guest and corporate network | Outdated software $(V_1)$. Network separation error $(V_2)$ |
| PC1 | Office computer with a Mail client | Corporate network | Incorrectly configured SPAM filter, antivirus $(V_4)$ |
| R2 | Router. Performs segmentation into corporate and technological domains | Corporate and technological networks | Outdated software; Network separation error $(V_2)$ |
| WS1 | Administrator workstation. Lack of a demilitarized zone | Corporate and technological networks | Dictionary password $(V_5)$ |
| WS2 | Engineer workstation with remote desktop. SCADA/HMI | Technological networks | Using saved remote connection sessions $(V_6)$ |
| S1 | SCADA server (electrical part) | Technological networks | Violation of update policy $(V_3)$ |
| S2 | SCADA server (thermo-technical part) | Technological networks | Violation of update policy $(V_3)$ |

**Cyber Threat Vector Modeling.** The presented example considers two cyber threats vectors. *Attack vector on the technology domain via guest WiFi.* The P1 router uses an outdated software version that allows one to gain a local admin access to the device $(V_1)(V_1)$ and performs network segmentation into guest and corporate domains at the logical level with an incorrectly configured firewall $(V_2)$. The exploitation of $V_1$ and $V_2$ vulnerabilities can lead to data interception with analysis of network traffic $(T_1)$ and gain access to the network map of the corporate domain $(T_5)$. Since the P2 router performs network segmentation into corporate and technological domains and also uses an outdated software version, it becomes possible to carry out similar actions for it and analyze network traffic in the technological domain of the network. Having gained access to the map of the technological network, it becomes possible to obtain sensitive information (IP, DNS) about the target assets, which are SCADA (Supervisory Control and Data Acquisition) servers S1 and S2. Since the target asset uses outdated software (update policy violation) $(V_3)$, the next stage is to gain a local admin access to it using an exploit. *Attacks vector on the technology network through phishing mailings.* The first stage is social engineering and gaining access to PC1 located in the corporate network by phishing mailing to the email client ($V_4$) and using malicious software. Further consolidation in the corporate network, obtaining a local admin access on PC1 ($T_2$), and increasing privileges by gaining a local admin access to the corporate domain ($T_4$). Whereas there is no demilitarized zone in the considered example, such access allows one to search for nodes in the technological network. The following stage is consolidation in the technology network ($T_7$) using a vulnerable version of the operating system of the engineer's workstation and attacks to target assets S1, S2 using saved remote connection sessions ($V_6$). *Target asset attacks.* Two options are considered as attacks on the target asset. There are i) deletion or encryption of SCADA server's data; ii) sending destructive control commands to the power unit (requires high skills and additional actions to deceive the protection mechanisms).

It is proposed to save, transfer and analyze knowledge about possible attack vectors in the form of production rules using an expert system. An example of a production rule describing penetration from a guest network to a corporate network via guest WiFi is shown in Table 3.

**Table 3.** An example of entering a production rule.

| An example of entering a production rule for describing penetration from a guest network to a corporate network via guest WiFi | | | |
|---|---|---|---|
| when | Asset(type == "Guest Corporate Wi-Fi router") ‖ Vulnerability (type == "Outdated version of software allowing access control") | then | set("Asset.type: Guest Wi-Fi router; CyberThreatList: Getting administrative access"); |
| end | | | |
| | | ... | |
| when | Asset(type == "Guest Corporate Wi-Fi router") ‖ CyberThreat(type= "Getting administrative access") | then | set("Asset.type: Guest Wi-Fi router; CyberThreatList: penetration to corporate network map"); |
| end | | | |

**Modeling Cyber Threat Vectors with Bayesian Belief Network.** The BBN model construction relies on the production rules of the expert system. The nodes of the graph are a reflection of vulnerabilities ($V$), threats ($T$, $W$), consequences ($C$), and arcs display the cause-and-effect relations as the sequence of the attack units. Information about the probabilistic characteristics of discrete random variables ($X$) corresponding to each of the nodes is determined with the expert's experience or the available statistical information. Figure 1 is for the probabilistic model of the present example. The posterior probabilities of consequences are determined by network reasoning and depending on the implemented attack scenario. The scenario offers a set of evidence ($E$) at the nodes corresponding to the vulnerabilities and threats.

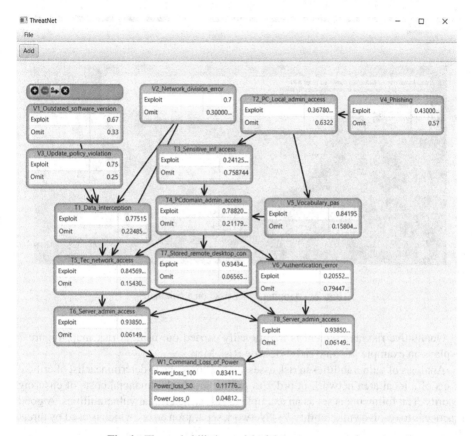

**Fig. 1.** The probabilistic model of the present example.

**Cyber threat Risk Assessment.** The risks are described and then calculated using the formula (1):

$$R_i = P\left(X_z^C | E = e\right) \times D_i, \tag{1}$$

where $R_i$ is risk of consequence $C_z$, $e$ stands for the set of values of the evidence $E$, $e \in [0, 1]$, $P\left(X_z^C | E = e\right)$ is posterior probability of consequence $C_z$, $D_i$ denotes damage from the consequence $C_z$. Provided that $D_i = \{d_{i1}, \ldots, d_{im}\}$, where $d_{i1}, \ldots, d_{im}$ are elements describing the structure of damage for consequence $C_z$.

The present example structure of damage for the consequences of "loss of power" can be as follows (2):

$$D_i = \{d_1, d_2, d_3, d_4, d_5, d_6\}, \tag{2}$$

where $d_1$ is lost profits; $d_2$ stands for losses related to various forfeits, fines, penalties, etc.; $d_3$ denotes losses associated with undersupply of energy; $d_4$ signifies production downtime losses; $d_5$ denotes accident investigation costs; $d_6$ is restoration cost.

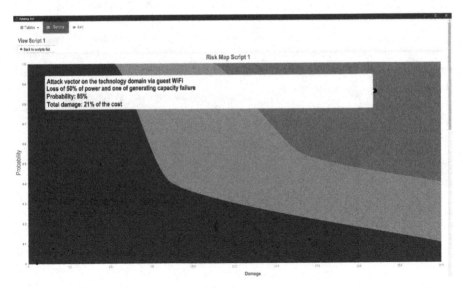

**Fig. 2.** Mapping cyber risks on a risk map.

Qualitative risk assessment is traditionally carried out using a risk map. Figure 2 displays an example of some risks view on Risk Map.

Analysis of vulnerabilities in risk assessment allows one to determine a list of critical assets of a local area network in order to further justify the financial costs of ensuring security. The following is set as an example of resolving critical vulnerabilities. A good practice is to resolve vulnerability $V_2$ "Network separation error" which is used by threat $T_5$ "Gain access to the network map" leading to unauthorized access to the technological domain in this sample. The resolving of this vulnerability is provided by the physical separation of the LAN into corporate and technological domains with locked out external access to the technological domain. Apart from that, it is recommended to create a demilitarized zone to locate services and tools required to work from a technological domain such as mail server or ERP system. Secure data link based on VPN increases the security both within the organization and possible unauthorized access to the endpoints of the corporate segment from outside. Threat $T_4$ "Gaining a local admin access to the corporate domain" results in critical threats like administrative access to the target asset. To overcome this, strengthen the protection of sensitive information during its transmission by file transfer protocols and improving password policy are proposed. Avoiding dictionary passwords, setting the password expiration date, limiting unsuccessful access attempts, logging user actions hold for the latter case.

Correcting the server's authentication errors by conformance of the update policy and limiting external processes are recommended to prevent the implementation of threats $T_6$ and $T_8$ "Administrative access to the target asset". It follows to organize an emulator of the software operating environment (Honeypot) for demilitarized zone. Such technique allows one to track intrusion attempt to LAN, study the behavior of attackers and malware they use. The general recommendation concerns the correct configuration of anti-virus protection, compliance with the software update policy, installation and configuration of firewalls to critical assets, increasing the computer literacy of the organization's employees and their awareness about possible types of cyber threats.

Cyber threat risk assessment in energy is a complex multi-stage task that requires the teamwork of specialists in the field of cybersecurity of critical infrastructures, energy security experts, and analysts.

## 4 Conclusion

The article reflects the author's approach to the risks assessment of cyber threats for the energy sector concerning cyber situational awareness. The integration of expert knowledge, information about cyber threats, assessments, and statistical information allows one to identify the most vulnerable elements in the system. The usage of probabilistic modeling tools and an expert system support both chains king of vulnerability-threat-consequence and probabilistic assessing, which is further used in risk assessment at facilities. Visualization of threat vectors is another advantage of the semantic modeling methods providing the approach described in the article. It allows one to highlight the most dangerous ones, which in turn helps in making decisions and issuing final recommendations.

Further work is planned to extend this approach to facilities such as large, complex cyber physical systems characterized by large volumes of information that significantly exceeds the data of a particular facility.

**Acknowledgments.** This work was executed within the framework of project on state task MESI SB RAS FWEU-2021-0007 № AAAA-A21-121012090007-7 using the resources of the High-Temperature Circuit Multi-Access Research Center (Ministry of Science and Higher Education of the Russian Federation, project no 13.CKP.21.0038). The studying of separated aspects was supported by RFBR grants No 19-07-00351, No 20-010-00204.

## References

1. Massel, L.V., Voropay, N.I., Senderov, S.M., Massel, A.G.: Cyber danger as one of the strategic threats to Russia's energy security. Cybersecur. Issues **4**(17), 2 (2016). https://doi.org/10.21681/2311-3456-2016-4-2-10. (in Russian)
2. APT attacks to Russian energy sector: review of tactics and techniques. Analytics of Positive Technologies. https://www.ptsecurity.com/ru-ru/research/analytics/apt-attacks-energy-2019. Accessed 23 Aug 2021. (in Russian)
3. Russian Federation Presidential Decree Energy security doctrine of the Russian Federation of 13th May 2019 No. 216. Collection of Legislation of the Russian Federation of 20th May 2019 No. 20, section 2421. (in Russian)
4. Rabinovich, M.A., Gaisner, A.D., Potapenko, S.P., Korotkov, V.A.: Application of the RETREN training complex for analysis of system accidents in the power industry. In: Voropai, N.I. (ed.) International Scientific Seminar. Yu.N. Rudenko 2018, Methodological issues in the study of the reliability of large energy systems, vol. 2, pp. 350–360. ESI SB RUS, Irkutsk (2018). (in Russian)

5. Endsley, M.R.: Situation awareness global assessment technique (SAGAT). In: Proceedings of the IEEE 1988 National Aerospace and Electronics Conference, pp. 789–795. Dayton, OH, USA (1988). https://doi.org/10.1109/NAECON.1988.195097

6. Endsley, M.R.: Theoretical underpinnings of situation awareness: a critical review. In: Endsley, M.R., Garland, D.J. (eds.) Situation Awareness Analysis and Measurement, pp. 3–32. LEA, Mahwah, NJ (2000)

7. Frank, U., Brynielsson, J.: Cyber situational awareness – a systematic review of literature. Comput. Secur. **46**, 18–31 (2014). https://doi.org/10.1016/j.cose.2014.06.008

8. The MITRE Corporation, Capabilities overviews. https://www.mitre.org/capabilities/cybersecurity/situation-awareness. Accessed 23 Aug 2021

9. Bazrafkan, M.H., Gharaee, H., Enayati, A.: National cyber situation awareness model. In: 2018 9th International Symposium on Telecommunications (IST), pp. 216–220 (2018). https://doi.org/10.1109/istel.2018.8660997

10. Poyhonen, J., Nuojua, V., Lehto, M., Rajamaki, J.: Cyber situational awareness and information sharing in critical infrastructure organizations. Inf. Secur. Int. J. **43**(2), 236–256 (2019)

11. Danyliw, R., Meijer, J., Demchenko, Y.: The Incident Object Description Exchange Format. RFC 5070 (Proposed Standard), Internet Engineering Task Force, 2007, updated by RFC 6685. http://www.ietf.org/rfc/rfc5070.txt. Accessed 23 Sept 2021

12. The MITRE Corporation, Common Event Expression (2014). http://cee.mitre. org/. Accessed 23 Sept 2021

13. IEEE SA - ICSG Malware Metadata Exchange Format Working Group. https://standards.ieee.org/industry-connections/icsg/mmdef.html. Accessed 23 Sept 2021

14. Curry, D., Debar, H.: Intrusion Detection Message Exchange Format Data Model and Extensible Markup Language (XML) Document Type Definition (2007). https://www.ietf.org/rfc/rfc4765.txt. Accessed 23 Sept 2021

15. Carroll, F., Chakof, A., Legg, P.: What makes for effective visualisation in cyber situational awareness for non-expert users? In: 2019 International Conference on Cyber Situational Awareness, Data Analytics and Assessment (Cyber SA). IEEE, Oxford, United Kingdom, pp. 1–8 (2019). https://doi.org/10.1109/cybersa.2019.8899440

16. Piatkova, N.I., Rabchuk, V.I., Senderov, S.M., Cheltsov, M.B.: Energy Security of Russia: Problems Solutions. SO RAN, Novosibirsk (2011). (in Russian)

17. Gaskova, D.A., Massel, A.G.: Modeling scenarios of extreme situations in the energy sector caused by cyber threats. In: International Conference of Young Scientists Energy Systems Research 2021. E3S Web of Conferences, Irkutsk, vol. 289, p. 03005 (2021). https://doi.org/10.1051/e3sconf/202128903005

18. Pappaterra, M.J., Flammini, F.: A review of intelligent cybersecurity with Bayesian networks. In: 2019 IEEE International Conference on Systems, Man and Cybernetics (SMC), pp. 445–452. IEEE, Italy (2019). https://doi.org/10.1109/smc.2019.8913864

19. Gaskova, D.A., Massel, A.G.: Technology of the intelligent system application for cyber threat analysis at energy facilities. In: Proceedings of the 21th International Workshop on Computer Science and Information Technologies (CSIT'2019), pp. 263–266. Atlantis Press, Austria, Vienna (2019). https://doi.org/10.2991/csit-19.2019.46

20. Aven, T.: Introduction to risk management and risk assessments. Challenges. In: Quantitative Risk Assessment: The Scientific Platform. Cambridge: Cambridge University Press, pp. 1–15 (2011). https://doi.org/10.1017/CBO9780511974120.002

21. Zhang, Q., Zhou, C., Xiong, N., Qin, Y., Li X., Huang S.: Multimodel-based incident prediction and risk assessment in dynamic cybersecurity protection for industrial control systems. IEEE Trans. Syst. Man Cybern. Syst. **46**(10), 1429–1444 (2016). https://doi.org/10.1109/TSMC.2015.2503399

22. Dantu, R., Kolan, P.: Risk management using behavior based Bayesian networks. In: Intelligence and Security Informatics, pp. 165–184 (2005)
23. Dantu, R., Kolan, P., Akl, R., Loper, K. Classification of attributes and behavior in risk management using Bayesian Networks. In: 2007 IEEE Intelligence and Security Informatics (2007). https://doi.org/10.1109/isi.2007.379536
24. Liu, Y., Man, H.: Network vulnerability assessment using Bayesian networks. In: Data Mining, Intrusion Detection, Information Assurance, and Data Networks Security (2005). https://doi.org/10.1117/12.604240
25. Glossary, NIST CSRC. https://csrc.nist.gov/glossary/term/cyber_risk. Accessed 01 Oct 2021
26. Carfora, M.F., Orlando, A.: Quantile based risk measures in cyber security. In: 2019 International Conference on Cyber Situational Awareness, Data Analytics and Assessment (Cyber SA). IEEE, Oxford, United Kingdom, pp. 1–4 (2019). https://doi.org/10.1109/CyberSA.2019.8899431.

# Features of Creating and Maintaining a Corporate Culture for Remote Workers Within the Organization in the Digital Transformation Era

Sofya Shatailova(✉) 📵 and Natalia Pulyavina 📵

Plekhanov Russian University of Economics, Moscow, Russia
Shatailova.S@edu.rea.ru, pulyavina.ns@rea.ru

**Abstract.** This paper is devoted to corporate culture created and maintained in a virtual environment for remote employees of organizations. The authors investigate how the trend of transition to a new format of the organization, the employees of which are geographically remote from each other and do not have access to the office of the company, influence corporate culture. The emergence and spread of this format is the result of digital transformation. Since the research is focused on a relatively new trend, its results provide useful knowledge for management and answer a number of questions that have not previously covered sufficiently, such as the role of a strong corporate culture in virtual organizations and ways of its maintenance and development. In this paper the specific features inherent in the corporate culture for remote workers were identified. A detailed analysis of methods and tools used to build and maintain a corporate culture in a virtual environment, such as cloud technologies, was conducted. The data for the study was obtained through a survey involving remote employees of various companies, as well as executives and HR managers of companies. The results of the study showed the importance of purposeful developing of corporate culture for remote employees, which increases the efficiency of their main work activities, their loyalty to the company and unleashes employees' potential.

**Keywords:** Corporate culture · Digital transformation · Remote work · Virtual organisation

## 1 Introduction

The development of information technologies and telecommunication systems has led to global changes in the forms of doing business in the 21st century and announced the era of digital transformation. The digitalization of business processes, changes in strategy, operations, and products lead to a change in the format of how employees work in the company. The format of remote work has become widespread worldwide. When working remotely the employer and his employees are at a distance from each other, communicate and perform official duties through modern means of communication, mainly through the global network—the Internet. According to a study conducted by OwlLabs, 62%

V. Taratukhin et al. (Eds.): ICID 2021, CCIS 1539, pp. 146–156, 2022.
https://doi.org/10.1007/978-3-030-95494-9_12

of employees aged 22 to 65 had remote work experience. At the same time, 16% of companies worldwide are already fully working remotely, and this percentage continues to increase [15]. According to WWR, the number of remote jobs is growing by an average of 30% annually [20]. However, the format of remote work has been particularly developed since 2020 during the global pandemic of the COVID-19 virus, during which the experience of mass implementation of the remote work in various companies was realized. Due to the need for quarantine measures, all work processes were performed remotely. These changes in the way organizations operate have revolutionized the concept of teleworking and its role in organizations. According to Kate Lister, president of Global Workplace Analytics, nearly 70% of the workforce will work remotely at least five days in a month by 2025 [8].

The processes of digital transformation of business ensure the spread of the transition to remote work, while modern technologies are being introduced into business processes. Not only does that change, but the corporate culture also changes during digital transformations. At the same time, managers and managers of companies faced the task of adapting and maintaining corporate culture in the conditions of remote work of employees. Older methods and tools either need to be adapted to the new format or replaced with something more efficient for remote work. Those companies that solved this problem before the pandemic and had a high degree of informatization of processes were able to successfully cope with the conditions in which the global spread of the virus put them. It is important to take into account the peculiarities of the corporate culture for remote workers, since the effectiveness of their activities also depends on how effectively communication and employee engagement are maintained, their adherence to the values of the company, which directly affects the effectiveness and profitability of the organization [12].

Speaking of how maintaining the corporate culture has changed over the last four years, it is especially important to focus on the conditions that the pandemic has faced us. If earlier it was possible to approach the formation and maintenance of culture gradually, now, abrupt changes in the situation, the transition to remote work, then going back to offline, have introduced the need for a more flexible, adaptive and rapidly changing system for maintaining a culture that would meet the current needs and trends of the environment. The last two years have started a number of trends that, although they originated earlier, have become widespread only now. Changes in employees' psychological state during a pandemic, such as increased stress, increased feelings of loneliness, lead to transformations in professional relationships and values. For example, digital etiquette, an understanding of the need to include remote employees in informal interaction, the formation of value and clear boundaries of work and personal time, care not only about the material, but also the psychological well-being of the employees, the openness of managers to communication and suggestions, the transparency of the company and highlighting the contribution of each employee, individual approach. Widespread of these trends requires all companies to follow them. In addition, in various programs, on platforms, etc. new functions for remote interaction appear, such as the appearance of 'rooms' in video conferences, the use of chatbots in the work environment. The culture of companies is now inextricably linked with technology. More than ever, companies need a corporate culture that helps to be open to new challenges and opportunities, innovation and readiness for change.

Previous studies do not always take into account the rapid and global changes that are brought about by the development of information and communication technologies. Every year there are more and more tools, methods, features of regulation for remote employees and their involvement in the corporate culture of the company. The attitude towards this form of employment is changing, moreover, new market needs appear in the form of, for example, new specialties or the digitalization of business processes associated with the era of digital transformation.

Studies of corporate culture for remote workers have been conducted since the end of the 20th century, when this trend was just emerging and when information technologies were just beginning their worldwide spread. At the same time, scientists foreshadowed the worldwide spread of the remote work format, which has been happening now since the beginning of the second decade of the 21st century [9]. At the moment, there are many relevant studies of recent years devoted to remote work, including related to the COVID-19 pandemic [5, 16, 17], as well as various statistical studies on this trend [1, 8, 15, 20], which take into account changes taking place in the economy, in approaches to human resource management, etc. In the course of highlighting the features of corporate culture and their causes, we also relied on a number of socio-psychological studies that describe the impact of remote communication on the psycho-emotional state of an employee [3, 19]. A number of scientists have conducted research in the field of the impact of the remote format on the organizational culture of the company, which is directly related to the topic of our study. Thus, a study by A. Spicer highlights the main groups of issues that managers in organizational culture have to face on the verge of changes in the work format [17], and Jennifer Howard-Grenville draws attention to the need to focus on key elements of corporate culture, and not on visual artifacts with the purpose of maintaining it in a remote format [10]. These theoretical data and scientific conclusions were used as a theoretical basis for this study, on the basis of which the characteristics of such corporate cultures were formulated.

## 2   Theoretical Background

According to one of the basic definitions, corporate culture is a unique set of norms, values, beliefs, patterns of behaviour, etc., which determine the way groups and individuals unite in an organization to achieve its goals [7]. Some scientists share the concepts of "corporate culture" and "organizational culture", in this paper both concepts are used as identical.

To substantiate the importance of forming and maintaining an organizational culture in the company, we have identified the following main reasons that most fully reveal the importance of corporate culture in the company:

1. Corporate culture ensures the unity of employees, increases motivation and efficiency of their work and commitment to the goals of the organization, job satisfaction and involvement in work;
2. Culture becomes the self-management mechanism of the organization, which ensures its smooth functioning;

3. A strong, consciously formed corporate culture ensures the achievement of organizational goals in the most effective, humane and socially acceptable way.

The term "organizational culture" is complex, many scientific papers are devoted to the study of its role, elements, and classification. In turn, for our research, as a basis, we took 10 elements of organizational culture, highlighted by F. Harris and R. Moran [9]:

1. Sense of self and space;
2. Communication and language;
3. Dress and appearance;
4. Food and feeding habits;
5. Time and time consciousness;
6. Relationships;
7. Values and norms;
8. Mental process and learning;
9. Work Habits and Practices;
10. Beliefs and attitudes.

Creation and maintenance of corporate culture is based on the creation and maintenance of its elements. Each of the elements was studied from the point of view of the remote format. A number of features inherent in the elements of culture in the remote format were highlighted, as well as recommendations for managing these elements. We considered companies that do not work in a remote format all the time, but have a part of remote employees on a par with on-site workers, or that were forced to reorganize into a remote work format during a pandemic. We also considered companies that do not have their own office and carry out all interactions with employees in a remote format on a regular basis, the so-called virtual organizations [13].

The data confirming and supplementing the theoretical studies were obtained through a survey and an interview. In order to collect the empirical base of the study, a questionnaire was developed and conducted, the respondents of which were people with remote work experience and on-site work experience in the same company. For example, when a company switches to a remote format during a pandemic. We received responses from 85 respondents. The questionnaire contained three-choice questions designed to reveal how the corporate culture of the company has changed with the transition of employees to a remote work format. Questions were formulated for each of the elements of culture with alternative answers that reflected the variable nature of the change.

Interviews were also conducted with 6 business representatives who had experience in managing remote employees, three of whom are managers of virtual organizations working completely remotely. They were asked open-ended questions about each of the elements of corporate culture in order to find out how the companies in which they work shape and maintain these elements in a remote work environment. And also to figure out their own methods and tools for involving employees in the culture of the company.

## 3 Features of a Corporate Culture for Remote Workers

According to authors' survey results, 64% of respondents rate the remote format of work positively for themselves, while only 40% of them noted changes in the culture of companies as positive. A relatively new format requires managers to competently manage elements of organizational culture, however, a company cannot always quickly rebuild its management models. Companies need time for this, as well as time for digital transformation, which is an integral part of the transition to a remote format. In such conditions, company executives and managers are not always ready for the consequences, for the features described below, which will dictate the transition to a remote format of work.

A feature for the remote format is the difficulty of tracking the employee's role in the organization and the role of other employees. The employee sees the monitor screen, and almost has no opportunity to observe what his colleagues are doing at this time, how effective they are. It is difficult for the employee to track what contribution his activities make to achieving common goals. It is more difficult for an employee to feel their value and relevance to the company and colleagues. 57% of respondents noted that the availability of information about processes in the company has become less. Therefore, when involving remote employees in the corporate culture of the company, transparency and accessibility of information in the company plays a special role. As methods of increasing the availability of information and transparency of processes, you can use newsletters about the company's activities, its current projects, or collect a common database on the company's activities and open access to it to all interested employees. Managers need to pay special attention to demonstrating the results of the employee and the team as a whole, announcing them at meetings, emphasizing what contribution each of the team members has made. The importance of feedback from managers has really increased with the transition to a remote format: 62% noted that the value of feedback on their activities in the company has increased for them, 32% shared that with an increase in the importance of feedback for them, its quantity from management has decreased, and 35% noted a deterioration in its quality.

Companies are faced with a new reality in which an individual attention to each employee becomes necessary for effective teamwork. 51% of respondents find it more difficult to contact their managers. The task of managers is to create a supportive environment in which each employee knows that he is not forgotten, that he can get in touch with his manager when he needs help. 75% of the interviewed managers introduced one-on-one online meetings with each of the employees as permanent, to discuss the employee's success, the current state of affairs, the difficulties he is facing and to discuss other issues of concern to the employee. Constant communication with managers is necessary for employees to feel like a part of the organization, the stronger this connection, the more the employee is involved in his activities [19].

The availability of management for communication is also an important part of such an element of corporate culture as the communication system and the language of communication. It is important for a business to have a clear policy on how to communicate to employees in the company, what programs to use, and define the communication style. With a remote work format, there is no face-to-face communication between company employees; all communications are completely transferred to telecommunication systems, such as platforms for video and audio conferencing. According to the survey, the use of such platforms in a company with a transition to a remote format compared to face-to-face. Increased by 44%, almost doubled. In the graph below, you can track how the use of technology in the company has changed with the transition to remote work (Fig. 1).

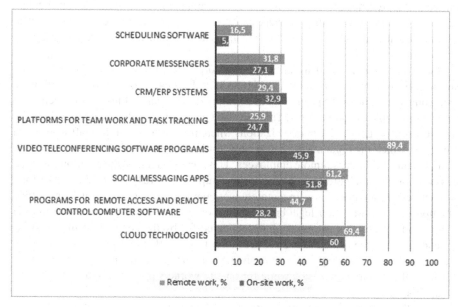

**Fig. 1.** The usage of technologies in face-to-face and remote work format.

Therefore, in order to create a virtual space in which employees can interact, managers must provide each employee with access to all systems, make sure that employees understand how to use them: conduct training, make clear instructions, create a channel where you can report a problem during use. communication systems. Since written interaction completely lacks an element of non-verbal communication, demonstration of emotions, body language, and with audio and video communication it is very limited, it can be difficult for people to understand each other, understand their tasks and perceive the transmitted information in general. 35% of a communication's understanding comes from the verbal part and 65% comes from the nonverbal part of the message [3]. A large proportion of companies underestimate the role of verbal communication; according to a survey, 75% of respondents do not have to communicate via video with their colleagues/managers during work. In connection with the importance of non-verbal communication, it is important to increase the number of video interactions between employees. The discussion of the most important issues should take place in such a format when people have the opportunity to convey information not only directly, but also using non-verbal communication. For written communication, it is recommended to develop regulations that formalize the language of communication, make it more specific and understandable, excluding the possibility of discrepancies.

In face-to-face work, such an element of culture as dress code plays a significant role in the identification of an employee as part of the company and the separation of work and personal [2]. When introducing certain dress code norms for remote employees, special attention should be paid to substantiating the value and importance of corporate dress code for employees. When employees interact with clients via video communication, the dress code was introduced in 5 out of 6 companies of managers whom we interviewed. In this case, the presence of a dress code during communication with external stakeholders provides the company with an image, and the absence of a dress code during intra-organizational communication allows creating an informal atmosphere and bringing people closer.

Food and feeding habits as an element of corporate culture essentially "falls out", since meals are not organized by a remote worker. However, companies can still make their contribution here: the promotion of a healthy diet, a healthy lifestyle in general, the personal example of team leaders. This can also become part of the corporate culture for remote workers. Especially in organizations where there is a part of the staff working in the office, which the organization can arrange for healthy meals, thereby supporting the idea of health care for its remote employees. This can also include the positioning in the company of the lunch break, as a time for the employee to rest, when he can take food and not deal with work issues. In a remote work environment, creating such an attitude of employees towards their legal break can not only relax the employee in the middle of the working day, but also make him feel cared for by the company. Unfortunately, not enough attention is paid to this element for remote workers, 5 out of 6 managers we interviewed noted that they in the company do not emphasize the importance of a quiet lunch break for remote workers, while for on-site workers this attitude to lunch breaks is encouraged.

With the transition to a remote format, the company's culture of attitude to time is changing. According to the survey, 36% noted an increase in the efficiency of their work and the efficiency of the use of working time. This is due to less distraction to formal and personal communication within the team, the employee's autonomy and self-organization increases. However, it is not always possible for an employee to cope with new conditions, 25% of respondents noted a decrease in work efficiency. There is a need on the part of managers to introduce various time management techniques and tools in order to organize the use of time by their employees when the ability to control them is reduced. At the same time, the requirements for meeting the work deadline are also growing, as well as the number of work tasks, 40% of respondents noted that the number of their work tasks has increased. In the culture of companies that work with remote workers, multitasking is expected, which can lead to decreased employee efficiency and increase employee stress levels.

Company values and norms are also being transformed. Of particular importance is the differentiation of working and personal time, the work-life balance. For a remote worker, the border of transition from work to free time is erased, in contrast to on-site workers, for whom such a border is the way home. In this regard, in order to maintain the efficiency of employees, it is necessary to pay special attention to the value of personal non-working time and the introduction of norms for the interaction of employees with each other during working hours. Set clear time frames for the working day, respect the free time and weekends of employees and not write/call them during this time. To do this, it is effective to use a common calendar, which displays the schedules of all employees and everyone can see at what time he can refer to another, without disturbing his personal time. Such a tool is provided by programs such as Google Calendar, Calendly, which can also be used to quickly schedule meetings: there is no need to specify with everyone the time when he can get in touch, just look at his calendar in the calendar.

In remote interaction, the value of the employee and his knowledge in the organization increases: companies depend solely on people, without whom technologies will not function. Remote work involves more mental activity, less mechanistic, so the formation of a knowledge management culture comes to the fore in the company.

With the transition to remote work, there is also an increase in the role of initiative, employee responsibility, and his ability to self-organize. Employees and their knowledge become the main value of the company; the manager does not just set a task and wait for its completion. Teamwork presupposes delegation and trust to employees, the possibility of their creative manifestation in solving problems. It is important for companies operating in this format to form the values of support and mutual assistance. In an environment where employees are already far from each other, and communication is limited, an atmosphere of care will allow the company to form a stable emotional environment in the team.

A feature in the creation and maintenance of a corporate culture of employee relationships with each other is that relationships for remote employees must be consciously formed, creating an environment for informal communication and discussions. In the office, this element is often not given close attention, because people themselves strive for communication and they have the opportunity to get to know each other, discuss the latest news, for example, when meeting at the beginning of the working day or at 'watercooler meetings'. Traditions between employees can be formed spontaneously in the face-to-face format. When employees do not have personal contact, it is necessary to purposefully create channels that will allow them to communicate with each other. The interviewed managers noted that such channels are used for informal, "live" communication, various thematic video meetings or discussions that unite people of interest, challenges for teams and individual employees, some also organize soft skills training, during which a favorable environment for networking is formed. The graph below shows the informal communication activities that were introduced by companies to involve remote employees (it was possible to choose several answers) (Fig. 2).

The loss of social and intimate interactions has resulted in loneliness and unmet social needs [12]. This is confirmed by our survey, in which 42% noted that they began to feel more lonely, and 54% began to feel less part of the work collective. Therefore, it is so important to create conditions for communication, for the formation of a sense of community, through activities and traditions. The creation of warm relations in the company, the formation of an understandable format and channels of interaction, which were mentioned above, also helps to prevent the development of conflicts. In a remote environment, conflicts mature longer, are more often associated with incorrect perception of information, and it is more difficult to resolve them without personal communication, which is noted by 35% of respondents. It is important for companies working with remote workers to collect feedback from employees in a timely manner, aiming at preventing conflict situations or their early resolution in order to avoid negative consequences and prolongation of the conflict. As mentioned above, remote employees are less emotionally involved in the activities of the company due to the lack of constant personal contacts,

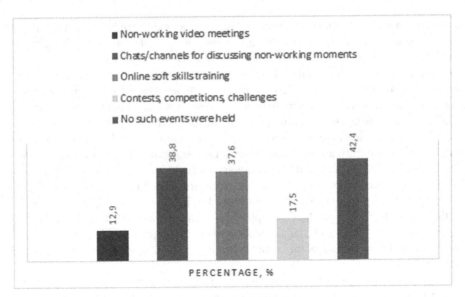

**Fig. 2.** Informal communication activities for remote employees

and the level of stress and feelings of loneliness can increase. In such conditions, it is necessary to maintain a sense of faith in employees, to provide support for their mental health. Creation of a psychological climate in which a person is comfortable, in which he knows that even if he is far from the team, this does not mean that there is no team or he is outside of it.

When a new remote worker comes to the company, he is not familiar with the employees, with the corporate culture, with his duties, it will not be possible to leave him alone with the computer in the hope that he will adapt. To maintain the corporate culture, it is necessary to develop a multilateral system of adaptation of new employees: it must be purposefully introduced to the team, help to master the programs used for communication, provide access to the necessary information, and ensure constant communication with a more experienced employee who will become his mentor during the adaptation. It should also be borne in mind that the onboarding of remote workers is slower and it is more difficult for them to master the culture of the company. The survey showed that 28% of companies do not help their employees in the transition to a remote format in any way: they do not conduct training, do not make sure they have the necessary equipment and do not provide it. This approach does not meet the current requirements for the culture of companies with remote workers. The remote format requires targeted involvement of employees, their continuous training in new technologies, new programs and tools. Whether employees will be able to perform their work tasks directly depends on this. The remote format also increases the importance of forming employees' readiness for technical problems and options for getting out of such problem situations for the smooth functioning of the company.

The interviewed managers note that there is practically no difference between the system of motivating a remote employee and a on-site employee. Motivation may decrease due to the fact that a remote employee loses contact with companies. Therefore, he needs more attention, he should establish his connection with the company so that he feels involved in something greater, then it will be possible to motivate him not only financially. In the survey, employees noted that the motivation to work in a company with the transition to a remote format did not change for 64%, and only for 21% the motivation decreased.

## 4   Conclusion and Future Work

It should be noted that the culture of each company is unique and requires its own approach, formed on the basis of the features highlighted above. The culture of the company is dynamic, not static, and companies need to be ready to adapt it to the emerging new conditions and technologies. In the Digital Transformation Era, it is necessary to purposefully manage the corporate culture in order to maintain a human attitude towards employees.

The limitations of this study include a high proportion of subjectivity in the responses of respondents, it is almost impossible to ensure the unity of opinions of employees regarding the corporate culture in their company. However, the data obtained made it possible to identify certain general trends and changes in culture associated with the transition to a remote format. The remote format of work also opens up opportunities for finding employees around the world, which exacerbates the importance of managing multinational teams and involving them in corporate culture. This aspect and its impact on the corporate culture of the organization are not covered in the paper, since the management of multinational teams is a separate large-scale area for study.

According to the study of Boston Consulting Group, companies that focused on culture were five times more likely to achieve breakthrough results in their digital transformation initiatives than those that didn't [1]. Cultural changes are the driving force behind digital transformation, which is also confirmed by our research. This study allowed us to formulate and structure the features of corporate culture for remote workers, to identify the main methods and tools used to adapt it to the features of the remote format, that could have practical implementation for companies with remote employees on staff, planning to transfer employees to a remote format or planning to develop digital transformation processes in the organization.

As prospects for further research, we can highlight such topics as: knowledge management in virtual organizations, a new work ethic for remote workers, the creation of a communication system for remote workers.

## References

1. BCG. How to Drive a Digital Transformation: Culture Is Key (2021). https://www.bcg.com/capabilities/digital-technology-data/digital-transformation/how-to-drive-digital-culture
2. Beno, M., Hvorecký, J., Caganova, D.: International journal of business and applied social science (IJBASS) an optimal e-working environment: online survey results. Int. J. Bus. Appl. Soc. Sci. 7(1) (2021). https://doi.org/10.33642/ijbass.v7n2p1
3. Birdwhistell, R.L.: Kinesics and Context: Essays on Body Motion Communication. University of Pennsylvania, USA (1970)

4. Coffey, R., Wolf, L. :The Challenge and Promise of Remote Work: A Brief Study of Remote Work and Best Practices. United States: N (2018)
5. Dvorak, J.: The impact on organization behavior in a telecommuting world and the potential impacts on the business. Dissertations, Theses, and Projects, 574 (2021)
6. Schein, E.H.: Organizational Culture and Leadership, 5th edn. Wiley, Hoboken, NJ (2017)
7. Eldridge, J., Crombie, A.: A Sociology of Organization. Allen & Unwin, London (1974)
8. Global Workplace Analytics. Remote Work Statistics: Navigating the New Normal (2021). https://www.flexjobs.com/blog/post/remote-work-statistics/
9. Harris, P.R.: Managing Cultural Differences/P.R Harris, R. T. Moran// Houston, TX: Gulf Publishing Co (1996)
10. Howard-Grenville, J.: How to sustain your organization's culture when everyone is remote. MIT Sloan Manag. Rev. **62**(1), 1–4 (2020)
11. Kniffin, K., Narayanan, J., Anseel, F., Antonakis, J., Ashford, S., Bakker, A., et al.: COVID-19 and the workplace: implications, issues, and insights for future research and action. Am. Psychol. **76**(1), 63 (2020). https://doi.org/10.31234/osf.io/gkwme
12. Kotter, J.P., Heskett, J.L.: Corporate Culture and Performance. - N.Y.: Free Press. -VIII, 214 p. – bibliogr, pp. 199–203 (1992)
13. Warner, M., Witzel, M.: Managing in Virtual Organizations, p. 224. International Thomson Business Press, London (2004)
14. Manko, B.: Considerations in the use of work-from-home (WFH) for post-pandemic planning and management. Management **25**, 118–140 (2021). https://doi.org/10.2478/manment-2019-0062
15. OwlLabs. State of Remote Work (2020). https://resources.owllabs.com/state-of-remote-work/2020
16. Pulyavina, N.: Possibilities of design thinking tools to improve electronic information and education systems. Econ. Entrepreneurship Law, **11**(2), 489–500 (2021). https://doi.org/10.18334/epp.11.2.111663
17. Spicer, A.: Organizational culture and COVID-19. J. Manage. Stud. **57**(8), 1737–1740 (2020). https://doi.org/10.1111/joms.12625
18. Staples, D.S.: An investigation of information technology-enabled remote management and remote work issues. Australas. J. Inf. Syst. **4**(2) (1996)
19. Staples, D.S., Hulland, J.S., Higgins, C.A.: A self-efficacy theory explanation for the management of remote workers in virtual organizations. J. Comput. Mediated Commun. **3**(4), JCMC342 (1998)
20. WWR, The Future of Remote Work (2019). https://nira.com/future-of-remote-work/

# Methodological Support of Analysis and Management Tools: Theoretical-Focused Research

# Simulation Modelling for Assessing the Adequacy of Decision Support Models with Choosing a Product Offer

Mikhail Matveev[1] (ID), Mikhail Shmelev[1](✉) (ID), and Andrey Budyakov[2] (ID)

[1] Voronezh State University, Voronezh, Russia
[2] JSC Gazprom Neft, Saint Petersburg, Russia

**Abstract.** Electronic commerce has become an effective tool for information integration of sellers and buyers. The significant effect can be achieved by automating decision-making processes that are based on mathematical information modelling. Practical application of such models in e-commerce markets is constrained by the lack of the possibility of checking the adequacy of these models. Adequacy verification is possible on a simulation model, which should take into account specific features of the product market. When simulating random demand, it is necessary to take into account the mutual influence of product characteristics.

The article discusses the problem of constructing a simulation model of demand. Special types of features can be displayed by the joint distribution of dependent characteristics and univariate distribution of independent characteristics. In the general case, the task is to construct a multidimensional distribution function of random characteristics that adequately reflects the features of their behavior. For interacting characteristics two-dimensional distribution is obtained on the basis of expert construction of a matrix of paired comparisons of values of these characteristics. It is shown that the eigenvector of the matrix of pairwise comparisons sets conditional probabilities of the distribution of one characteristic for given values of other. For an independent characteristic the expert indicates the required table distribution. With obtained discrete distributions, random values of characteristics are generated by the method of inverse functions.

For given example, a simulation model of the generation of customer demand was built, which reflects features of the behavior of product characteristics.

**Keywords:** Product characteristics · Eigenvector · Inverse functions · Demand

## 1 Introduction and Statement of the Problem

Electronic commerce has become an effective tool for obtaining and integrating information about sellers and buyers. This provided the opportunity to automate trade business processes and digital management of these processes based on actively developing information technologies. This development is observed most successfully in the processes of the document flow automation and the formation of analytical information for management [1]. The greatest effect can be achieved by automating decision-making processes

© Springer Nature Switzerland AG 2022
V. Taratukhin et al. (Eds.): ICID 2021, CCIS 1539, pp. 159–166, 2022.
https://doi.org/10.1007/978-3-030-95494-9_13

that are based on mathematical and information modelling. Such models are increasingly appearing in scientific publications, for example, [2–4], but their practical application in e-commerce markets is constrained by the lack of the possibility of checking the adequacy of these models before the start of their expensive testing on real trading objects. Adequacy verification is possible on a simulation model, which should take into account the specifics of a particular type of the product market. When simulating random consumer demand, it is necessary to take into account the generalized behavioral type of buyers, for example, their attitude to the ratio of price and quality of the desired product, the mutual influence of product characteristics and other features of the behavior of buyers.

Information about the behavior of buyers is very important for the organization of market processes. Methods of obtaining, presenting and using information about customer behavior are detailed, for example, in works [5, 6]. The obtained information is used to manage processes of organizing sales [5]. Our research aims to address other challenges:

- How to obtain a formalized model of behavior that can be represented by program code based on the available qualitative information about behavior of customers?
- How to build a simulation model of this behavior on the basis of a customer behavior model, designed to test the adequacy of certain mathematical models for managing sales organization processes?

The solution of these problems is extremely difficult. Some simplification can be obtained by considering only a homogeneous product, the various types of which are interchangeable. In this case, the formalized description of consumer demand can be presented as a set of unified vectors of characteristic parameters of the desired product, where each type has its own characteristic values [7, 8]. We will consider the formation of consumer demand as a set of random events for the implementation of a set of characteristic values. Then we can use a probabilistic-theoretical approach to solve the first problem, which is in good agreement with methods of simulation modelling. As a model of a consumer demand, it is proposed to use a multivariate distribution function of values of vector components of characteristic parameters for a homogeneous product.

## 2   Simulation of Customer Demand Behaviour

Before building the model, we will make a number of simplifying assumptions. These assumptions do not fundamentally change the proposed modelling methodology, but they make possible to significantly simplify the presentation of the methodology and provide a clear interpretation of its main provisions.

Suppose that the vector of characteristic parameters of the desired product has three components: $g = (g^1; g^2; g^3)$. Let for the convenience of interpretation, shoes are a homogeneous product. Then components of vector $g$ will be interpreted as "price", "quality", "size". It is natural to assume that price $(g^1)$ and quality $(g^3)$ are interacting characteristics, and size $(g^2)$ is independent. All components are measurable. Let size is measurable on a discrete numerical scale, and price and quality are measurable on a rank scale.

The distribution of shoes size values for consumer demand should be viewed as a normal distribution with a thickest value close to the most demanded average size. The situation is more complicated with the price-quality pair. Here it is necessary take into account preferences of the buyer in the local market to the price level and the ratio $g^1/g^3$. Suppose we know of some verbal statements about this. For example, buyers in the local market tend to buy shoes at a low price, but not of the lowest quality. Another example, the most of local market buyers prefer to buy shoes of medium or high quality and are willing to pay at the average price.

To formalize a verbal statement about needs, the procedure of paired comparisons [9] seems to be the most adequate, which is included in the arsenal of modern methods of social research [10]. Consider the first example of a need: buy shoes at a low price, but a good quality. For illustrating the methodology, it is convenient for price and quality to select rank scales with three values: low, medium and high. Let's construct a scale for the ratio $w = g^1/g^3$ as follows:

$$w(i = j) = 1; \quad w(i < j) = 2; \quad w(i << j) = 3; \quad w(i > j) = \frac{1}{2}; \quad w(i >> j) = \frac{1}{3}$$
(1)

Formula (1) indicates: $i \in (1; 2; 3)$– scale grade numbers for price; $j \in (1; 2; 3)$– scale grade numbers for quality. The matrix of pairwise comparisons of price and quality will take the form:

$$
A = \begin{array}{c|ccc}
 & low & medium & high \\
\hline
low & w_{11} = 1 & w_{12} = 2 & w_{13} = 3 \\
medium & w_{21} = \frac{1}{2} & w_{22} = 1 & w_{23} = 2 \\
high & w_{31} = \frac{1}{3} & w_{32} = \frac{1}{2} & w_{33} = 1
\end{array}
$$
(2)

Matrix $A$ – strictly positive antisymmetric matrix. According to [11], this matrix has a single eigenvector $x = (x_1; x_2; x_3)$, corresponding to the maximum eigenvalue. The numerical value of the eigenvector is defined as follows [11]:

$$x = \frac{A \cdot e}{e^T \cdot A \cdot e} = (0.53; \ 0.31; \ 0.16),$$
(3)

Where $e^T = (1; \ 1; \ 1)$.

The corresponding eigenvalue is calculated by determining the arithmetic mean of values $\lambda_i = \frac{A \times x}{x_i}$ [11], where $i$ – number of vector component (3). The resulting value $\lambda = 3,01$ allows calculating the matrix consistency index $A$ [11]:

$$I = \frac{\lambda - n}{n - 1} = \frac{3.01 - 3}{2} = 0.005;$$
(4)

where $n$ – dimension of matrix $A$. The computed value of the consistency index indicates satisfactory, but not ideal consistency between the paired comparison results.

Let's perform following transformations of elements $w_{ij}$ of matrix (2):

$$p_{ij} = \frac{w_{ij}}{\sum_i \sum_j w_{ij}} \in (0; 1)$$

The elements obtained after the transformation can be interpreted as probabilities of joint events $i \wedge j$. Then transformed matrix $A^*(p_{ij})$ can be considered as a discrete two-dimensional distribution of probabilities of these events. Conditional distributions $p(i/j)$ и $p(j/i)$ are calculated by the formulas:

$$p(i/j) = \frac{p_{ij}}{\sum_i p_{ij}}; \quad i = 1, 2, 3; \tag{5}$$

$$p(j/i) = \frac{p_{ij}}{\sum_j p_{ij}}; \quad j = 1, 2, 3. \tag{6}$$

The calculated values of the conditional distributions (column vectors) are shown in Table 1 and Table 2.

**Table 1.** Conditional distributions of price values in customer demand

| Conditional distribution of price $i$ with low quality ($j = low$) | Conditional distribution of price $i$ with medium quality ($j = medium$) | Conditional distribution of price $i$ with high quality ($j = high$) |
|---|---|---|
| $p(1/1) = 0.55$ | $p(1/2) = 0.57$ | $p(1/3) = 0.50$ |
| $p(2/1) = 0.27$ | $p(2/2) = 0.29$ | $p(2/3) = 0.33$ |
| $p(3/1) = 0.18$ | $p(3/2) = 0.14$ | $p(3/3) = 0.17$ |

**Table 2.** Conditional distributions of quality values in customer demand

| Conditional distribution of quality $j$ with low price ($i = low$) | Conditional distribution of quality $j$ with medium price ($i = medium$) | Conditional distribution of quality $j$ with high price ($i = high$) |
|---|---|---|
| $p(1/1) = 0.17$ | $p(1/2) = 0.14$ | $p(1/3) = 0.16$ |
| $p(2/1) = 0.33$ | $p(2/2) = 0.29$ | $p(2/3) = 0.30$ |
| $p(3/1) = 0.50$ | $p(3/2) = 0.57$ | $p(3/3) = 0.54$ |

The data shows that buyers tend to buy cheaper goods, but at the same time want high quality. It is easy to see that distributions of price values are similar for different quality values. Also distributions of quality are similar for different price values. This result is expected since the buyer's behavioral trend (see matrix $A$) does not change, but only shifts along the corresponding scale. The obtained result indicates that in this case buyer does not have a relationship between the price and the quality of the desired product. In principle, the existence of such dependence is possible for a reasonable buyer, but understanding the existence of dependence is more inherent for the seller. In Tables 1 and 2, some difference in values of vectors of conditional distributions is explained by the non-ideal consistency of the matrix $A$. In this case, it is more reasonable to find the arithmetic mean of components of corresponding vectors, which will better match the absence of dependence of one variable on another.

The average probabilities of the distribution of desired price and quality values among buyers are equal, respectively:

$$p(i/\forall j) = (0.54; \ 0.30; \ 0.16); \quad p(j/\forall i) = (0.16; \ 0.30; \ 0.54)$$

The probability distributions of desired prices and product quality are represented by the same values arranged for prices in descending order, and for quality in ascending order, in accordance with the structure of the matrix. Moreover, it should be noted that averaged probability distributions practically coincide with the eigenvector (3) of the matrix $A$. This justifies the need to prove the relationship of the eigenvector of pairwise comparisons matrix with the distributions of conditional probabilities. The proof is beyond the scope of this study, but it is very relevant, since in the case of a successful solution, it provides a sharp reduction in computational labor costs.

For the considered example, the initial data for simulating the characteristics of consumer demand are three one-dimensional discrete distributions:

- distribution of shoe sizes desired by buyers, approximated by a normal distribution;
- distribution of prices desired by buyers, $p(i/\forall j) = (0.54; \ 0.30; \ 0.16)$;
- distribution of qualities desired by buyers, $p(j/\forall i) = (0.16; \ 0.30; \ 0.54)$.

## 3  Imitations Algorithms Based on the Method of Inverse Functions

Simulation modelling of stochastic phenomena has a sufficiently developed theoretical basis, for example [12, 13]. This modelling assumes some source of randomness. As a source of randomness, we will consider the generator of uniformly distributed on [0; 1] pseudo-random numbers $\xi$.

Let a random variable $X$ (in our case price, size and quality) is given by a table distribution:

$$F(x) = \begin{pmatrix} x_1, \dots, x_k, \dots, x_n \\ p_1, \dots, p_k, \dots, p_n \end{pmatrix}, \tag{7}$$

where $\sum\limits_{k} p_k = 1$.

Simulation modelling involves the implementation of a specific value of a random variable $X$ endowed with a probabilistic measure. Formally, this corresponds to the solution with respect to $X$ of the equation:

$$\xi = F(x) \tag{8}$$

Also we introduce $s_i = p_1 + p_2 + \ldots + p_i$ at $i = 0;\ 1;\ \ldots;\ n$. The resulting numbers are commonly referred to as the cumulative probability. The imitation is carried out using the method of inverse functions [14]:

$$F^{-1}(\xi) = x_k, \quad \text{if } s_{i-1} \leq \xi \leq s_i, \quad i = 1, 2, \ldots, n \tag{9}$$

Consider a simulation of customer behavior for the example from Sect. 1. We concretize the discrete distribution of values of the characteristic "size". Let the distribution for the size characteristic is close to normal, e.g.:

$$F_{size}(x) = \begin{array}{c} x \\ p \end{array}\left( \begin{array}{ccccccc} 39 & 40 & 41 & 42 & 43 & 44 & 45 \\ 0.05 & 0.1 & 0.2 & 0.3 & 0.2 & 0.1 & 0.05 \end{array} \right)$$

Let we show the previously obtained distributions of values of characteristics "price" and "quality":

$$F_{price}(x) = \begin{array}{c} x \\ p \end{array}\left( \begin{array}{ccc} low & medium & high \\ 0.54 & 0.30 & 0.16 \end{array} \right);$$

$$F_{quality}(x) = \begin{array}{c} x \\ p \end{array}\left( \begin{array}{ccc} low & medium & high \\ 0.16 & 0.30 & 0.54 \end{array} \right).$$

Simulation modelling was performed in environment Simulink of MATLAB.

A fragment of simulated values of vector components is $g = (g^1; g^2; g^3)$ presented below (Table 3):

**Table 3.** Fragment of simulated values of characteristics "price", "size" and "quality"

| № | price | size | quality | № | price | size | quality |
|---|---|---|---|---|---|---|---|
| 1 | low | 42 | medium | 11 | low | 41 | medium |
| 2 | low | 41 | medium | 12 | high | 41 | high |
| 3 | medium | 43 | medium | 13 | low | 41 | high |
| 4 | medium | 42 | high | 14 | low | 39 | high |
| 5 | medium | 42 | high | 15 | medium | 42 | high |
| 6 | low | 40 | medium | 16 | high | 42 | medium |
| 7 | low | 43 | medium | 17 | medium | 40 | low |
| 8 | high | 44 | high | 18 | low | 42 | high |
| 9 | low | 42 | high | 19 | high | 44 | low |
| 10 | medium | 42 | high | 20 | low | 43 | high |

A relatively small fragment shows the dominance in consumer demand of low prices and high quality with a normal distribution of shoe size.

## 4 Validating Characteristics Aggregation Model Using a Simulation Model

One of the tasks of decision support is to assess the efficiency of the seller's offer of a homogeneous product on the marketplace. Let characteristics of a product are determined by analogy with characteristics of consumer demand, i.e. the values of price, size and quality of shoes are set. We denote the corresponding vector: $q = (q_1; q_2; q_3)$. We introduce the concept of generalized consumer demand by analogy with [15]. We will assume that the componentwise correspondence of vector $q$ and generalized consumer demand is determined by finding the probability value in the generalized demand for a specific component value of vector $q$.

A weighted sum is used to obtain the generalized demand. To aggregate local correspondences, the densities of the fuzzy measure of local correspondence were found in an expert way: $den = (0.64; 0.58; 0.4)$. As in [15], the Sugeno fuzzy measure and the Choquet integral are used to find the probability of a trade for a particular product.

Bernoulli's formula is used to calculate the number of possible transactions in the market. The formula is as follows:

$$\bar{n} = p \cdot m,$$

where $\bar{n}$ – average number of transactions, $p$ – the probability of a transaction taking into account the generalized demand, $m$ – number of buyers in the market.

As a result, it turned out that the highest probability of a transaction is obtained for a product with the following characteristics: low price, 42 shoe size, high quality. This probability is $p \approx 0.501$; number of purchases for $m = 1001$ is equal to 501.

For example, the probability of a transaction for a product with medium price, 43 shoe size and medium quality is $p \approx 0.286$.

Thus, the result obtained for the simulation model correlates with our idea of a specific market.

## 5 Conclusion

Imitation market modelling was carried out for a specific product with 3 characteristics: price, size and quality. The following example was considered: buy shoes at a low price, but a good quality. When simulating random demand, we took into account the mutual influence of product characteristics, in our case, the interaction of price and quality. For given example, a simulation model of the generation of customer demand was built, which reflects features of the behavior of product characteristics.

For the "price-quality" pair, a matrix of paired comparisons was used as a modern mathematical tool. For interacting characteristics (price and quality) two-dimensional distribution was obtained on the basis of expert construction of a matrix of paired comparisons of values of these characteristics. Also with obtained discrete distributions, random values of characteristics are generated by the method of inverse functions.

The model of aggregation is checked using a simulation model. First, the concept of generalized consumer demand was introduced. Also, the Sugeno fuzzy measure and the Choquet integral were used to find the probability of a trade for a particular product.

As a result, it turned out that the highest probability of a transaction is obtained for a product with low price and high quality. The validation obtained for the simulation model showed that it matched our perception of a particular market.

Further research will focus on the types of buyers present in the market in more details. In this paper, only two possible types of buyers are shown.

## References

1. Arora, S., Frieze, A., Kaplan, H.: A new rounding procedure for the assignment problem with applications to dense graph arrangement problems. Math. Prog. **92**, 1–36 (2002). https://doi.org/10.1007/s101070100271
2. Amin, S.H., Zhang, G.: An integrated model for closed-loop supply chain configuration and supplier selection: multi-objective approach. Expert Syst. Appl. **39**(8), 6782–6791 (2012)
3. Mendoza, A., Ventura, J.A.: Analytical models for supplier selection and order quantity allocation. Appl. Math. Model. **36**(8), 3826–3835 (2012)
4. Budyakov, A.N., Getmanova, K.G., Matveev, M.G.: Resources and their supplier's selection problem solving within contradictory technical and commercial requirements. In: Proceedings of Voronezh State University. Series: Systems Analysis and Information Technologies, no. 2, pp. 66–71 (2017)
5. Blackwell, R.D., Miniard, P.W., Engel, J.F.: Consumer Behaviour, 10th edn., p. 944. St. Petersburg, Publ, Piter (2007)
6. Gunter, B., Furnham, A.: Consumer Profiles: An Introduction to Psychographics, p. 300. St. Petersburg, Publ, Piter (2007)
7. Matveev, M.G.: Information technologies for supply creation on e-trading platform with marketplace technology. Econ. Math. Meth. **1**, 114–121 (2021)
8. Budyakov, A.N.: The formalization of requirements of the technical policy in the automation of the selection of equipment oil and gas company based of fuzzy logic. Control Syst. Inf. Technol. **4**, 29–33 (2017)
9. Thurstone, L.L.: The method of paired comparisons for social values. J. Abnorm. Soc. Psychol. **21**(4), 384–400 (1927)
10. Gorshkov, M.K., Sheregi, F.E., Doktorov, B.Z.: Prikladnaya sotsiologiya [Applied sociology], 3rd edn., p. 334. Moscow, Publ, Urait (2019)
11. Saati, T.L.: Prinyatie resheniy. Metod analiza ierarhiy [Making decisions. Hierarchy analysis method] Moscow, Publ. Radio i svyaz, p. 278 (1993)
12. Ermakov, S.M., Mikhailov, G.A.: Statisticheskoe modelirovanie [Statistical modelling] Moscow, Fizmatlit, p. 296 (1982)
13. Gentle, J.E.: Random Number Generation and Monte Carlo Methods (Statistics and Computing), 2nd ed., p. 381. Springer, New York (2005). https://doi.org/10.1007/b97336
14. Zadorozhnyi, V.N.: Imitatsionnoe i statisticheskoe modelirovanie [Simulation and statistical modelling], 2nd edn., p. 136. Omsk, Publ, OmSTU (2013)
15. Matveev, M.G., Shmelev, M.A., Aleynikova, N.A.: Information technologies for formation of services on e-trading platform. In: Proceedings of Voronezh State University. Series: Systems Analysis and Information Technologies, no. 1, 63–73 (2021). https://doi.org/10.17308/sait.2021.1/3371

# The Study of Fuzzy Quantifiers in Multi-criteria Decision-Making

Mikhail Matveev[1] , Natalya Alejnikova[1]([✉]) , Vladislav Safonov[1] ,
and Lyudmila Korobova[2]

[1] Voronezh State University, Voronezh 394018, Russia
{alejnikova_n_a,vladislavsf}@sc.vsu.ru
[2] Voronezh State University of Engineering Technologies, Voronezh 394036, Russia

**Abstract.** The paper examines fuzzy quantifiers, which serve to formalize human reasoning. In this paper, quantifiers are considered in relation to the problems of decision-making on a set of alternatives based on a combination of criteria. Using fuzzy quantifiers and OWA aggregation operators, in which quantifiers are used to calculate weights, it is possible to implement basic decision-making strategies. In this paper, we study various quantifiers that are most often used when choosing the best alternative, such as "Most", "The more, the better", "At least k%". As a result of the study of these quantifiers, the boundaries of the values of the parameters of the membership functions were found, at which the OWA operator will have compensatory properties. The paper also points out the disadvantages of the most commonly used quantifiers when they are used in the OWA operator. The presence of "insensitivity zones" of the quantifier with piecewise linear functions of belonging to the change in the values of the components of the criteria vector is established. It is shown that this problem is solved when passing to a continuous membership function in the form of an s-shaped (logistic) curve. A modification of the OWA operator is proposed in the form of a superposition of partial estimates and a membership function of the fuzzy concept of "Good correspondence". This modification ensures that when comparing alternatives, not only the number of private assessments that meet the criteria is taken into account, but also the quality of compliance.

**Keywords:** OWA operator · Multi-criteria · Fuzzy quantifiers · Decision-making

## 1 Introduction

In the tasks of making a decision on a multiple alternatives based on a set of criteria, information about the acceptable form of compromise between estimates according to different criteria (private estimates) plays an important role. The tolerance level is a subjective, fuzzy concept that can be defined by a fuzzy quantifier [1, 2]. Fuzzy quantifiers are an extension of the classical quantifiers of generality and existence, they serve to formalize human reasoning. A certain quantifier is fuzzy if it is possible to construct a function of belonging to the corresponding fuzzy set for it [3]. Examples of fuzzy quantifiers are such concepts as "about half", "in general", "most", etc. In the theory of

V. Taratukhin et al. (Eds.): ICID 2021, CCIS 1539, pp. 167–179, 2022.
https://doi.org/10.1007/978-3-030-95494-9_14

decision-making, a fuzzy quantifier is a fuzzy statement about the acceptable form of compromise between particular estimates, reflecting the intuitive idea of the decision-maker about the preference of decisions. In particular, fuzzy quantifiers are used in OWA aggregation operators [4]. A special feature of these operators is to obtain a weighted average ordered by the magnitude of the partial estimates.

When comparing alternatives with each other with the help of OWA, it is necessary to take into account how large the values of partial estimates for one criteria compensate for small values for other criteria. This property of the OWA operator is called compensation [5]. It is important to be able to manage these compensatory properties, which depend on the form of the quantifier. In this paper, we study various quantifiers that are most often used when ordering and choosing the best alternative. Among the quantifiers, it is necessary to choose those that most adequately reflect the form of compromise and take into account the opinion of the decision-maker. The conditions under which the aggregating operator will have compensatory properties are considered. The disadvantages of the most commonly used quantifiers when they are used in the OWA operator are indicated. It is proposed that when comparing alternatives, not only to take a decision on a set of alternatives based on a set of criteria into account the number of private assessments that meet the criteria, but also the quality of compliance. All of the above is illustrated by examples.

## 2  Formalization of the Decision-Making Problem on a Set of Alternatives

Let's consider the problem of making a decision on a set of alternatives based on a set of criteria $C = \{c_1, c_2, \ldots, c_J\}$. Such a multi-criteria problem can be represented by the following tuple [6]

$$<X, G, P, D>, \tag{1}$$

where $X = \{x_1, x_2, \ldots, x_m\}$ – the set of alternatives; $G = \{g_1(x), g_2(x), \ldots, g_n(x)\}$ - the vector evaluation of the alternative $x \in X$, where the partial estimates $g_j(x) : R \rightarrow [0, 1]$ determine the degree of compliance $x \in X$ with the criteria $c_j \in C$, $(j = \overline{1, n})$; $P$ – the system of preferences of the decision-maker; $D$ - the decisive rule.

With the help of the preference system $P$, it is possible to determine a strategy for comparing partial estimates of alternatives and build a decisive rule $D$. It sets the procedure (algorithm) for performing the required action on a set of alternatives. The specified action may consist in ordering alternatives by preference, distributing them by classes of solutions, or choosing the optimal alternative [6].

The system of preferences of the decision-maker $P$ can be represented in the form

$$\left\langle \underset{x \in X}{agg}(g_1(x), g_2(x), \ldots, g_n(x)) \rightarrow \max, \quad x \in X; \quad \Lambda, Q \right\rangle, \tag{2}$$

where $agg$ is the aggregation operator; $\Lambda$ is information about the relative importance of criteria, usually given as a set of weights $\lambda_j \geq 0$, giving 1 in total; Q is information about the acceptable form of compromise between estimates for different criteria.

Further, instead of $g_j(x)$, we will write simply $g_j$.

Let's take a closer look at the concept of an aggregation operator. The aggregation operator is a function of $n$ variables (criteria, partial estimates) $agg : \bigcup_{j \in n} [0, 1]^j \to [0, 1]$ that satisfies a number of mandatory conditions [7, 8] on a set of arbitrary $x, y \in [0; 1]$:

1)  $agg(x) = x$;
2)  $agg(0, \ldots, 0) = 0$ and $agg(1, \ldots, 1) = 1$;
3)  $agg(x_1, \ldots, x_n) \le agg(y_1, \ldots, y_n)$ if $(x_1, \ldots, x_n) \le (y_1, \ldots, y_n)$.

The aggregation operator allows us to obtain a generalized (complex) assessment that characterizes the object as a whole according to all criteria [5]. At the same time, three main strategies can be implemented [9]:

a)  a conjunctive strategy, according to which a generalized estimate cannot be better than the worst of the partial estimates; in this case, the degree to which the alternative $x \in X$ meets all the criteria at once is defined as

$$agg(g_1, \ldots, g_n) = \min(g_1, \ldots, g_n); \qquad (3)$$

b)  a disjunctive strategy, according to which the generalized estimate is determined by the best of the partial estimates. The degree to which it meets at least one of the criteria is defined as

$$agg(g_1, \ldots, g_n) = \max(g_1, \ldots, g_n); \qquad (4)$$

c)  a compromise strategy, according to which the generalized estimate occupies an intermediate position between the private estimates involved in the aggregation:

$$\min(g_1, \ldots, g_n) \le agg(g_1, \ldots, g_n) \le \max(g_1, \ldots, g_n). \qquad (5)$$

The disjunctive strategy is characteristic of the optimistic position of the decision-maker, while the pessimistic decision-maker's tends to rely on the worst properties of objects in its judgments, and, consequently, on the conjunctive strategy.

In cases where the importance of the values of particular estimates is primary, the ordinal weighted aggregation operators [5–7, 10, 11] are used, OWA operators that aggregate the components of the vector estimate ordered in a certain way:

$$agg(g_1, \ldots, g_n) = OWA(g_1, \ldots, g_n) = \sum_{j=1}^{n} w_j g_{\sigma(j)}, \qquad (6)$$

where $\sigma$ is the index of ordering by the magnitude of the elements, such $g_{\sigma(1)} \ge g_{\sigma(2)} \ge \ldots \ge g_{\sigma(n)}$, $w = (w_1, \ldots, w_n)^T$ is the vector of weights, such that $\sum_{j=1}^{n} w_j = 1$, $w_j \ge 0$, $j = \overline{1, n}$.

In this operator $w_j$, the weight is associated not with a specific element of the vector G, but with its comparative value relative to other objects (the largest element gets the weight $w_1$, the next one after it $w_2$, etc.).

At the same time, by assigning certain values of weights, it is possible to implement disjunctive, conjunctive and compromise strategies for aggregation.

For example, when $w = (1, 0, \ldots, 0)^T$, we get

$$OWA* = OWA(g_1, \ldots, g_n) = 1 \cdot g_{\sigma(1)} = \max(g_1, \ldots, g_n), \tag{7}$$

that is, the OWA* operator implements a disjunctive strategy. Therefore, the aggregation takes into account the best property (correspondence) of the object.

When $w = (0, 0, \ldots, 1)^T$,

$$OWA_* = OWA(g_1, \ldots, g_n) = 1 \cdot g_{\sigma(n)} = \min(g_1, \ldots, g_n), \tag{8}$$

that is, the operator $OWA_*$ implements a conjunctive strategy, only the worst property of the object is taken into account. Therefore, a generalized estimate cannot be better than the worst of the partial estimates.

For $w = \left(\frac{1}{n}, \frac{1}{n}, \ldots, \frac{1}{n}\right)^T$,

$$OWA(g_1, \ldots, g_n) = \frac{1}{n} \sum_{j=1}^{n} g_{\sigma(j)} \tag{9}$$

is the arithmetic mean.

It is assumed that the aggregation operator has a compensatory property if small values of partial estimates for one indicator are compensated by large values of estimates for other indicators. Operator (3) does not have compensatory properties, operator (4) implements full compensation.

The indicator that characterizes the presence of a compensatory property in OWA with a particular set of weights is calculated using the formula

$$orness(w) = \frac{1}{n-1} \sum_{j=1}^{n} (n-j)w_j \tag{10}$$

If, for a given set of weights $orness(w) > 0,5$, the OWA operator has compensatory properties and implements a strategy close to disjunctive. If $orness(w) < 0,5$, then to the conjunctive.

Another indicator associated with $orness(w)$ and the inverse of it in value

$$andness(w) = \frac{1}{n-1} \sum_{j=1}^{n} (j-1)w_j = 1 - orness(w) \tag{11}$$

For example, for (7), $orness(w) = 1$, $andness(w) = 0$, for (8), $orness(w) = 0$, $andness(w) = 1$, for (9), $orness(w) = 0,5$.

## 3  Investigation of the Properties of Quantifiers in Relation to the Problem of Decision-Making on a Set of Alternatives

Let's return to the system of preferences (2). To formalize the information Q, we use the concept of a fuzzy quantifier [4]. Fuzzy quantifiers are an extension of the classical set of logical quantifiers, which includes the quantifiers $\exists$ ("exists") and $\forall$ ("for all"), by introduction of fuzzy concepts "almost for everyone", "about half", etc.

Let's consider the case when we are talking about ordering alternatives by preference and the criteria are of equal importance, that is $\lambda_j = const$. If we consider the behavior of the decision-maker, then the natural reasoning associated with the choice of the most preferred alternative will be those that are based on the assumption that the more criteria the alternative meets, the better. Another type of reasoning related to determining the quality of an alternative is based on the fact that the alternative must meet the majority of criteria or at least k% of the criteria. Such arguments can be formalized using quantifiers. The quantifier determines an approximate estimate of the number of aggregated values that greatly affect the value of the generalized estimate [5]. The quantifier is a fuzzy variable, the carrier of which is the fraction of partial estimates of r, defined on the segment [0; 1]. The membership function of the quantifier Q (r) corresponds to the degree of preference of an alternative that satisfies the fraction r of the entire set of criteria. For example, for Q(0,6), the specified percentage corresponds to 60%. In the future, we will consider quantifiers whose membership function satisfies the conditions:

$$1. \; Q(0) = 0, \quad Q(1) = 1; \tag{12}$$

$$2. \; Q(r_1) \leq Q(r_2), \; \text{at } r_1 < r_2; \tag{13}$$

$$3. \; Q(r) \; - \; \text{piecewise continuous function.} \tag{14}$$

Then, using such a quantifier, we can find the weights $w_j$ of descending-ordered partial estimates in the OWA operator by the formula

$$w_j = Q\left(\frac{j}{n}\right) - Q\left(\frac{j-1}{n}\right). \tag{15}$$

That is, the greater the rise in the value of the quantifier gives an increase in the share of the set of ordered partial estimates due to the $j$-th estimate, the greater its weight. The geometric meaning of the weights found by the formula (15) is shown in Fig. 1.

Suppose that, according to the information $\Lambda$, all partial estimates have the same importance. Then the decisive rule D in the model (1)–(2) will have the form

$$D(x) = \underset{x \in X}{OWA}(g_1(x), \, g_2(x), \, \ldots, \, g_J(x)) \rightarrow \max, \tag{16}$$

where the weights of the OWA operator are given by the formula (15).

Let's consider the types of quantifiers that satisfy the conditions (12)–(14) that are most often used in decision-making.

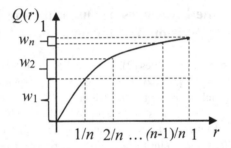

**Fig. 1.** The geometric meaning of the weights found by the formula (15)

The quantifier "For all" is determined by the formula (Fig. 2)

$$Q_\forall(r) = \begin{cases} 0, & r < 1, \\ 1, & r = 1. \end{cases}$$

The weights obtained by the formula (15) will be equal to

$$w_j = \begin{cases} 0, & j < n, \\ 1, & i = n. \end{cases}$$

The membership function of the quantifier is shown in Fig. 2.

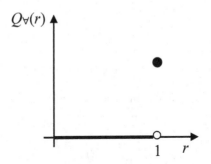

**Fig. 2.** The "For All" quantifier

When substituting these weights in (6), we get the operator $OWA*$. The quantifier "Exists" (Fig. 3) is determined by the formula

$$Q_\exists(r) = \begin{cases} 0, & r = 0, \\ 1, & r \le 1. \end{cases}$$

The weights obtained by the formula (12) will be equal to

$$w_j = \begin{cases} 1, & i = 1, \\ 0, & i < n. \end{cases}$$

In this case, we get the operator $OWA*$.

The quantifier "The more, the better" or "For as many as possible" (Fig. 4) can be determined by the formula

$$Q(r) = r \tag{17}$$

The weights obtained by the formula (12) will be equal to

$$w_j = \frac{1}{n}$$

and in this case we get the operator (9).

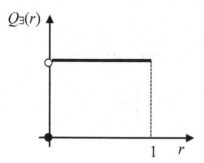

**Fig. 3.** The quantifier "Exists"

In order to determine to what extent the OWA operator, whose weights are found by formula (15), can implement conjunctive or disjunctive strategies, use the formula

$$orness(Q) = \int_0^1 Q(r)dr.$$

Consider a family of quantifiers whose membership function depends on the parameter $\alpha > 0$

$$Q(r) = r^\alpha \tag{18}$$

We investigate the influence of the parameter on the compensation properties, for this we define

$$orness(Q) = \int_0^1 r^\alpha dr = \frac{1}{\alpha + 1}. \tag{19}$$

The operator will have a compensation property if $\frac{1}{\alpha+1} > 0{,}5$ or $\alpha < 1$.

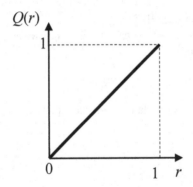

**Fig. 4.** The quantifier "The more, the better"

Let's consider a quantifier that can be used to express the concept of "For the majority", an example of such a quantifier is given in [6] (Fig. 5)

$$Q(r) = \begin{cases} 0, & 0 \le r \le a, \\ \frac{x-a}{b-a}, & a < r \le b, \\ 1, & b < r \le 1. \end{cases} \tag{20}$$

We'll find it

$$orness(Q) = \int\limits_0^1 Q(r)dr = \frac{b-a}{2} + (1-b) = 1 - \frac{a+b}{2}.$$

$$orness(Q) > 0.5, \text{ at } a + b < 1$$

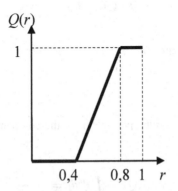

**Fig. 5.** Graphical representation of the quantifier "For the majority"

Consider a family of quantifiers that depend on two parameters, of the form

$$Q(r) = \begin{cases} 0, & r = 0, \\ \frac{1}{1+e^{-a(r-b)}}, & 0 < r < 1, \\ 1, & r = 1, \end{cases}$$

$$a > 1, \quad 0 < b < 1. \tag{21}$$

This function is continuous, monotonically increasing, $a > 1$ and has an s-shape and one inflection point with coordinates $x = b$, $y = 0,5$. The higher the value of $a$, the faster the transition from the shape of the curve convex down to convex up.

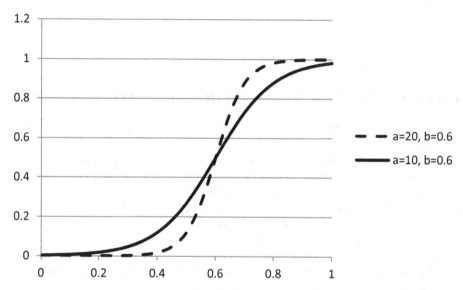

**Fig. 6.** Dependence of the curve shape of the function (21) on the parameter a

With the help of such a family of quantifiers, it is convenient to express concepts such as "At least for k%". The quantifier "For the majority" can be considered as a special case of this quantifier. Because "the majority" can be interpreted as "at least for 50%". We examine this quantifier for the presence of compensatory properties:

$$orness(Q) = \int_{0}^{1} \frac{1}{1 + e^{-a(r-b)}} dr = \frac{1}{a} \ln\left(e^a - 1\right) - b. \tag{22}$$

Let's determine at what values of parameters $a$ and $b$ it will have compensatory properties, that is

$$\frac{1}{a} \ln\left(e^a - 1\right) - b > 0,5 \text{ or } \frac{1}{a} \ln\left(e^a - 1\right) > b + 0,5. \tag{23}$$

Since $a > 1$, the function $f(a) = \frac{1}{a}\ln(e^a - 1)$ is monotonically increasing and $\lim_{a \to \infty} \frac{1}{a}\ln(e^a - 1) = 1$, then this inequality will be satisfied only when $b < 0,5$. That is, the function will have compensatory properties if the abscissa of the inflection point is less than 0,5.

In order for the indicator $orness(Q)$ to reach a certain value, which we denote by $\alpha^*$, it is necessary to solve the equation

$$\frac{1}{a}\ln(e^a - 1) - b = \alpha^*. \tag{24}$$

Let's fix $a$, in this case $b = \frac{1}{a}\ln(e^a - 1) - \alpha^*$.

If $b$ is fixed, then it is necessary to solve the equation with respect to $a$:

$$e^{a(\alpha^*+b)} - e^a + 1 = 0, \quad a > 1. \tag{25}$$

This equation has a single root, provided $\alpha^* + b < 1$.

## 4 Problems that Arise When Using Quantifiers in OWA Operators

We note a number of problems that may arise when using certain quantifiers in the OWA operator.

When using the "for the majority" quantifier in the form (20) with parameters $a = 0,4$, $b = 0,8$ (Fig. 5), the OWA operator becomes insensitive to the first values $g_{\sigma(j)}$ for which their total share does not exceed 0.4. The weights of these estimates will be zero. For example, if we are talking about comparing two alternatives, whose partial estimates are based on a set of criteria (already ordered in descending order):

(0,9; 0,6; 0,4; 0,3; 0,1) and (0,7; 0,6; 0,4; 0,3; 0,1),

then they turn out to be equivalent. For both options, OWA = 0,38. Although the first alternative is slightly preferable to the second due to the larger value of the first estimate.

Or, for example, for alternatives with private estimates (0,9; 0,6; 0,4; 0,3; 0,1), (0,8; 0,8; 0,4; 0,3; 0,1) and (0,9; 0,7; 0,4; 0,3; 0,1), OWA will also be equal to 0,38. Between the two alternatives with private estimates (0,4; 0,3; 0,3; 0,2; 0,2) and (0,9; 0,8; 0,3; 0,2; 0,2) the OWA operator will also not allow you to choose the best one, for both alternatives it will be equal to 0,29.

This is due to the fact that for the first ranked values of private estimates, their share is small and until it reaches 40%, the values of these estimates will not be taken into account. To avoid this problem, we can use the sigmoid function (21) as the membership function for the "for the majority" quantifier. With the help of the parameter b (the abscissa of the inflection point), it is possible to influence the proportion of criteria reached, as the concept of majority becomes more pronounced, with the value of the membership function greater than 0,5.

For the above cases, the OWA operator with the use of the quantifier "most" in the form of the function (21) will give the following results:

for (0,9; 0,6; 0,4; 0.3; 0,1) OWA = 0,3501 for (0,7; 0,6; 0,4; 0,3; 0,1) OWA = 0,3500, that is, the first alternative is slightly preferred;

for the alternative with private estimates (0,9; 0,6; 0,4; 0,3; 0,1) OWA = 0,3501 for (0,8; 0,8; 0,4; 0,3; 0,1) OWA = 0,3536 and (0,9; 0,7; 0,4; 0,3; 0,1), OWA = 0,3518;

for (0,4; 0,3; 0,3; 0,2; 0,2) OWA = 0,24997 and (0,9; 0,8; 0,3; 0,2; 0,2), OWA = 0,25896.

The OWA operator using the quantifier "The more, the better" in the form (17) may have problems when comparing alternatives if the sums of the partial estimates are the same. This is due to the fact that the weights $w_j$ obtained using this quantifier are the same. As a result,

$$OWA(g_1, \ldots, g_n) = \sum_{j=1}^{n} w_j g_{\sigma(j)} = w \sum_{j=1}^{n} g_{\sigma(j)} \qquad (26)$$

and, for example, alternatives with partial estimates (0,4; 0,35; 0,3; 0,03; 0,02) and (0,3; 0,2; 0,2; 0,2; 0,2) they turn out to be equivalent, OWA = 0,23.

## 5  Modification of the OWA Operator

To fine-tune the mechanism for comparing alternatives with each other using the OWA operator, it is necessary not only to take into account the share of private estimates of $r$, but also how well this or that private estimate involved in the formation of this share corresponds to the representation of the decision-maker about the degree of achievability of compliance with the criterion. It is proposed to supplement the OWA operator with a fuzzy function $h(g_{\sigma(j)}) : [0, 1] \rightarrow [0, 1]$ that allows describing the fuzzy concept of "Good matching". In this case, the rule Q for the quantifier Q1 = "for the majority" can be formulated as follows: "A GOOD match must be achieved for most criteria". For the quantifier Q2 = "the more, the better", the rule will take the form: "The more criteria a GOOD match is achieved, the better". The membership function "GOOD match" can be set:

$$h(g_{\sigma(j)}) = \begin{cases} 0, & g_{\sigma(j)} = 0, \\ \frac{1}{1+e^{-a(g_{\sigma(j)}-0,5)}}, & 0 < g_{\sigma(j)} < 1, \\ 1, & g_{\sigma(j)} = 1. \end{cases}$$
$$a > 1, \qquad (27)$$

where the abscissa of the inflection point is 0,5.

Note that each criterion can have its own degree of reachability of a good match, then the function $h(g_{\sigma(j)})$ for each criterion must be set separately.

Then the OWA operator will take the form

$$OWA(g_1, \ldots, g_J) = \sum_{j=1}^{J} w_j h(g_{\sigma(j)}) \qquad (28)$$

We will use the operator (28) with the quantifier "for the majority" in the form (21) and the rule "A GOOD match must be achieved for most criteria" with the membership function (27) ($a = 10$), which is the same for all criteria for ranking alternatives with characteristics:

(0,9; 0,6; 0,4; 0,3; 0,1) and (0,7; 0,6; 0,4; 0,3; 0,1). In the first case, OWA = 0,20064, in the second case, OWA = 0,20061;

(0,9; 0,6; 0,4; 0,3; 0,1), (0,8; 0,8; 0,4; 0,3; 0,1) and (0,9; 0,7; 0,4; 0,3; 0,1), in the first case, OWA = 0,20064, in the second, OWA = 0,20454, in the third, OWA = 0,20328;

(0,4; 0,3; 0,3; 0,2; 0,2) and (0,9; 0,8; 0,3; 0,2; 0,2). OWA1 = 0,0833, OWA2 = 0,0983.

When ranking alternatives with characteristics (0,4; 0,35; 0,3; 0,03; 0,02) and (0,3; 0,2; 0,2; 0,2; 0,2) using the quantifier "the more, the better" and the rule "The more criteria a GOOD match is achieved, the better", the OWA operator will allow them to be ranked, in the first case OWA = 0,1175, in the second case OWA = 0,062.

## 6 Conclusion

As a result of the study of various fuzzy quantifiers, the boundaries of the values of the parameters of the membership functions were found, at which the OWA operator will have compensatory properties. The presence of "insensitivity zones" of the quantifier with piecewise linear functions of belonging to the change in the values of the components of the criteria vector is also established. It is shown that this problem is solved when passing to a continuous membership function in the form of an s-shaped (logistic) curve.

A modification of the OWA operator is proposed in the form of a superposition of partial estimates and a membership function of the fuzzy concept of "Good correspondence". This modification ensures that when comparing alternatives, not only the number of private assessments that meet the criteria is taken into account, but also the quality of compliance. At the same time, for each criterion, the degree of reachability of a good match can be set by its own membership function.

## References

1. Zadeh, L.A.: A computational theory of dispositions. In: Turksen, I.B., Asai, K., Ulusoy, G. (eds.) Computer Integrated Manufacturing. NATO ASI Series (Series F: Computer and Systems Sciences), vol. 49. Springer, Heidelberg (1988). https://doi.org/10.1007/978-3-642-83590-2_9
2. Dubois, D., Prade, H.: Possibility Theory: An Approach to Computerized Processing of Uncertainty. Plenum, New York (1988)
3. Pospelov, D.A.: The modeling of reasoning. Experience in the analysis of mental acts. Radio and communication, Moscow (1989). 184 p.
4. Yager, R.R.: On ordered weighted averaging aggregation operators in multicriteria decision making. IEEE Trans. Syst. Man. Cybern. **18**, 183–190 (1988)
5. Ledeneva, T.M.: Aggregation of information in evaluation systems. In: Ledeneva, T.M., Podvalny, S.L. (eds.) Proceedings of Voronezh State University, Voronezh. Series: Systems Analysis and Information Technologies, no. 4, pp. 155–164 (2016)
6. Averchenkov, V.I., Lagerev, A.V., Podvesovskiy, A.G.: Presentation and proceedings of fuzzy information in multi-criteria decision models for the problems of socioeconomic systems management. Herald Bryansk State Tech. Univ. **2**(34), 97–104 (2012)

7. Detyniecki, M.: Mathematical aggregation operators and their application to video querying. M. Detyniecki Ph.D. dissertation, Docteur de l'Universite, Paris (2000). 185 p.

8. Mesiar, R., Komornikova, M.: Aggregation operators. In: XI Conference on Applied Mathematics "PRIM 1996", Proceedings, Novi Sad, pp. 193–211 (1997)

9. Ledeneva, T.M.: Models and methods of decision-making: a tutorial. Voronezh State Technical University, Voronezh (2004). 189 p.

10. Fuller, R.: OWA operators in decision making. In: Carlsson, C. (ed.) Exploring the Limits of Support Systems, pp. 85–104. TUCS General Publications, No. 3, Turku Centre for Computer Science, Abo (1996). ISBN 951-650-947-9, ISSN 1239-1905

11. Kouchakinejad, F., Šipošov6, A.: Ordered weighted averaging operators and their generalizations with applications in decision making. Iran. J. Oper. Res. **8**(2), 48–57 (2017)

# The Dialectic Interrelation Between Digital Organization Architecture Elements

Vladimir V. Godin[(✉)] [ID] and Anna E. Terekhova[ID]

State University of Management, 99 Ryazanskii Prospect, 109542 Moscow, Russia
godin@guu.ru

**Abstract.** The paper examines the mutual influence of business architecture and system architecture on enterprises of different types: traditional enterprises, enterprises advanced in terms of use of information and communication technologies, digital business enterprises, digital platform enterprises, and digital ecosystems. Enterprise types are aligned with architecture patterns that describe business architecture, information architecture, application architecture, and technology architecture. The paper describes a general methodological model of diverse elements dynamics of enterprise business architecture and elements of system architecture. The developed methodological model allows to predict emergence of libraries of enterprise architecture templates for multiple use.

**Keywords:** Enterprise architecture and system architecture · Enterprise architecture templates · Automation · Informatization · Digitalization · Digital enterprises · Digital platforms and ecosystems · SAP solutions

## 1 Introduction. Problem Statement

The role of information and communication technologies (ICT) and information systems (IS) plays in modern day enterprises and the economy can hardly be overestimated. Technologies development changes business environment and creates new opportunities, which forces enterprises to change their strategies, business models, operating activities, and which ultimately form the demand on the ICT and IS change. Therefore, there is a mutual influence of the economy (and enterprises), relevant technologies and IS. This mutual influence is so critical that IS management including interaction of business and technologies has taken shape as a separate knowledge field, with its own theory and concepts, methodologies, methods, and tools. This management activity is now represented by special departments within companies which are involved among other things in monitoring of interrelations of enterprises and IS used by them.

This interrelation varies over time and manifests itself in a number of processes. Based on internal and external reasons enterprises make decisions about development, use and liquidation of information systems (IS). At the same time, enterprises depend on changes in the business environment, including those caused by ICT and IS evolution. On the other hand, enterprise properties manifest themselves in the properties of created

© Springer Nature Switzerland AG 2022
V. Taratukhin et al. (Eds.): ICID 2021, CCIS 1539, pp. 180–190, 2022.
https://doi.org/10.1007/978-3-030-95494-9_15

IS. Involvement of certain ICTs and IS into the enterprise work practice leads to transformation of the enterprise itself. This transformation can be seen at all levels in terms of ways of functioning, organizational structure, employment, activities, formalization, etc.

The mutual influence of enterprise elements in the era of accelerating digitalization requires a concept that would allow enterprises to use existing transformation experience to their advantage. Development of such a concept not only would ease decision making, but also allows to allocate resources in the more appropriate or urgent elements of the enterprise architecture in order to facilitate and speed up transformation. The aim of the study is to link into a general methodological model the diverse dynamics of economic changes under the influence of ICTs with the ICTs and IS as a technology, enterprises, and their chosen IS. The concept of the enterprise architecture ("Enterprise Architecture") considered at several levels of abstraction is proposed as a tool. Within the framework of the enterprise architecture, it describes what and how an enterprise does (mission, goals, strategy, main functions), what it consists of (properties of elements), where the elements are placed (the structure of the enterprise), how the parts of the enterprise (elements) relate and on what principles they use to interact (the relationship of elements).

There are many studies in the field of enterprise architecture, for example, publications by Jung J., Fraunholz B. [1] devoted to the management of enterprise architecture. Benade S.J., Pretorius L. [2] apply the system architecture and the processes associated with it to a typical (physical) system, enterprise and project. This leads to the concepts of system architecture, enterprise architecture, and project architecture, respectively. The authors explore the similarities and relationships between these architectures and the corresponding methodologies in search of a better interaction between them. Pérez-Castillo R., Ruiz F., Piattini M., Ebert C. [3] show how, in the context of digital transformation, the concept of enterprise architecture is used as a technological, continuous process of change for companies, which allows companies to model IT and assess the needs for changes in IT, business processes, cloud services and distributed systems. Portugal T., Barata J. [4] discuss the phenomenon of unplanned and sometimes uncontrollable changes in elements and interactions related to architecture (applications, business entities, processes, strategy, motivation and physical infrastructure). Kleehouse M., Matthes F. [5] propose an approach to creating an enterprise architecture model based on a connected enterprise knowledge graph. This makes it possible to ensure the relevance of information about the enterprise architecture by automatically creating and updating static models of the enterprise architecture and information about their execution environment.

## 2   Setting the Framework. The Research Methodology Description

A "framework" is a set of principles and approaches to describe the interaction of components of an enterprise architecture. There are numerous models for architecture description (IEEE POSIX 1003.23 [6], TOGAF [7], Gartner methods [8, 9], META Group methods [10], Giga [11], etc. [12]). The purpose of this study is best met by using an integrated concept of enterprise architecture [13], in which the enterprise architecture contains four domains (business architecture, information architecture, application architecture, technological architecture), and is considered at five levels of abstraction (context level, conceptual level, logical level, physical level, implementation level).

At the context level, the business architecture domain, necessary organizational structure elements, sales channels, and functional model of organization are determined based on analysis of external environment, chosen driving forces and factors, vision, mission, goals and strategy. The information architecture domain defines a list of business entities and relationships between them. In the application architecture domain, there is a list of business processes supported by IS. The technological architecture contains a list of business locations.

The concept level in a corresponding domain will contain a description of the business model, key business processes and data, services, and the relationship of services.

The logical level of the architecture contains a description of the main functional components and their interrelationships with each other without technical implementation details – classes of application systems, technologies and data that must be supported (workflow models, business events, process locations, role definition; logical data models, data schemas, document specifications; definition of services and relationships between them; logical server types, geographical distribution of servers, hosted software).

The physical level of abstraction describes the physical structure model of the implemented system. It sets design principles, standards and rules, and deployment models.

At the implementation level, the IS architecture model is defined in terms of certain products and technologies.

The proposed integrated enterprise model is a certain simplification, but it is sufficient and detailed enough to consider the elements interaction dynamics. The integrated enterprise model is based on the mutual influence of business architecture and system architecture (which consists of information architecture, application architecture and technological architecture).

## 3   Implementation of the Framework. Analysis of the Mutual Influence of Enterprise Architecture and System Architecture

### 3.1   Hypothesis and Classification of the Enterprise Types for the Purpose of the Study

Enterprise architecture review is the practice of constant and endless analysis, design, planning and implementation using an integrated approach for the successful development and implementation of the strategy.

The following assumptions can be made in order to consider this practice.

1. Private mutual influence. Changes at any level in the business architecture elements lead to changes in data architecture domains, applications and technologies. And vice versa. Any change in the system architecture element forms transformation opportunity for the business architecture in terms of strategy, business model, structure and operations.

2. General mutual influence. Any system architecture (information architecture, application architecture, and technological architecture) is a consequence and an environment for implementing the business architecture. The business architecture "forces" the system architecture to change. The system architecture supports the forms of business models. At certain moments it creates opportunities for generating changes in the business architecture.

3. Enterprise architecture components have different dynamics of changes – changes are occurring at a different pace.

Considering the impact of the system architecture development on the business architecture, a significant factor to mention is the role of ICT (system architecture) impact, measured at different business levels (operational, structural and strategic). When assessing such an impact, one can talk about the auxiliary role of ICT, ICT as a restructuring resource and ICT as technologies of digital transformation. At the same time, processes of using ICT are associated with the implementation of automation, informatization and digitalization of an enterprise.

In this context automation is understood as the use of self-regulating technical and software tools that fully or partially exempt a person from participating in production processes. Informatization is the process when information use efficiency is enhanced by implementation of ICT. Informatization is impossible without automation and represents the next step in the use of ICT (after automation). The phenomenon that encompasses automation, and then informatization as subsets, is digitalization. Digitalization is the process of applying digital transformation technologies that allows forming a production scope of goods and services in the form of cyber-physical systems which are an integral interaction between the virtual and real parts of the world.

It is possible to classify manifestation of the system architecture influence on business architecture in a form of five classes (types) of enterprises (forms of business architectures), within which the processes of automation, informatization and digitalization are implemented at a different rate:

- traditional enterprises (automation and limited informatization);
- advanced enterprises in terms of use of ICT (automation and full-fledged informatization);
- digital business enterprises (digitalization);
- digital platform enterprises (digitalization of the enterprise itself and digital services for other enterprises);
- enterprises – digital ecosystems (digitalization of the enterprise and digital services for other enterprises, replication of these solutions to non-core markets).

For the all listed above types of enterprises, one can specify recurring tasks and solutions used to form the architecture (design solutions and logical models as templates). Therefore, it becomes possible to create a library of enterprise architecture templates for reiterated/multiple use (business templates, design templates, application templates, etc.).

## 3.2   Traditional Enterprises

In this case, the role of ICT is auxiliary and partly restructuring. Most often this role is manifested in the automation of daily activities of employees. For a traditional enterprise, it is possible to form the system architecture description and the nature of the formation of its domains at the implementation levels and at the physical level. The corresponding IS architectures are formed from the variants of the information architecture domains, application architecture and technological architecture within the framework of combinations of three dimensions: the presentation layer (user interface) - business logic (application algorithms) - the data access layer (storage, selection, modification and deletion of data). This will manifest itself in the form of various IS architectures (file-server architecture, client-server architecture, three-level client-server architecture, multi-tier client-server architectures). As well in the form of classes of applied IS corresponding to the forms of information (data, information, knowledge), levels of the management hierarchy (automated process control systems (automated process control systems), MES (manufacturing execution system) – systems, ERP (Enterprise Resource Planning), BPMs (Business Process Management System), EPM (Enterprise Project Management)/PPM (Project Portfolio Management) – systems, Knowledge Management class systems, decision support systems (DSS), BI (Business Intelligence)/BW (Business Warehouse) – systems, ECM (Enterprise content management) - systems) and the nature of the tasks to be solved (large amounts of information, many users in real time, complex calculations, etc.).

## 3.3   Enterprises Advanced in Terms of ICT Use

The ICT development at a certain stage has turned them into a restructuring resource for the economy, business strategies and business models of organizations. At this level of ICT impact on the economy as a whole, the following changes can be noted. The electronic economy has formed with the corresponding models: B2B, B2C, C2C, G2C, etc. Which brought along such new concepts as information economy (B. Gates [14]), information age (M.Castels [15]), knowledge economy.

There was a shift of business model: the idea of the "network enterprise" appeared on the basis of the "traditional enterprise" concept, which carried out the transition from mass production to flexible production, value-added communities, meta-markets, strategic alliances. It is important to note the change of business priorities: from the era of production to the era of quality, and then to the era of the consumer [16].

All these changes in the business architecture were accompanied by changes in the domains of the system architecture at various levels of abstraction. It was manifested in the development and implementation of the appropriate models of data architectures, application architectures and technological architectures.

## 3.4   Digital Business Enterprises

The next stage of ICT development has formed the digital economy. First it was manifested in the technological possibility of creating cyber-physical systems in all spheres of human activity, where life changing convenient products and services are being built.

The cyber-physical system is the information technology concept that implies the integration of computing resources into physical entities of any kind. The new concept of "digitalization" involves inclusion of the previous phases of software and hardware use in the form of automation and informatization. A digital business enterprise is a company, which products and services, as well as processes (both customer-oriented and internal) have been digitized and received digital interfaces. There is the three-level digital company architecture concept (Gartner):

- Accounting support system (System of Record) - ERP/CRM.
- Support system for unique processes (System of Differentiation) - Process Management, BPMS (Business Process Management Suite).
- Support system for innovative projects (System of Innovation). The concept of case management and SW of the ACM (Adaptive Case Management) class.

The following properties are typical for digital business companies:

- Presence of a single technological platform that provides support for all types of collaboration: processes, projects, cases, assignments.
- Interoperability (the ability, for example, to call processes from cases and vice versa).
- Unified architecture (a complete map of business capabilities, translated into specific functions of the ERP system, processes, projects or cases).
- Event-dependent systems - constant recognition and use of opportunities, which requires continuous registration of business events, such as delivery contract execution, aircraft landing, etc. Events can be recognized faster and analyzed in depth with the help of event brokers, the Internet of Things technologies, clouds, blockchain, smart contracts, data analysis in memory and artificial intelligence. Decisions can be made in real time, which means more adequate respond to events.
- Unified environment of social interaction at work.
- Adaptive security architecture. Continuous adaptive risk and reliability assessment.

All the emerging forms of digital companies can be described as a combination of a digital core (for example, SAP S/4 HANA) and the digital transformation technologies. Although there are different definitions of the digital core, the definition formed by SAP is best suited [17] for the purposes of this work. The digital core is understood as a platform that combines main data and basic business processes. For example, the use of Big Data technologies based on the digital core will lead to the formation of a Data-Driven Company. The digital core, blockchain, smart contracts and cryptocurrencies will lead to the emergence of decentralized autonomous organizations. The Digital core, the industrial Internet of Things, cloud computing, digital twins, etc. create new industrial platforms (Industry 4.0 - Smart Factory, Maker Economy, Robotic Process Automation). Also new economic forms have appeared, such as sharing economy, maker economy, smart grid, smart cities, new generation automation, connected healthcare, etc. [16].

Obviously, under the influence of digital transformation technologies, we have to consider the architecture of an enterprise in the context of cloud computing, the Internet of Things, Big data, mobility and collaboration networks, etc. An illustration of such changes in business architecture and system architecture under the influence of digital transformation technologies can be the work of [18]. It is devoted to the analysis of the impact of infrastructures and components of the Internet of Things on the architecture of enterprises. The paper proposes an approach for the integration of architectural objects of the Internet of Things, which are semi-automatically combined into an integrated environment of the architecture of a digital enterprise.

### 3.5   Digital Platform Enterprises

Platform economy is the economic activity based on platforms [19, 20]. The concept of a platform, in architectural terms, is described as an online system that provides complex standard solutions for interaction between users, including commercial transactions. Platforms provide the ability to use specific software solutions and related services without the need for independent development. The key task of the platform is to serve as a basis for direct interaction between participants.

The transition to platform economy meant several fundamental changes in the nature of business:

- transition from resources control to coordination of resources;
- shift from internal optimization to interaction with external parties;
- focus change from consumer value to network interaction value;
- reorientation of information technologies and systems from organizational management to social networks and consumer community networks.

Analysis of the platform economy development allows us to record the rapid growth of companies that have created platforms, a sharp increase in sales, the emergence of additional ways to create added value, and other economic effects. Such companies have essentially become digital companies, with digital platforms, digital processes, digital sales and service channels, with a specific corporate culture that supports the use of analytics, flexible adaptation and change. Due to the need to trust a platform, constant development of security systems (platform and information protection systems) is required. Depending on the type of the digital platform, it is possible to formulate the properties of the system architecture that should ensure the existence of such a platform. In any case, the system architecture will provide the following: algorithmization of relationships between platform participants, high centralization of IT solutions, ease of registration, clear interfaces, ease of downloading content, a simple process of joining the platform, finding partners, exchanging information, availability 24 h 7 days a week, constant development of security systems in terms of protecting the platform itself and the information used in it. The implementation of a significant part of the platform requirements is based on the possibility of data exchange between the platform's IP and the participants' IP based on open APIs (Application Programming Interface). The functioning of the platform as a form of business allows to accumulate and analyze data (Big Data, artificial intelligence and machine learning), and make decisions based on such analysis.

## 3.6   Enterprises - Digital Ecosystems

The next step of ICT development in the framework of digitalization is the transition from the platform economy (platform as a business model) to the ecosystem economy (digital ecosystem as a business model) [21–24]. Digital platform companies have accumulated experience in operating platforms to ensure interaction between various companies (they have formed an understanding of businesses and their customers based on data analysis). Companies have a sufficient number of technological solutions that could be considered as prototypes of IS. This ensures the availability of technologies to generate new businesses in the fields of non-core markets and products. Implementation of these processes of the platform economy development and its transition to the ecosystem economy have ensured the existence of companies themselves and satisfaction of the market needs. For instance, obtaining high-quality products and services quickly and with minimal effort using convenient digital channels, providing personal data for obtaining targeted and personalized products and services, making decisions based on data analysis. There are certain properties which are necessary for a company in order to be able to organize an ecosystem:

- company must be attractive to potential participants of the ecosystem (have a well-known brand and business reputation, own a large-scale customer base and extensive relationships with customers, ensure mutually beneficial coexistence of participants);
- company must have the resources to build a digital ecosystem platform (the properties of the system architecture of digital ecosystems are openness, flexible IT infrastructure, microservice architecture, providing integrative and flexible interaction with partners and customers through API, staffing, analytics-competencies for collecting and analyzing client and other data, modular structure support and strengthening the redundancy of the ecosystem organizer components to preserve its stability);
- company must be able to solve organizational issues of building and operating an ecosystem based on the principles of self-organization.

Now at least three types of ecosystems can be observed: an ecosystem as a platform for trading and providing services, an ecosystem as an association of participants in the value chain (added value community) and an ecosystem as a self-developing organization.

## 3.7   Enterprise Architecture Templates and Architectural Styles

Any existing enterprise can be referred to one of the enterprise classes described above. This allows to describe the architecture of such an enterprise in the "as is" and "to be" terms and match them with an appropriate information architecture, application architecture and technological architecture. The conceptual level for the business architecture contains top-level business process models which describes primary and secondary activities. The models are built using the VAD (value added chain diagram) and eEPC (event-driven process chain) notations [25]. Information, application, and technology architectures are the consequence and implementation environment of business architecture. Therefore, the consequence of building the business process models are the requirements for the environment of their implementation – in other words, the system architecture. Semantic models, relationship models, and Entity-relationship models are to be built for the information architecture. Within the framework of the application architecture, processes are divided into services/subsystems/modules, etc. The logical level for the business architecture contains the description of business event models, detailed process models (workflows), etc. As a presentation tool for the business architecture one can choose the eEPC model or any other model of a similar class. For information architecture – logical data models, data schemas, document specifications. For the technological architecture – logical types of servers (databases, mail, transactional, etc.), geographical distribution of servers, hosted software, etc. Process specifications, process integration models, description of manual procedures, quality standards will be set at the physical level and the business architecture implementation level. For information architecture – physical data models, data reference books. For application architecture – program code, interface description, process schedule, workflow code.

For all levels of the architecture description one can specify recurring tasks and their solutions in the formation of architectures (these are logical models of technologies in the form of design ideas and their embodiments/realization). Each of the general solutions to any recurring problem in a certain context is a template. Templates description can be performed with varying degrees of details according to the architecture level. Templates are certain typical elements of the architecture.

An enterprise has an opportunity to create a library (repository) of enterprise architecture templates for multiple use (business templates, design templates, application templates, etc.). The main idea of creating such a library is to determine rules for applying the templates for the specific architecture of enterprise information systems in the future. Ultimately, this means creating a methodology as a tool for forming a wide range of different architectures. Its tasks are to define a common dictionary of terms used; to set a tool for describing architecture elements at different levels of abstraction; to form methods for designing architecture in terms of using certain templates and linking them to each other; to describe recommended standards and compatible products that can be used to build various elements of the enterprise architecture.

Various business processes are supported by specific applications, information, and technologies, which corresponds to a certain set of architecture templates. It is possible to allocate business process classes (for example, real-time operations, analytics, mass transaction processing, etc.) to similar architectures. Such a built-up set of architecture templates, put in accordance with the classes of business processes, is a certain architectural style. Since an enterprise cannot develop a system architecture in all directions, it is important to analyze the existing and future classes of business processes, their corresponding architectural style. This approach which will allow to predict the system architecture requirements.

## 4   Conclusion

Business architecture properties at the contextual, conceptual, logical, physical and implementation levels determine system architecture requirements. Enterprises, depending on the role of ICT for them, can be allocated to different classes of business architectures: traditional enterprises (automation and limited informatization); enterprises advanced in terms of the ICT use (automation and full-fledged informatization); digital business enterprises (digitalization); digital platform enterprises (digitalization of the enterprise and the offer of digitalization services for other enterprises); enterprises–digital ecosystems (digitalization of the enterprise and the offer of digitalization services for other enterprises, replication of these solutions to non-core markets). Within each class of business architecture, it is possible to distinguish repetitive implementations with corresponding business models, business processes, organizational structures, and operational activities. Such repetitive implementations are the patterns (templates) of business architectures. They are characterized by the corresponding patterns of information architecture, application architecture and technological architecture. Therefore, there is an opportunity to create an architecture templates library (business templates, design templates, application templates, etc.) and specify the architectural styles for their repeated use in the current practice of enterprise development management and forecasting of the system architecture requirements.

To test the theoretical approach discussed in the article, the authors applied an architectural approach to the process of digital transformation of an educational institution - a university. From the point of view of the impact of ICT on the university and on the educational process carried out in it, the university falls into the same classes of organizations that were discussed above. At the same time, changes at the university occur in three directions: pedagogy, technology and organization.

Firstly, the university may belong to the class of traditional educational institutions (traditional enterprise). The role of ICT is auxiliary. We are talking about automating the operational activities of teachers, staff and students. The dominant technologies are computer base training (CBT) and web base training (WBT).

Secondly, the university can belong to the class of advanced enterprises in terms of the use of ICT. ICT forces the transformation of the university's strategy, structure and operations. The system architecture supports any form of Distance Learning.

Thirdly, the university may have turned into a digital business company. ICTs support the implementation of the concepts of e-Learning, smart education. The personal educational trajectory, the student's digital twin, the widespread use of the ideas of Big data, artificial intelligence, etc. at least turn the university into a data driven company. Templates of types of interactions (teacher - student, student - study material, student - student, student with himself), phases of the educational process (presentation of educational material, its consolidation, control), educational process management processes, educational institution management processes allow you to build appropriate standard solutions in terms of system architecture as described in the paper.

# References

1. Jung, J., Fraunholz, B.: Masterclass Enterprise Architecture Management. Springer International Publishing, Cham (2021). https://doi.org/10.1007/978-3-030-78495-9
2. Benade, S.J., Pretorius, L.: System architecture and enterprise architecture: a juxta position? S. Afr. J. Ind. Eng. **23**(2), 29–46 (2013). https://doi.org/10.7166/23-2-328
3. Pérez-Castillo, R., Ruiz, F., Piattini, M., Ebert, C.: Enterprise architecture. IEEE Softw. **36**(4), 12–19 (2019). https://doi.org/10.1109/MS.2019.2909329

4. Portugal, T., Barata, J.: Enterprise architecture erosion: a definition and research framework. In: 27th Americas Conference on Information Systems (AMCIS) (2021)
5. Kleehaus, M., Matthes, F.: Automated enterprise architecture model maintenance via runtime IT discovery. In: Zimmermann, A., Schmidt, R., Jain, L.C. (eds.) Architecting the Digital Transformation. ISRL, vol. 188, pp. 247–263. Springer, Cham (2021). https://doi.org/10.1007/978-3-030-49640-1_13
6. IEEE 1003.23-1998 - IEEE Guide for Developing User Open System Environment (OSE) Profiles. https://standards.ieee.org/standard/1003_23-1998.html, Accessed 22 Nov 2021
7. The Open Group. TOGAF® Version 9 - Download. Architecture Forum. http://www.opengroup.org/architecture/togaf9/downloads.htm, Accessed 22 11 2021
8. Drobik, A., et al.: The Gartner definition for the Real-Time enterprise. Gartner Research Note COM-18-3057 (2002)
9. Application Architecture Standards/Guidelines: Case Study. Gartner (2003)
10. Meta Group. Enterprise Architecture Desk Reference (2002)
11. Basic Elements of Enterprise Architecture Methodology. Giga (2003)
12. Wout, J., Waage, M., Hartman, H., Stahlecker, M., Hofman, A.: The Integrated Architecture Framework Explained: Why, What, How. Springer, Heidelberg (2010). https://doi.org/10.1001/978-3-642-11518-9
13. Danilin, A., Slyusarenko, A.: Architecture and strategy. "Yin" and "Yang" of information technologies of the enterprise. Internet-Un-t Inform. Technologies, p. 504 (2005)
14. Gates, Bill. Business @ the Speed of Thought. Grand Cantral Publishing (2009)
15. Castells, Manuel. Information Age: Economy, Society and Culture. Vol. I-III. Oxford: Blackwell Publishers (1996–1998)
16. Godin, V.V.: Transformation of business under the influence of information technologies. In: Materials of the II International Scientific Forum "Step Into the Future: Artificial Intelligence and Digital Economy Revolution in Management: A New Digital Economy or a New World of Machines, GUU, no. 2, pp. 434–443 (2018)
17. Digital core. https://news.sap.com/2016/06/sap-s4hana-the-digital-core, Accessed 22 Nov 2021
18. Zimmermann, A., Schmidt, R., Sandkuhl, K., Wißotzki, M., Jugel, D., Möhring, M.: Digital enterprise architecture - transformation for the internet of things. In: Conference: IEEE International Enterprise Distributed Object Computing Conference (EDOC 2015), SoEA4EE Workshop, Adelaide, Australia, vol. 19 (2015)
19. Evans, P.C., Gawer, A.: The rise of the platform enterprise: a global survey. The Center for Global Enterprise (2016)
20. Choudhary, S.P., Van Alstyne, M.V., Parker, G.: Platform revolution. How network markets are changing economies - and how to make them work for you, p. 304 (2017)
21. Martin, R., Levin, S., Daichi, U.: The biology of corporate survival (2016). https://hbr-russia.ru/biznes-i-obshchestvo/fenomeny/a17381, Accessed 30 July 2021
22. Godin, V.V., Terekhova, A.E.: Digital ecosystems as a form of modern business transformation. In: Proceedings of the 1st International Conference of Information Systems and Design, Moscow, Russia, 5 December 2019, CEUR-WS.org (2019). http://ceur-ws.org/Vol-2570/paper19.pdf, Accessed 22 Nov 2021
23. Mark, J.G., Wei, W.: Business Ecosystems in China, Paperback (2017)
24. Rong, K., Shi, Y.: Business Ecosystems: Constructs, Configurations, and the Nurturing Process (2014)
25. Scheer, A.-W.: Business Process Engineering Study Edition. Springer, Heidelberg (1998). https://doi.org/10.1007/978-3-662-03615-0

# New Areas of Blockchain Technology Applications During the COVID-19 Pandemic

Vladimir V. Godin$^{(\boxtimes)}$ (ID), Anna E. Terekhova (ID), and Sofia V. Kalamagina (ID)

State University of Management, 99, Ryazanskii Proskpekt, 109542 Moscow, Russia
godin@guu.ru

**Abstract.** The article is devoted to the analysis of new directions of application of blockchain technology that have emerged in the conditions of the Covid-19 pandemic, generalization of existing experience in using blockchain and definition of tasks for blockchain technology development. The paper analyzes the properties of blockchain technology, manifested in the created systems, which can solve several management problems in the process of combating the Covid-19 pandemic. The article describes the Covid-19 pandemic fight and their possible actions. The paper provides analysis and systematization of new areas of blockchain application, such as tracking contact with infected people, verification of vaccine authenticity, maintaining of the clinical trial data integrity, identifying counterfeit medicines and personal protective equipment, identifying foci of infection. The authors describe examples of specific solutions, showed that the blockchain technology can play an innovative role in solving a few problems that appeared during the pandemic. The existing limitations are revealed both in the blockchain technology itself and in the conditions of its application, which do not allow to fully realize the full potential of the technology. The description of new areas of application of blockchain technology is important both for the organizers of the COVID-19 pandemic response, and for researchers and developers of systems using blockchain technology.

**Keywords:** Digitalization · Digital technologies · Blockchain · Pandemic · Covid-19 · Healthcare

## 1 Introduction

Today, the issue of digitalization is widely discussed both in theoretical and practical aspects. The Russian Federation has approved a list of technologies related to end-to-end digital technologies (the program called "Digital Economy of the Russian Federation"). These are technologies such as artificial intelligence, virtual and augmented reality, quantum and neurotechnology, big data, distributed registry systems [1]. Distributed registry technologies, in particular, blockchain, are one of the promising tools of digitalization. The technology is used in such areas as digital identity, user authentication, secure bilateral transactions without a third-party guarantee, cryptocurrencies, etc. [2]. The pandemic, which began in 2019, has made its own changes in all spheres of life, including redefining the use areas of this technology.

© Springer Nature Switzerland AG 2022
V. Taratukhin et al. (Eds.): ICID 2021, CCIS 1539, pp. 191–202, 2022.
https://doi.org/10.1007/978-3-030-95494-9_16

Large amount of publications have been devoted to the topic of using various digital technologies to combat the COVID-19 pandemic or eliminate its consequence. E.g., [3] is considering the application of artificial intelligence and big data, [4] - the Internet of things, unmanned aerial vehicles and 5G, [5] - artificial intelligence (AI). Sufficient amount of works is considering specific areas of digital technologies, the blockchain technology application. For example, [6] presents specific algorithms used to track those who were in contact with the infected of COVID-19 and their test results. Experience of the Charity Wall application which uses blockchain technology to ensure secure donations is discussed in [7].

The purpose of this paper is to analyze new areas of application of blockchain technology that have emerged in the context of the Covid-19 pandemic, highlight the main directions, and demonstrate examples; identify limitations that do not allow to fully use the potential of the blockchain technology. From a practical point of view, results of the study will allow to determine strategic guidelines for developers and consumers of blockchain technology.

Therefore, the paper is mainly focused on showing how the properties of the blockchain are manifested in the capabilities of the systems being created to combat the COVID-19 epidemic, generalizing existing experience and defining objectives of technology development.

## 2   Research Methodology. Properties of the Research Object

Numerous scientific researches on the theoretical and methodological aspects of the construction and use of blockchain technology were observed as a theoretical basis for this paper. Reports from the World Economic Forum, broadcasts of Digital Planet (BBC World Service), the Information Age portal articles, etc. were used as information materials of the study. The methodological basis of this paper is the general scientific research method that is used to study reality.

According to Gartner's definition, "a blockchain is a regularly updated list of cryptographically signed, irrevocable transaction records that all network participants have access to".

The author of the book "Blockchain for Business", William Mougayar, suggested such a distribution of blockchain functions: cryptocurrency, computing infrastructure, transaction platform, decentralized database, distributed account registry, development platform, open source software, financial services market, peer-to-peer network, trust services level [8]. To provide such a wide set of functions, the blockchain has a number of properties. The inherent properties of the blockchain, such as openness, transparency, security, and others, ensure the universality of its application and the ability to reduce transaction costs. For example, the article "The Blockchain Practice", posted on the official Deloitte page, describes the following areas: digital assets (crowdfunding, exchange trading, audit, supply chain management, etc.), digital identification (proof of ownership, chronology of documents, insurance, automatic deduction schemes, etc.), smart contracts (notarial service, voting, distributed trading, compliance, etc.), applications (financial asset management, automatic tax collection, etc.) [9]. The approximate size of the phenomenon called "blockchain application" can be estimated as follows.

Decentralized applications in the form of cryptocurrencies (issuance, transfers, payments, exchanges) – billions of users, digital assets and digital identification (shares, property rights, debt obligations, crowdfunding, crowdlending) – hundreds of thousands of millions of users, smart contracts (family trusts, escrow, derivatives, leasing, insurance, labor contracts) – hundreds of millions of users, applications in the form of decentralized autonomous organizations and systems (decentralized autonomous organizations, industrial platforms, transportation, medicine, decentralized exchanges (commodity, stock) – hundreds of thousands of users.

The properties and capabilities of the blockchain in these areas of use show their potential in the COVID-19 pandemic combat. At least blockchain allows to create a coordinated general view of information and facilitate the exchange of it. As a maximum, it allows to create specific applications. The difference between such applications and centralized solutions is the lack of problems specific for all centralized solutions: unreliability, the existence of a point of destruction, vulnerability to data manipulation, problems of access to personal data, restrictions on integration and data exchange, the inability to track and ensure data transparency, etc.

## 3 Results of the Study

### 3.1 Blockchain Technology in the Fight Against the COVID-19 Pandemic

COVID-19 is a respiratory infection that has affected various sectors around the world, such as the economy, healthcare, transport, education and many others. COVID-19 is estimated to have reduced global economic growth by 3–6% in 2020. COVID-19 has spread beyond 213 countries and independent territories, where at the moment of writing this paper more than 187 million people have been infected, and the number of deaths has reached 4 million. Health institutions, such as the World Health Organization (WHO), have recommended several protective measures to respond immediately and limit the unprecedented global spread of COVID-19. Preventing the adverse consequences of the spread of COVID-19 requires significant efforts, in particular, application of digital technologies to solve arising problem and minimize their negative impact.

At first glance, the two phenomena under consideration-blockchain and Covid-19- have never been associated together. However, preventing the adverse consequences of the spread of COVID-19 requires coordinated actions of many people and organizations, which can be facilitated by the use of blockchain technology.

Let's consider possible areas of application of blockchain technology. To do this, we will describe the stakeholders and their actions in the fight against the COVID-19 epidemic. Here, due to the limited volume, we will do this at the layer level, but the same actions of stakeholders can be described as a model of top-level processes, for example, in the form of a Value Added Chain diagram, and in more detail (by decoding the diagrams of added quality) as models of the event - driven process chain (extended Event-driven Process Chain).

In general, as part of the fight against the COVID-19 pandemic, stakeholders include international organizations, state bodies, local governments, research centers and scientists, the medical industry, the management of medical institutions, doctors and medical personnel, mass media, employers, and the population. As an example, actions these stakeholders are taken or being a subject to can include: management (coordination, financing, certification), prevention (lockdown, work at home, restrictions on movement, propaganda, combating fakes, vaccination), research (basic and applied research, vaccine development, forecasting), treatment (localization and tracking of patients, ambulance, home treatment, hospital treatment, training of doctors and staff), supply of medical drugs and equipment, food supply, supply of goods, etc.

**Table 1.** Systematization of blockchain application areas during the Covid-19 pandemic

| Action | Examples of blockchain use | Blockchain possibilities |
|---|---|---|
| Prevention<br>• Lockdown control<br>• Remote work organization<br>• Restriction on movement<br>• Vaccination | Ex3. Personal data<br>Ex4. Digital contact tracing<br>Ex5. Health monitoring<br>Ex8. Immunization<br>Ex9. Smart contracts | Contact tracing (exposed to an infected person)<br>Traceability, transparency, and immutability of data related to Covid-19<br>Ensuring personal data confidentiality and security |
| Research<br>• New drugs development and treatment protocols<br>• Vaccine creation<br>• Forecasting of the epidemiological trends | Ex8. Immunization<br>Ex9. Smart contracts<br>Ex11. Forecasting of epidemiological trends<br>Ex12. Authorities' decisions | Vaccine authenticity verification<br>Clinical data integrity<br>Identification of non-compliant personal protective equipment (PPE)<br>Detection of foci of the infection |
| Treatment<br>• Patients tracking<br>• Provision of the first aid<br>• Management of the medical institutions<br>• Medical personnel training | Ex4. Digital contact tracing<br>Ex10. Medication identification | Clinical data confidentiality<br>Contact tracing (exposed to an infected person)<br>Transparent exchange of medical resources |
| Supply<br>• Drug and medical equipment supply<br>• Food supply<br>• Supply of goods | Ex1. Global supply chains<br>Ex2. Medication via air<br>Ex9. Smart contracts | Audit of the operation of aircraft delivering medicines<br>Guaranteed execution of the supply chain |
| Regular activities<br>• Mass events management<br>• Public catering<br>• Public transportation<br>• Combating fake news and disinformation | Ex3. Personal data<br>Ex4. Digital contact tracing<br>Ex6. Fake news<br>Ex7. Donations<br>Ex9. Smart contracts | Evidence of the health status while maintaining confidentiality and data security<br>Tracking of people in real time<br>Donation control (tracking and further use)<br>Tracing of the fake news sources |

To systematize the results of the study, a table is constructed (see Table 1), the 1st column of which contains grouped actions performed by stakeholders in the framework of the fight against the COVID-19 pandemic. In the detailed version, the contents of this column will be a model of actions in the form of an event chain of the process. In fact, such models will contain an action algorithm, input and output information, organizational units of interested parties and applied information systems. Blockchain, due to its properties, will serve as the basis for the formation of a number of applications in these information systems. What systems can blockchain help to create and how do they differ from existing centralized and local solutions? The following columns of the table provide examples and the main possibilities of using the blockchain technology within the framework of these actions.

The rows of the table describe the role and examples of using blockchain technology. Such systematization will allow us to highlight the advantages and existing problems of using blockchain, as well as to formulate promising research tasks.

## 3.2 Examples of Blockchain Technology Application in the Fight Against the Covid-19 Pandemic

### Ex1. Global Supply Chain

The first obvious problem that appeared in the pandemic is the global supply chain vulnerability. Covid-19 has exposed weaknesses in global supply chains with countless reports of problems with PPE (personal protective equipment), food shortage in poor regions, business disruption even in areas with remained constant demand. Possible solutions to this problem using blockchain technology are considered, for example, in [10, 11]. Blockchain-based applications, due to their decentralized nature, create a transparent ecosystem, which can provide an instant overview of all supply chains to highlight problems as soon as they appear.

Moreover, it is possible to make fault-tolerant systems a reality with the help of smart contracts, which can ensure guaranteed supply chain and required trust. Thus, thanks to blockchain technology, it is possible to control all stages of supply chains and reduce the number of errors.

The World Economic Forum has developed a blockchain deployment toolkit – a set of high-level guidelines that help companies implement best practices within blockchain projects, especially those that help solve supply chain problems. They worked with more than 100 organizations for a year, exploring 40 different use cases of blockchain, including traceability and automation, to help organizations in their efforts to solve real problems in supply chains using blockchain [12].

### Ex2. Medicine Delivery

In 2020, China decided to conduct an experiment: it was decided to deliver medicines from one city to another using drones. Blockchain technology can help track the location of a drone, check the level and quality of services provided, and calculate the reputation rating of drones based on its performance in a reliable, accountable and transparent way. For example, such application of blockchain technology is described in [13].

Thanks to the introduction of access control and identity management protocols, blockchain technology minimizes the possibility of attacks from competing vehicles. The technology stores in the database instructions which are given to air vehicles (for command non-compliance audit) by the controller, as well as actions of cleaning the areas heavily infected with the virus, detection of human interactions and their control. The idea is to use several autonomous aircrafts (drones) that work together to achieve a common goal. Blockchain technology can be used to make a reliable global decision by making transactions securely.

For example, with the help of a blockchain-based analysis system, mobile devices will be able to determine the most densely populated public places for further disinfection [14, 15].

### Ex3. Personal Data

Another issue is the re-opening of objects with large concentration of people (such as clubs, stadiums, restaurants). There is an economic and social need to re-organize social infrastructure that is safe for everyone. Most of the currently used or proposed solutions focus on personal data collection and therefore have fatal problems related to compliance (general data protection regulation, security issues, data leakage etc.) The blockchain old cryptographic concept, called zero-knowledge proof, is extremely relevant for such cases of using proof of health status while maintaining data confidentiality and security.

### Ex4. Digital Contract Tracing

Blockchain can also become an assistant in digital tracing of contacts. As an example of research on this topic can be named [16, 17]. Digital contact tracking constantly monitors infected people in order to quickly and effectively identify all social interactions that occurred during the incubation period of infected COVID-19 patients. GPS or Bluetooth are mainly used to obtain proximity data to identify social interaction with a person infected with the virus. Ensuring the confidentiality of a person's personal and medical data, along with the occurrence of a minimal risk of obtaining a false positive result of COVID-19, use of blockchain is the key for digital contact tracking solutions [18].

### Ex5. Healthcare Monitoring

Many programs and applications have been recently developed around the world to assist the authorities in monitoring public health in the fight against the COVID-19 pandemic. These solutions should have access to personal data, such as the location of a person and the results of tests for COVID-19, to determine the rate of spread of the virus and predict the distribution areas. Most of the developed applications have a centralized architecture for data storage [19].

For example, Singapore has developed a solution for tracking contacts called Trace-Together. The application uses Bluetooth technology to detect potential exposure to the virus through an infected person. Since the solution is centrally managed, service providers can access user data, violating their confidentiality. Similarly, data records and transactions in centralized systems are vulnerable to changes, fraud, or deletion. In addition, centralized systems are less reliable, since they are subject to all the shortcomings of a single point of failure. Centralized systems are not able to ensure full traceability, transparency and immutability of data stored and exchanged during various operational processes, in particular, processes related to the elimination of the consequences of COVID - 19. Blockchain technology can play a vital role in solving all of these issues: an immutable record of transactions stored in a distributed network of nodes in geographically distributed locations; high security and reliability of data stored in the blockchain, without the possibility of an attack from a single point of failure. The record of transactions and data stored in the blockchain is transparent for each participant of the network.

**Ex6. Fake News and Disinformation**
The next promising area of blockchain application is the fight against fake news. During the pandemic the Internet became full of unreliable news about Covid-19 itself and about other important issues. Ernst & Young has developed and implemented blockchain called the "ANSAcheck". This development allows readers to see where the information they read was taken from, trace the entire history of its publication and conclude whether it is reliable and therefore can be trusted or not; the hash code or digital fingerprint of any material posted on the network is stored in the blockchain.

**Ex7. Donations**
Another issue which has limited number of possible solutions apart from blockchain is the problem of tracking donations. We also need to take into account the problem of tracking donations. During the pandemic, the WHO opened an aid fund, and now everyone can contribute to the fight against the virus. Since the possibility of fraud is one of the main problems in donations, the blockchain has a solution for this problem. There is always a concern that millions of dollars donated are not being used where it is needed the most. With the help of blockchain capabilities, financial donors can see where funds are most urgently needed and can track their donations to make sure that their contributions were used for their intended purpose. Blockchain can make this process transparent for the general public, so that not only a specific interested person, but also the community as a whole understands how and for what donations were used [14].

As an example of the practical implementation, we can refer to the blockchain project of the Chinese Hangzhou Quilan Technology and China Xiongan Group. They have developed a platform for tracking donations - the "Shanzong", which keeps a transparent record of each donation made, the amount, and accurate information about the distribution of the resources received to people in need.

**Ex8. Immunization**
Another problem area in the fight against the spread of COVID-19 is the immunization of people against the virus, including their vaccination. Conducting clinical trials for the development of a vaccine against COVID-19 is a complex, time-consuming and expensive process. This requires close coordination and cooperation between organizations involved in clinical trials of the vaccine, despite the fact that they are often located in geographically dispersed locations. Researchers, organizations, donors and pharmaceutical companies are examples of organizations that are actively involved in clinical trials for the successful development of a vaccine.

Traditional centralized clinical trial data management systems face a number of problems, which are mainly related to the registration of subjects, limited productivity and non-compliance with the requirements for clinical trials. In particular, there are problems of ensuring data confidentiality, compliance with the rules of clinical trials to ensure the safety of the health of their participants, the integrity of clinical trial data. In addition, the organizations participating in the tests can create several versions of information repositories, as a result of which the data is duplicated and, in case of a violation of consistency, several versions of the data about these tests may appear. The centralization of clinical trial data storage makes them vulnerable to changes by external hackers or participants [20].

Blockchain technology allows pharmaceutical companies and research institutes to maintain the integrity of clinical trial data during vaccine development [21]. The blockchain will be able to guarantee a unified and synchronized representation of data available to all authorized organizations. There is also an opportunity not to face such problems as duplication and fragmentation of data due to the inconsistency of existing centralized clinical trial management systems.

### Ex9. Smart Contracts

Blockchain technology can use smart contracts to automate business processes and resolve disputes between employees of organizations. A smart contract is a program that monitors and ensures the proper fulfillment of the obligations of the contract [22]. For example, in a blockchain-based system that is used to control the logistics of deliveries of COVID-19 polymerase chain reaction (PCR) testing kits, smart contracts can play a key role in such processes as: tracking the location of containers of testing kits; identifying defective testing kits; providing government officials with access to data to analyze the demand for kits in various geographical areas.

In addition, smart contracts could simplify registration and management processes related to tracking vaccine trials, tracking COVID-19 outbreaks and ensuring the confidentiality of user data using registration services of ready-made blockchain platforms. Smart contracts can verify an organization's access rights before allowing it to use clinical trial data to guarantee the confidentiality and security of such data. For example, to meet the requirements of clinical trials, smart contracts can verify that authorized participants in clinical trials have signed a consent form in digital form before launching a transaction to read or write health data to the card. Thus, anonymized data collection and verifiable consent management can allow participants to share their medical histories with authorized organizations without revealing their identity [23].

To attract more participants in a clinical trial, medical companies usually offer participants a reward in the form of cash or gift cards. Smart contracts can help speed up the payment process by providing an automated, transparent, and accountable mode for the transfer of cryptocurrencies. Transparency and accountability also ensure that the data can only be used for the purpose for which it is collected, thereby increasing user confidence and trust.

### Ex10. Medication Identification

At the time of writing this article, many people have already get vaccinated with either first or both components of vaccines. There already have been several cases of spread of fake medications and fake vaccines and risk remains remarkably high. Using blockchain technology in hospitals medical workers and other responsible people would be able to access data related to all stages of the life cycle of medicines to identify, track and verify data on vaccines before their introduction. An example of such usage of blockchain technology is analyzed in [24].

**Ex11. Epidemiological Trends Forecasting**
Another function of the blockchain that emerged during the pandemic is forecasting epidemiological trends. Particularly, article [25] is devoted to this problem. At the end of March, the World Health Organization, and companies such as IBM, Oracle launched their platform to combat Covid-19. Their developed "MiPasa system" should identify asymptomatic carriers of the disease and inform residents about the location of dangerous areas of infection. Access to such important information can help the authorities develop policies to prevent the further spread of the virus.

"MiPasa" is a blockchain-oriented platform that integrates, processes and shares information related to the spread of the COVID-19 virus from several verifiable sources, such as the WHO, registered health organizations and authorities. This helps authorities to identify both human errors and incorrect reporting, thereby allowing data researchers and public health officials to develop solutions to limit the spread of the virus. For example, using analytics based on reliable and verified blockchain-based data, "MiPasa" can help government agencies identify COVID-19 carriers and hotspots in a timely and secure manner.

"MiPasa" is a completely private system implemented on the basis of the Hyperledger Fabric environment. Through web interfaces, individuals and representatives of public health can use "MiPasa" to upload the location of infected persons. In response, it ischecked using data provided by the WHO and ECDC to make sure that the new data corresponds to the original. At the next stage, the new verified data is transmitted to the state authorities and health care institutions specified by countries.

**Ex12. Authorities' Decisions**
Blockchain technology can serve authorities. The advantages of blockchain technology in terms of significant trust, security, traceability and transparency can greatly help the authorities in developing solutions to combat the COVID-19 pandemic. Using a distributed registry, local authorities can make decisions faster and more efficiently and offer citizen a customized action plan. Each citizen identification number is compared with the records available in the register, which allows, for example, to determine a safe time for shopping for each participant of the platform, etc. The unchanged data associated with the outbreak of COVID-19 in the city can be used by authorities to correctly identify the foci of infection. Access to such important information can help authorities develop policies to prevent the further spread of the virus [10].

## 4    Conclusion

The results of the conducted research prove that the epidemiological situation that has arisen in the world, among other things, has determined new goals and directions for the use of blockchain technology. History shows that a serious crisis is always a fertile ground for innovations. Structuring new directions of using blockchain technology, in addition to determination of their existence, creates strategic guidelines for developers and consumers.

Briefly the mentioned above possibilities of using blockchain technology to combat the consequences of the pandemic can be summarized as follows:

- organization of efficient and trouble-free supply chain;
- audit of delivery drones (including those involved in medication delivery);
- proof of the state of health, immune status, etc. in compliance with the security conditions for the storage and transfer of personal data;
- digital tracing of contacts with infected people;
- preventing the spread of fake news;
- ensuring a safe donation collection process;
- maintaining the confidentiality of hospital data;
- forecasting of epidemiological trends.

Despite a sufficient number of examples of the use of blockchain technology in the context of the Covid-19 pandemic, the study revealed a number of limitations both in the blockchain technology itself and in the conditions of its application. To overcome these limitations, which do not allow to fully realize the potential capabilities of the technology, it is necessary to solve the following problems and research challenges:

1. Development and use of closed systems based on blockchain. These systems have a limited number of nodes, a fixed number of miners, are able to provide secure work with personal data, they are easier to integrate with other applications, they are less demanding on the infrastructure, they are less expensive.
2. Use of ledger database management systems instead of open blockchain systems, as closed local solutions for enterprises.
3. Development of UX (User Experience)/UI (User Interface) technologies supporting blockchain.
4. Development of the idea of "seamless" compatibility and integration of all elements of the blockchain technology application system: the platform itself, the smart contract implementation system, the system of interfaces, languages, protocols.
5. Offering solutions for information and functional compatibility, integration and interaction of blockchain platforms. The "ability to interact with other platforms" should be added to the currently typical elements available in each individual blockchain platform-distribution, encryption and immutability.
6. Overcoming the security problems of smart contracts and trust in external data.
7. Creating incentives for using data from decentralized sources (in particular, blockchain platforms).
8. The problem of blockchain scalability. We are talking not only about the speed of writing and bandwidth in a dramatically enlarged system, but also about the block size in the blockchain. In the basic blockchain, it is limited to 1 MB. Now there are already solutions with a block of 5 MB, which allows you to create decentralized data warehouses with images and video files.
9. Implementation of blockchain is hindered by the legal uncertainty of using the system in different jurisdictions. Coordinated actions of supranational and national regulators, business and expert community are required.

10. The formation of a single database of medical records of patients based on the blockchain allows, in addition to obtaining new information, for example, about the real effectiveness of medicines, to develop effective methods of decision-making based on the patterns obtained from the data. This is the so-called thinking patterns and knowledge dynamics.

# References

1. Tadvisor. https://www.tadviser.ru/index.php/Статья:Блокчейн_(Blockchain). Accessed 22 Nov 2021
2. Alblooshi, M., Salah, K., Alhammadi, Y.: Blockchain-based ownership management for medical IoT (MIoT) devices. In: International Conference on Innovations in Information Technology (IIT), pp. 151–156. IEEE (2018)
3. Pham, Q., Nguyen, D., Hwang, W., Pathirana, P.: Artificial intelligence (AI) and big data for coronavirus (COVID-19) pandemic: a survey on the state-of-the-arts. IEEE Access **8**, 130820–130839 (2020)
4. Chamola, V., Hassija, V., Gupta, V., Guizani, M.: A comprehensive review of the COVID-19 pandemic and the role of IoT, drones, AI, blockchain, and 5G in managing its impact. IEEE Access **8**, 90225–90265 (2020)
5. Ghimire, A., Thapa, S., Adhikari, S.: AI and IoT solutions for tackling COVID-19 pandemic. In: Proceedings of the 4th International Conference on Electronics, Communication and Aerospace Technology, pp. 1083–1092. https://doi.org/10.1109/ICECA49313.2020.9297454
6. Xu, H., Zhang, L., Onireti, O., Fang, Y., Buchanan, W.J., Imran, M.: BeepTrace: blockchain-enabled privacy-preserving contact tracing for COVID-19 pandemic and beyond. IEEE Internet Things J. **8**(5), 3915–3939 (2021)
7. Rangone, A., Busolli, L.: Managing charity 4.0 with blockchain: a case study at the time of Covid-19. Int. Rev. Public Nonprofit Mark. **18**, 491–521 (2021). https://doi.org/10.1007/s12 208-021-00281-8
8. Mougayar, W.: The Business Blockchain: Promise, Practice, and Application of the Next Internet Technology. Bombora, Moscow (2018)
9. Deloitte. https://www2.deloitte.com/uk/en/pages/innovation/solutions/deloitte-blockchain-practice.html. Accessed 22 Nov 2021
10. Kamalakshi, N., Naganna: Role of blockchain in tackling and boosting the supply chain management economy post COVID-19. In: Gururaj, H.L., Ravi Kumar, V., Goundar, S., Elngar, A.A., Swathi, B.H. (eds.) Convergence of Internet of Things and Blockchain Technologies. EAISICC, pp. 193–205. Springer, Cham (2022). https://doi.org/10.1007/978-3-030-76216-2_12
11. Nabipour, M.: On deploying blockchain technologies in supply chain strategies and the COVID-19 pandemic: a systematic literature review and research outlook. Sustainability **13**(19), 10566 (2021). https://doi.org/10.3390/su131910566
12. World Economic Forum. http://www3.weforum.org/docs/WEF_Building_Value_with_Bloc kchain.pdf. Accessed 22 Nov 2021
13. Singh, M., Bali, R., Singh, A.: Blockchain-enabled secure communication for drone delivery: a case study in COVID-like scenarios. In: Proceedings of the 2nd ACM MobiCom Workshop on Drone Assisted Wireless Communications for 5G and Beyond, pp. 25–30 (2020). https://doi.org/10.1145/3414045.3415937
14. Ahmad, R., Salah, K., Jayaraman, R., Yaqoob, I., Ellahham, S., Oma, M.: Blockchain and COVID-19 pandemic: applications and challenges. TechRxiv. https://doi.org/10.36227/tec hrxiv.12936572.v1. Accessed 22 Nov 2021

15. Kalla, A., Hewa, T., Mishra, R., Ylianttila, M., Liyanage, M.: The role of blockchain to fight against COVID-19. IEEE Eng. Manag. Rev. **48**(3), 85–96 (2020)
16. Hasan, H., Salah, K., Yaqoob, I.: COVID-19 contact tracing using blockchain. IEEE Access **9**, 62956–62971 (2021). https://doi.org/10.1109/ACCESS.2021.3074753
17. Platt, M., Hasselgren, A., Román-Belmonte, J.: Test, trace, and put on the blockchain?: A viewpoint evaluating the use of decentralized systems for algorithmic contact tracing to combat a global pandemic. JMIR Public Health Surveill. **7**(4), e26460 (2021). https://doi.org/10.2196/26460
18. World Health Organization. https://www.who.int/publications/i/item/contact-tracing-in-the-context-of-covid-19, last accessed 2021/11/22
19. Choudhury, Y., Goswami, B., Gurung, S.: CovidChain: an anonymity preserving blockchain based framework for protection against Covid-19. Cornell University publication. https://arxiv.org/pdf/2005.10607v1.pdf. Accessed 22 Nov 2021
20. ResearchGate publication. https://www.researchgate.net/publication/343093194_A_Survey_of_COVID-19_Contact_Tracing_Apps. Accessed 22 Nov 2021
21. Rotbi, M., Motahhir, S., Ghzizal, A.: Blockchain technology for a safe and transparent Covid-19 vaccination. arXiv preprint arXiv:2104.05428
22. Godin, V., Terekhova, A.: Blockchain: philosophy, technology, applications and risks. Vestnik Universiteta (9), 54–61 (2019). https://doi.org/10.26425/1816-4277-2019-9-54-61
23. Information Age. https://www.information-age.com/2020-year-blockchain-came-of-age-123490520/. Accessed 22 Nov 2021
24. National Library of Medicine. https://pubmed.ncbi.nlm.nih.gov/28357041/. Accessed 22 Nov 2021
25. Torky, M., Hassanien, A.E.: COVID-19 blockchain framework: innovative approach. arXiv preprint arXiv:2004.06081 (2020)

# Development of Special Software for Solving Large-Dimensional Transport Problems with a Modified Genetic Algorithm Using Multithreading Properties

S. L. Podvalny[1,2] ⓘ, D. A. Vdovin[1] ⓘ, and Y. A. Zolotukhina[2(✉)] ⓘ

[1] Voronezh State Technical University, Voronezh, Russia
[2] The Russian Presidential Academy of National Economy and Public Administration, Moscow, Russia

**Abstract.** The article solves the problem of developing special software for solving transport problems of large dimension using genetic algorithm.

Meaningful setting of the task: on the terrain map (for example. Region of several regions) in different places there are a number of customers, as well as a centralized warehouse where a given number of goods is stored. Several vehicles are engaged in transportation of goods, for each of them it is necessary to organize their route. Limitations of task are, in addition to traditional balance ratios, necessity of taking into account interval of consumer service time and necessity of taking into account driver rest after certain travel time.

Criterion selected: minimization of service time at resource limitations. Even in its simplified production (commuter problem), it belongs to the class of intractable problems and it is most often solved using evolutionary genetic algorithms. Standard optimization procedures using genetic algorithms with increasing dimension (more than 50 customers) require large computing resources exponentially growing from the number of customers.

Compared to traditional genetic algorithms, it was required to create some version of modified genetic algorithm, which uses the second operator of mutation, -to reduce probability of getting into local extremum.

Detailed description of the algorithm of the software that allows: to make initial settings of genetic operators (population size, number of iterations) and basic elements of evolutionary search (selection, inheritance, crossover, mutation)- and what is especially important when solving large-size problems, to preserve the routes obtained when the vehicle is re-solved by customers in combination with the database used in a timely manner - as for customers(for example, lower and upper limits of service time), and on vehicles.

The software was developed by the author in the Microsoft Visual Studio 2017 programming environment, as programming language used with sharp. Presented interface software. To increase efficiency of processor usage and in particular to increase speed when solving problems of large dimension. For this purpose elements of parallelisation of computational process were used, in which method of multithreading is selected. The work shows that due to such approach search time decreased by more than 5 times and is comparable with search time by other methods, for example, SWARM.

V. Taratukhin et al. (Eds.): ICID 2021, CCIS 1539, pp. 203–214, 2022.
https://doi.org/10.1007/978-3-030-95494-9_17

**Keywords:** Algorithm · Software · Transportation task · Multithreading ·
Heuristic algorithm

# 1 Introduction

Transportation of various quantities of car-go is a topical topic today in the life of modern man. Solving such problems increases the number of organizations, and minimizes the cost of transporting goods. The task is completed if a certain amount of cargo has reached the consumer body under certain restrictions. To in-crease the speed and efficiency of finding a better solution, algorithms are used that run on a computer. The development of such algorithms should take into account the specifics of the computer hardware.

Transport tasks (TT) can be divided into the following types:

- TT to limit the cost of transportation.
- TT for time limitation.
- TT for search of shortest distance by specified points.

This article will discuss the second type of task. The criterion for the completion of the TT will be considered the minimum time spent on transporting goods to customers.

The restrictions are the following conditions:

- any consumer has its own time interval for maintenance;
- any vehicle starts and finishes its journey in the warehouse;
- the total amount of cargo transported should not exceed the capacity of all vehicles;
- customer service is performed once and only by one vehicle;
- the driver of each vehicle must rest after a certain number of time units.
- Taking into account these circumstances, the generalized mathematical formulation of the problem is presented below:

$$\sum_{a \in A} \sum_{(i,j) \in S} c_{ij} W_{ij}^a, \tag{1}$$

$$\sum_{a \in A} \sum_{j \in L} W_{ij}^a = 1, \forall i \in K, \tag{2}$$

$$\sum_{i \in K} d_i \sum_{j \in L} W_{ij}^a \leq g, \forall a \in K, \tag{3}$$

$$\sum_{j \in L} W_{0j}^a = 1, \forall a \in A, \tag{4}$$

$$\sum_{i \in L} W_{ir}^a - \sum_{j \in L} W_{rj}^a = 0, \forall r \in K, \forall a \in A, \tag{5}$$

$$\sum_{j \in L} W_{i,n+1}^a = 1, \forall a \in A, \tag{6}$$

$$\sum_{j \in L} W_{i,j}^a \left( T_i^a + t_{ij} - T_j^a \right) \le 0, \forall (i,j) \in S, \forall a \in A, \tag{7}$$

$$n_i \le T_i^a \le m_i, \forall i \in L, \forall a \in A, \tag{8}$$

$$W_{ij}^a \in \{0, 1\}, \forall (i,j) \in S, \forall a \in A, \tag{9}$$

A meaningful statement of the problem: a certain number of customers are located on the map in different places, and there is also a warehouse where a certain amount of goods is stored. Cargo transportation is carried out by one or more vehicles. For each vehicle, it is necessary to organize its own route along which the delivery of goods and customer service will be performed. There is a certain limit for each vehicle on the route.

Regardless of the type of criterion and constraints, such discrete optimization problems belong to the class of NP-hard problems and are solved using approximate methods that allow finding a solution close to the optimal result with an increase in the dimension of the search space, when the volume of necessary calculations increases exponentially and at the same time the speed of calculations increases linearly [1].

One of the methods for solving optimization problems is genetic algorithms (GA), based on the principles of natural selection of C. Darwin. A genetic algorithm is a heuristic search algorithm used to solve optimization problems and model the random selection, combination and variation of the desired parameters using mechanisms resembling biological evolution with simulating living process schemes: crosses, crossovers, mutations and breeding-to ensure convergence to a suboptimal solution [2, 3].

The article discusses the algorithm of the software tool, which implements the search for the best solution using a modified GA. The need for such a technique is due to the fact that in each specific case, the search for the settings of the standard genetic algorithm results in the creation of a specific external contour of adaptive tuning. They are different in each individual case and it is impossible to find a universal best algorithm for the entire class of problems, as the no free lunch theorem confirms [4]. The method proposed in this paper differs from the standard algorithm in that it uses the second mutation operator. Thanks to him, the search for the optimal path has been reduced several times, and the probability of getting into the local minimum has also decreased.

The software tool (ST) uses the threading property. This method allows you to reduce the execution time of the algorithm by distributing the CPU load on individual threads.

This article is devoted to the development of a software tool for solving TT with time limitation using a modified genetic algorithm.

## 2 Algorithm of Software Operation

The modified GA for the software is shown in Fig. 1.

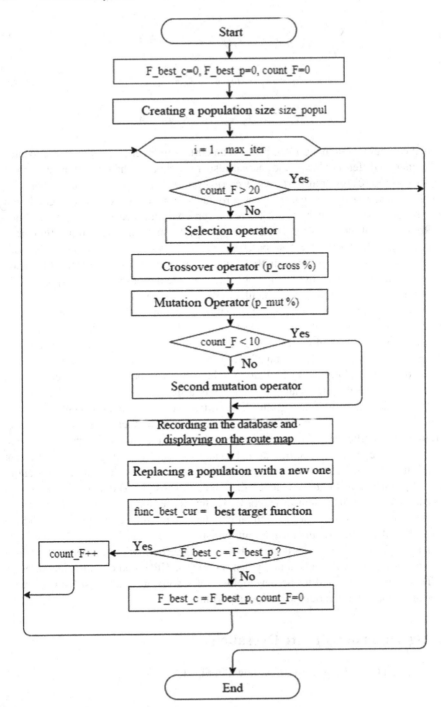

**Fig. 1.** Algorithm of software operation.

The encoding of the solution occurs as it is shown in Fig. 2.

$$0[0;0;300;0;0] \rightarrow 8[32;59;240;32;17] \rightarrow 6[101;133;180;74;39]$$
$$\rightarrow 2[168;182;80;109;34] \rightarrow 0[230;230;20;109;34]$$

**Fig. 2.** Displaying routes

When displaying the route, the information about the customer is as follows:

- the lower limit of the time of arrival of the vehicle;
- upper limit of the departure time of the vehicle;
- the variable denotes the load of the vehicle after the cargo has been delivered to this customer;
- the total path traveled by all vehicles on the map;
- the total time, including the delay of the vehicle on the way.

To explain the operation of the modified genetic algorithm in the software, the following designations must be entered:

- «max_iter» – a variable indicating the maximum number of steps performed by the algorithm;
- «size_popul» – shows the number of routes in the population;
- «p_cross» – a variable that allows a crosser operator in a modified GA with a cer-tain probability;
- «p_mut» – a variable, a value that indicates the probability of using a mutation opera-tor;
- «F_best_c» – Best Objective Function (OF) in this best route search step
- «F_best_p» – OF of the best solution in the population on the previous iteration of the algorithm;
- «count_F» – the number of steps of the fixed values of the objective function
- «i» – is the current iteration step of the algorithm.

During algorithm startup, the following variables must be reset: «F_best_c», «F_best_p» and «count_F». Next, an initialization statement is applied, that is, an array of solutions (a set of routes) is created in a number equal to the variable «size_popul». On the following step there is a comparison of a step of «i» with the maximum possible iteration of «max_iter». If the first variable exceeds the second, then the execution of the algorithm ends, otherwise we proceed to the next step. Further, if the number of unchangeable «count» solutions during ten iterations is the same, then we complete the algorithm process, otherwise we use the selection operator. With a given probability «p_cross,» the crosser operator is used.

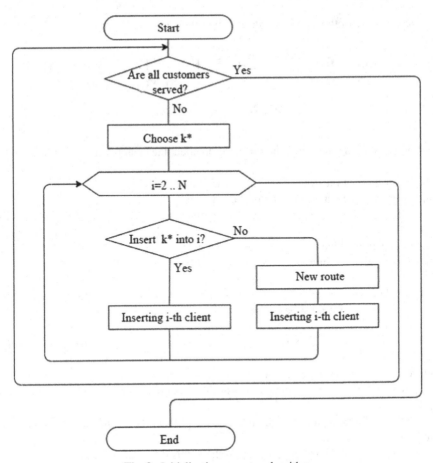

**Fig. 3.** Initialization operator algorithm.

The algorithms of the initialization operator and the crossing operator are shown in Figs. 3 and 4.

Similarly, the operator is the first mutation operator (MO1). After, if the number of unchanged solutions exceeds five, then we apply the second operator of the MO2 mutation), otherwise the operator is skipped. The operator of MO2 is capable to diversify population with new routes, thereby providing an exit from a local minimum with which the operator of MO1 doesn't cope. Next, you need to write the solution to the table «History» in the database, for further reporting, and also display the current route on the map. Afterwards we change the old decision for new, and we appropriate the best criterion function of the «F_best_»c variable. If the previous best CF «F_best_pr» is equal current, then we add «count» on unit, otherwise we will equate them and we nullify «count_F». Then the algorithm will be executed until either the iterations end or the exit occurs.

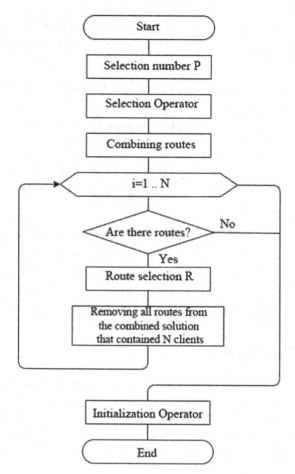

**Fig. 4.** Crossover operator algorithm.

After a short downtime, the solution uses the second mutation operator after OM 1. The number M2 is set, which will correspond to the number of customers being deleted. In this algorithm, the number M2 = [number of customers]/4, where the brackets "[]" mean the whole part. Next, we remove M2 consumers from the population. Customers are selected probabilistically for deletion. We insert remote consumers into the solution using the initialization operator, after which we get a ready-made solution.

The algorithm of the second mutation operator is shown in Fig. 5.

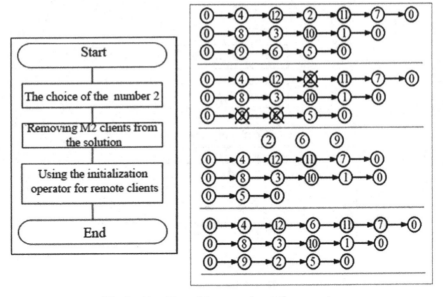

**Fig. 5.** Algorithm of the second mutation operator.

# 3   Software Tool Interface

The software allows you to clearly display customers (points on the map) and the movement of the vehicle (V) from one consumer to another (line between two points). The program is written in C #, development environment MicrosoftVisualStudio 2017. The program is designed to run under operating systems Windows10 and above. The main window of the program "Solution of TT using MGA" is shown in Fig. 6.

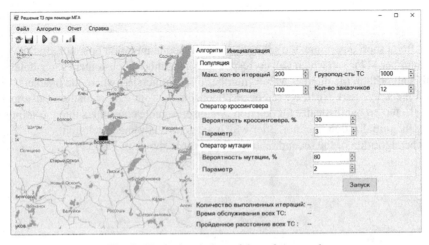

**Fig. 6.** The main window of the software tool.

The right part of the ST provides the opportunity to configure various parameters of the genetic algorithm, thereby improving the quality of the resulting solution. The superstructure of various parameters is divided into three stages: initialization, crosser operator and mutations.

The size of the population allows you to create the desired number of initial solutions (gene or chromosomes). A large number of choice of solutions increases the probabilities of choosing the best set of routes. When the maximum number of iterations specified in the field is reached, the algorithm displays the last route processed. The value of lifting capacity is relevant for each vehicle leaving the warehouse. The Number of Customers option creates the specified number of rows to initialize the data for each consumer, and it is the maximum number that can be applied to the work zone.

In the field "Crossing over operator", you can configure two parameters "Probability of crossing over" and "Parameter". The first field shows the probability of using this operator in the modified algorithm. The value of the second field allows you to select the specified number of routes from the generation for further application of the crossing over operator on them. Mutation operator and fields: "mutation probability" and "Parameter" are similar to the crossing-over operator.

The "Initialization" tab allows you to fill in the initial data about customers, namely: customer number; upper and lower border dispatch the vehicle from the customer; the requested weight of the goods; coordinates "X" and "Y" of the customer; the time at which the current consumer can be served on the vehicle.

The left side of the program is the work area. It is necessary to install clients on the map, as well as to display the routes of the vehicle from the warehouse. The map is updated at each iteration of the algorithm. At the bottom of the working area, the following are displayed: the number of the last iteration, the total service time for all customers and the distance traveled by all vehicles.

## 4   Development of a Multithreaded Algorithm

Parallel execution of the code of modern software is realized on the one hand by the hardware capabilities of modern personal computers, and on the other by the capabilities of operating systems, such as process and thread multitasking [6]. A process is an object that is created and controlled by the operating system. It has its own address space and execution priority at the operating system level. Processes can communicate through the services of the operating system: sockets, pipes, and sending messages. It should be noted, however, that there are relatively high data transfer overheads.

Thread multitasking relies on the concept of a thread - an isolated sequence of commands that are executed within a process. Streams are controlled both by the program that generates them and by the operating system. Multithreading allows a process to split its algorithm of work into separate parallel executable groups of commands.

GA is one of the best algorithms that is able to search for the best target on parallel processes because it contains elements of parallelism [5]. The separation of the processor load is carried out by implementing data processing in different processes. A thread (in the C# programming language, it is a "Thread" library) is an independent sequential execution of a set of commands and functions in a software tool [5].

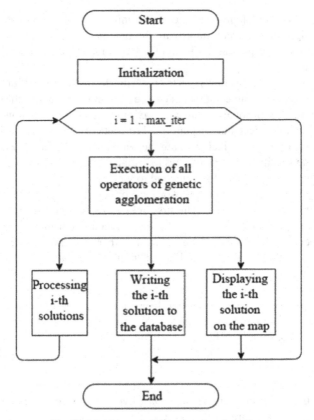

**Fig. 7.** Multithreaded PS operation algorithm.

The algorithm for parallelizing the above operations is shown in Fig. 7.

Additional threads are launched after receiving the route set. Further, the program is subdivided into the main thread and two additional ones. In the first, processing of route data and estimation of the objective function continues. In an additional stream, the resulting population is written to the database table, and in the second stream, the population is graphically plotted at the current iteration, i.e. displaying the movement of the vehicle on the map. Auxiliary streams operate independently of the main stream, thereby reducing the time to find the best solution.

In VisualStudio, to use the multithreading function, the "System.Threading" library is connected. To initialize an additional thread, the "Thread" class is used.

## 5 Results

In the article [4], a comparison of heuristic algorithms for finding the optimal route was carried out. Modeling took place in a simulation environment, namely AnyLogic.

For comparison, the results of the standard genetic algorithm were taken, and the results of the multithreaded execution of the software tool were also added, but with minor transformations. The changes were as follows: the integration with the SQL database was removed and the number of vehicles was reduced to one. These transformations were necessary because of the initial conditions in the work [7], as well as so that the results were close to the standard genetic algorithm.

The graph of the dependences of the route calculation time on the number of customers for each of the algorithms is shown in Fig. 8.

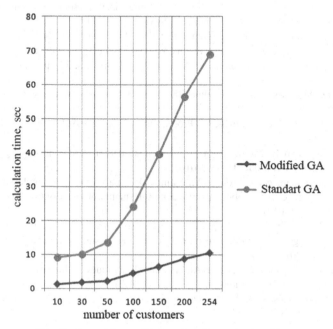

Fig. 8. Comparison of the results of genetic algorithms.

Based on the results of the comparison [8] of the standard and modified, we can say that the developed algorithm reduces the execution time of the search for the best path by about 5 times.

## 6   Conclusion

This article discusses a modified GA for solving transport problems with time constraints.

The article presents an algorithm for the operation of a software tool for solving transport problems with a time limit. At the stage of developing the interface of the software tool, integration with the database was performed.

The developed PS algorithm makes it possible to search for the best solution with a large dimension, and also prevents hitting the local minimum.

To increase the speed of searching for the optimal path of movement of vehicles, as well as recording in the database and displaying routes, a multithreaded method was used. During comparison, this mode showed a reduced program runtime by about 5 times, compared to the standard genetic algorithm.

Note that the advantages of multithreading in other tasks may, of course, be different, but in any case, multithreading should be used as widely as possible to improve overall performance by parallelizing processor calculations and I/O operations [9, 10].

# References

1. Batishchev, D., Nejmark, E., Starostin, N.: Solving discrete problems using evolutionary genetic algorithms, p. 199. NGGU, Nizhniy Novgorod (2011)
2. Podvalny, S., Kremer, O.: Software implementation of the solution of optimization problems by the method of genetic algorithm. Vestnik Voronezh State Tech. Univ. 5(3), 21–24 (2012)
3. Gladkov, L., Kurejchik, V., Kurejchik, V.: Genetic Algorithms, 320 p. Fizmatlit, Moscow (2006)
4. Wolpert, D., Macredy W.: No free lunch theorems. Sanmta fe institute technical report 95-02-010 (1995)
5. Deniskin, A.: Multithreading in the C # Programming Language, vol. 2, pp. 21–24. Academy, Ivanovo (2017)
6. Sobolev, P., Filenko, E.: Multithreading Implementation in Modern Programming Languages. Mod. Trends Dev. Sci. Educ. 14–19 (2016)
7. Semenov, S., Pedan, A., Volovikov, V., Klimov, I.: Analysis of the complexity of various algorithmic approaches for solving the traveling salesman problem. Control Commun. Secur. Syst. 1, 116–131 (2017)
8. Podvalny, S., Vdovin, D.: Development of special software for solving transport problems with a modified genetic algorithm using multithreading. Vestnik Voronezh State Tech. Univ. 16(4), 7–12 (2020)
9. Olukotun, K.: Chip Multiprocessor Architecture-Techniques to Improve Throughput and Latency, pp. 115–127. MC Publishers, Princeton (2007)
10. Nemirovsky, M., Tulsen, D.: Multithreading Architecture, vol. 6, pp. 134–145. MC Publishers, Princeton (2013)

# Analysis of the Estimates Quality in the Problem of Parametric Identification of Distributed Dynamic Processes Based on a Modified OLS

M. G. Matveev and E. A. Sirota(⊠)

Voronezh State University, 1 Universitetskaya pl., Voronezh 394018, Russia

**Abstract.** The control problem is now a fairly common one in various technical, economic and social applications. In order to obtain good control, it is necessary to construct a model that meets the conditions of adequacy and accuracy. For this purpose it is necessary to obtain a solution to the identification problem of the constructed model.

When modeling objects of such applications we usually use multidimensional autoregressive equations with regressors located at adjacent nodes of spatial coordinates. Obviously, in this case there is usually significant correlation dependence between regressors. There appears the effect of quasi-multicollinearity resulting in overestimation of autoregressive parameter estimates standard error and bias of obtained parameter estimates. Thus the presence of two error components makes it necessary to find a trade-off between bias and variance, which is well known in machine learning. We focus on multiple autoregressive equations based on approximation of homogeneous partial differential equations with constant parameters by difference equations with conservativity property. A difference scheme is called conservative if it preserves the same conservation laws on the grid as in the original differential problem. A comparative analysis of the solution of the parametric identification problem using ordinary least squares method (LSM), Ridge Regression and two author's methods of dimensionality reduction has been carried out in the frame of this paper. Both positive and negative parameters have been considered. A comparative analysis of the application of the investigated identification methods to parameter estimation has shown a significant dependence of the estimation quality on the observation noise intensity. At low noise all methods successfully cope with the identification problem. The author's method of dimensionality reduction based on taking into account the conservatism property of the difference scheme displays satisfactory efficiency when the noise intensity increases, in the case of positive coefficients. In the case of negative coefficients there is no positive effect of the method in question.

**Keywords:** Autoregressive model · Finite difference equations · Identification · OLS estimates · Biased estimates of model parameters · The reduction of the dimensionality

© Springer Nature Switzerland AG 2022
V. Taratukhin et al. (Eds.): ICID 2021, CCIS 1539, pp. 215–223, 2022.
https://doi.org/10.1007/978-3-030-95494-9_18

# 1 Research Models and Methods

The task of management today is a fairly common task in various areas of technical, economic and social applications. In order to get good control, it is necessary to build a model that meets the conditions of adequacy and accuracy. To do this, it is necessary to solve the problem of identifying the constructed model.

Identification of parameters of distributed dynamic systems is one of the most important applied tasks. When modeling objects of these applications, passive identification methods are often used, based on statistical processing of the observed values of variables of a functioning object. In this case, the original linear differential equation is replaced by the corresponding difference scheme [1, 2]. The difference scheme allows us to switch to a regular grid of spatial and temporal coordinates at the nodes of which variables are observed. In this case, for modeling, it is advisable to use the equations of multidimensional autoregression with regressors located in adjacent nodes of spatial coordinates [3–5]. It is evident that in this case there is usually a significant correlation between the regressors. There is a quasimulticollinearity effect, the consequence of which is an overestimated value of the standard error of the estimates of the regression parameters [6]. The second source of model identification errors is the correlation of random interference of observation of variables of the non-stationary process with regressors, the consequence of which is the bias of the obtained parameter estimates [6]. Among the various methods of improving the quality of statistical estimates, there are many that reduce the standard error and increase the bias, and vice versa. For example, reducing the size of the autoregression feature space, using regularization methods reduces the standard error [13–15], but may increase the bias. The method of instrumental variables [15] is effective in combating bias, but it can increase the standard error of the estimate.

Thus, the presence of two components of the error generates the need to find a compromise of "bias and spread", which is well known in machine learning [13], a special case of which is the identification problem under consideration.

The object of our research is the equations of multiple autoregression, obtained on the basis of approximation of homogeneous partial differential equations with constant parameters, different equations with the conservativity property. A difference scheme is called conservative if it preserves the same conservation laws on the grid as in the original differential problem [9]. The reduced difference scheme is considered

$$y_i^{k+1} = a_1 y_{i-1}^k + a_2 y_i^k + a_3 y_{i+1}^k \tag{1}$$

with the specified initial and boundary conditions:

$$y_i^0 = c_i, y_{i-1}^k = b_{i-1}^k, y_{i+1}^k = b_{i+1}^k, \forall k \tag{2}$$

where $i$ - discrete values of the spatial coordinate, $k$ discrete time; $a_1 + a_2 + a_3 = 1$ this ensures that the scheme is conservative. Indeed, the stationary mode in all coordinates can be provided only when the right part of the expression (1) is a convex linear combination. The values of the variable $y_i^k$ are measured at each node $i$ with an error $\xi_i^k$ formed by a random process of the "white noise" type. We will denote the measured value $x_i^k = y_i^k + \xi_i^k$. Then expressions (1) and (2) can be written in the form of an autoregressive dependence describing, generally speaking, a non-stationary time series:

$$
\begin{aligned}
x_i^{k+1} &= a_1\left(x_{i-1}^k - \xi_{i-1}^k\right) + a_2\left(x_i^k - \xi_i^k\right) \\
&+ a_3\left(x_{i+1}^k - \xi_{i+1}^k\right) + \xi_i^{k+1} = a^T \cdot x^K + \omega_i ,
\end{aligned}
\tag{3}
$$

where $\omega_i = \xi_i^{k+1} - a^T \cdot \xi^k$, T – here and then the transpose sign. The fundamental difference between the difference scheme (1) and autoregression (3) is that in the right part of the latter, the variables $x_i^k$, $\forall i, k$ are measured in the grid nodes. The values of the initial and boundary conditions are also determined by the measurement results.

Within the framework of this work, a comparative analysis of the solution of the parametric identification problem (3) is carried out in the case of applying the usual least squares method (OLS), ridge regression and two author's methods of dimension reduction [10, 11] to it.

The estimates of the OLS of the parameters (3) have the following form [6]

$$
\hat{a} = (X^T X)^{-1} X^T (Xa + \omega).
\tag{4}
$$

The presence of an offset is checked by applying the expectation operator to the expression (4):

$$
a \neq a + \left(X^T X\right)^{-1} X^T \omega;
$$

under the conditions of correlation of random interference, observations of variables of a non-stationary process with regressors. It should be noted such a property of OLS estimates as efficiency, i.e. the minimum variance of parameter estimates.

As an alternative to OLS, we consider the estimates of the parameters of the model (3) obtained using ridge regression. In this case, regularization is added to the OLS criterion [14]

$$
Q_p(a) = \|x - y(a)\|^2 + \tau \|a\|^2 \to \min_a ,
\tag{5}
$$

where $\tau$ is the regularization coefficient.

The regularized OLS estimate is obtained in the form

$$
\hat{a}_p = \left(X^T X + \tau I\right)^{-1} X^T y .
\tag{6}
$$

The matrices $X^T X$ and $X^T X + \tau I$ the eigenvectors coincide, and the eigenvalues differ by $\tau$. Therefore, the condition number for the matrix $X^T X + \tau I$ is

$$
\mu(X^T X + \tau I) = \frac{\lambda_{\max} + \tau}{\lambda_{\min} + \tau}.
$$

It turns out that the more $\tau$, the smaller the number of conditionality. With growth, the stability of the problem increases, but the bias of estimates increases.

We proposed methods for reducing the dimension by taking into account the conservativeness property of the difference scheme [10, 11]. So in [10], taking into account the correlation of time series in adjacent grid nodes, it was proposed to change the level of the series in the $i$-th node to the expression

$$x_i^k = \beta_1 x_{i-1}^k + \beta_2 x_{i+1}^k + \xi^k . \tag{7}$$

Expression (7) allows us to expect to obtain OLS estimates $\hat{\beta}$ with a smaller standard error both due to a decrease in the dimension and due to a decrease in the correlation of time series in non-adjacent nodes.

Substituting the expression (7) in (3) we get

$$
\begin{aligned}
x_i^k &= (a_1 + a_2\beta_1)x_{i-1}^k + (a_2\beta_2 + a_3)x_{i-1}^k \\
&+ \xi_i^{k+1} - \xi_{\Sigma}^k = \theta_1 x_{i-1}^k + \theta_2 x_{i+1}^k + \Delta\xi ,
\end{aligned}
\tag{8}
$$

where $\xi_{\Sigma}^k$ is a convex linear combination of interference $\xi_{i-1}^k$, $\xi_i^k$, $\xi_{i+1}^k$.

The expression (8) with the obtained estimates $\hat{\beta}$ allows us to construct a system of linear equations with respect to the parameters $\hat{\beta}$:

$$
\begin{aligned}
a_1 + a_2\hat{\beta}_1 &= \theta_1, \\
a_2\hat{\beta}_2 + a_3 &= \theta_2, \\
a_1 + a_2 + a_3 &= 1.
\end{aligned}
\tag{9}
$$

The determinant of the system (9) $\hat{\beta}_1 + \hat{\beta}_2 - 1$ is different from zero in the case when $\hat{\beta}_1 + \hat{\beta}_2 \neq 1$.

Thus, the estimates are calculated in two stages: first, the estimates $\hat{\beta}$ of the expression model (7) are calculated, then the estimates $\hat{\theta}$ are calculated, and the system (9) is solved.

The conservativeness condition can be used directly [11]. Expressing, for example, $a_2 = 1 - a_1 - a_3$ we get the following modification of the expression model (3)

$$x_i^{k+1} - x_i^k = \hat{a}_1(x_{i-1}^k - x_i^k) + \hat{a}_3(x_{i+1}^k - x_i^k) \tag{10}$$

Here, as in the previous approach, we can expect to obtain OLS estimates of parameters with a smaller standard error.

Further, the above four methods of obtaining estimates will be denoted as follows: OLS – ordinary OLS; PP-ridge-regression; TSOLS-two-stage OLS; MOLS-modified OLS.

## 2   Experimental Study of the Quality of Assessments

The study was carried out on the example of a one-dimensional convective diffusion equation with respect to the spatial coordinate

$$\frac{\partial y}{\partial \tau} = D\frac{\partial^2 y}{\partial z^2} - v\frac{\partial y}{\partial z}$$
$$y(0, z) = \varphi(z) \qquad\qquad (11)$$
$$y(t, z^{\min}) = f_1(t), \quad y(t, z^{\max}) = f_2(t),$$

where $D \geq 0$ is the diffusion coefficient, $v \geq 0$ is the convection velocity, $z$ is the spatial coordinate.

The sample statistics required for the model study were obtained using an analytical solution equations (11) with the specified values of the parameters $D$ and $v$, initial and boundary conditions.

The analytical solution of the problem has the form [12]

$$x(t, 1) = \exp\left(\frac{v}{2D}\left(l - \frac{vt}{2}\right)\right)(\exp(-Dt)\sin l$$
$$+ \exp(-4Dt)\sin 2l + \exp(-9Dt)\sin 3l$$

The differential equation under consideration can be represented using the reduced difference scheme (1).

$$y_i^{k+1} = a_1 y_{i+1}^k + a_2 y_i^k + a_3 y_{i-1}^k,$$

where

$$a_1 = \left(\frac{D\Delta t}{\Delta z^2} - \frac{v\Delta t}{2\Delta z}\right) = 0.2583;$$
$$a_2 = \left(1 - \frac{2D\Delta t}{\Delta z^2}\right) = 0.5;$$
$$a_3 = \left(\frac{D\Delta t}{\Delta z^2} + \frac{v\Delta t}{2\Delta z}\right) = 0.2417.$$

It is not difficult to verify that $a_1 + a_2 + a_3 = 1$.

An additive observation noise $\xi_i^k$ was added to the values of the variable $y_i$ in the corresponding grid nodes, obtained using a Gaussian-type independent random number generator with zero mathematical expectation and unit variance $\sigma_\xi^2$. The intensity of the interference $c\sigma_\xi^2$ was set at the levels: $c = 0.1, c = 0.01, c = 0.001$.

To obtain reliable results of estimates of the bias and standard error, allowing comparisons of numerical values, the experiments in all modes were repeated 1000 times and their results were averaged.

For each of the four methods, estimates of the parameters of the initial identification problem were obtained, as well as estimates of the average value of the bias and the standard error. In the case of ridge regression, different values of the parameterization parameter at each noise level were considered.

Table 1 shows the influence of the value of the ridge regression regularization parameter $a_1$ on the quality of parameter estimation with interference intensity $c = 0.01$. For the remaining parameters, the results are qualitatively analogous.

Table 1 clearly shows that with the growth of the regularization parameter, the standard error of the estimate decreases, and the bias increases. For a comparative analysis of ridge regression with other methods, the value of the regularization parameter $\tau = 12.456$ was chosen, at which the standard error of the parameter $a_1$ is commensurate with the corresponding value of the OLS estimate.

Table 2 shows the results of the implementation of the methods of OLS, RR, DOLS and MOLS: estimates of the standard error and offset when identifying the parameters of the difference scheme (1) with different interference intensity $c$. Data from Table 2 show that for low interference ($c = 0.001$), all identification methods provide a satisfactory estimate of the parameters of the difference scheme (1) - an error of no more than 1% of the absolute value of the parameter. For large ($c = 0.1$) and medium ($c = 0.01$) interferences, only the MOLS can more or less cope with identification, providing a standard error and a laugh of estimates up to 10% for large interferences and up to 4% for medium interferences. Other methods demonstrate practically unacceptable values of the standard error and offset. At an average noise level, the standard error of the estimation by the PP method is slightly lower than the corresponding estimate of the OLS, which confirms the theoretical expectations.

**Table 1.** Standard error and parameter offset in the case of ridge regression

| Regularization parameter | Standard error | Offset |
| --- | --- | --- |
| 2.0498 | 0.09207 | 0.129 |
| 4.4976 | 0.09201 | 0.138 |
| 8.9851 | 0.0909 | 0.145 |
| 12.456 | 0.09 | 0.152 |
| 14.876 | 0.0806 | 0.201 |
| 16.976 | 0.08 | 0.399 |

The conducted studies of the model example in the case of problem (1–2) for non-negative values of the coefficients confirmed the effectiveness of the modified OLS. However, modeling of multidimensional time series based on the results of practical observations showed that some coefficients of the reduced difference model can take negative values. At the same time, the sum of all coefficients remains equal to one. A study of the quality of estimates obtained using a modified OLS was conducted, which showed the absence of a positive effect. For the data of the model experiment, in the case

**Table 2.** Standard error and parameter bias in the case of OLS, RR, DOLS, MOLS

| Parameters | C | OLS | RR | DOLS | MOLS |
|---|---|---|---|---|---|
| The average value of the offset | | | | | |
| a1 = 0,2583 | 0,1 | 0,1375 | 0,678 | 0,1321 | 0,0300 |
| | 0,01 | 0,0937 | 0,152 | 0,0900 | 0,0110 |
| | 0,001 | 0,0030 | 0,0357 | 0,0010 | 0,0009 |
| a2 = 0,5 | 0,1 | 0,1707 | 0,5680 | 0,1690 | 0,0286 |
| | 0,01 | 0,1625 | 0,1986 | 0,1601 | 0,0160 |
| | 0,001 | 0,0049 | 0,0456 | 0,0008 | 0,0004 |
| a3 = 0,2417 | 0,1 | 0,0244 | 0,6450 | 0,0369 | 0,0020 |
| | 0,01 | 0,0683 | 0,1745 | 0,0701 | 0,0010 |
| | 0,001 | −0,0017 | −0,0299 | 0,0002 | 0,0001 |
| The average value of the standard deviation | | | | | |
| a1 = 0,2583 | 0,1 | 0,1351 | 0,1532 | 0,1320 | 0,0270 |
| | 0,01 | 0,0922 | 0,0900 | 0,0900 | 0,0100 |
| | 0,001 | 0,0021 | 0,0091 | 0,0009 | 0,0008 |
| a2 = 0,5 | 0,1 | 0,1699 | 0,1622 | 0,1689 | 0,0271 |
| | 0,01 | 0,1608 | 0,1501 | 0,1600 | 0,0156 |
| | 0,001 | 0,0024 | 0,0020 | 0,0008 | 0,0004 |
| a3 = 0,2417 | 0,1 | 0,0207 | 0,0201 | 0,0360 | 0,0019 |
| | 0,01 | 0,0665 | 0,0602 | 0,0643 | 0,0008 |
| | 0,001 | 0,0001 | 0,0001 | 0,0001 | 0,0001 |

when one of the coefficients is negative (for example, a1 = 1.8252, a2 = −2.1720, a3 = 1.3433), a comparison of the usual OLS and the modified one is presented in Table 3.

**Table 3.** Standard error and parameter offset

| | Offset | | | Standard deviation | | |
|---|---|---|---|---|---|---|
| | a1 | a2 | a3 | a1 | a2 | a3 |
| OLS | 0.00123 | 0.00097 | 0.00087 | 0.00145 | 0.002 | 0.00079 |
| OLS with the replacement of the equation by 1 | 0.0035 | 0.0026 | 0.0078 | 0.0028 | 0.0025 | 0.0018 |

# 3 Conclusion

A comparative analysis of the application of the studied identification methods to the assessment of the parameters of the conservative difference scheme (1) showed a significant dependence of the quality of the assessment on the intensity of observation interference. At low interference ($c = 0.01$ and $c = 0.1$), all the considered methods successfully cope with the identification task. With an increase in the intensity of the method, only the modified least squares method, based on taking into account the conservativeness property of the difference scheme, demonstrates satisfactory performance, but only if we have positive parameters of the model. Thus, the modified OLS can be recommended for the identification of models of distributed dynamical systems, the different representation of which satisfies the given property of conservativeness $a_1 + a_2 + a_3 = 1$ in the case of non-negative coefficients. It should be expected that in problems with a higher dimension of spatial coordinates - ($R^2$ and $R^3$), the use of the conservativeness property will also have a positive effect.

# References

1. Egorshin, A.O.: Piecewise-linear identification and differential approximation on a uniform grid. In: Proceedings of the XII all-Russian Conference on Problems of Management 2014, pp. 2807–2822 Moscow (2014)
2. Bezruchko, B.P., Smirnov, D.A.: Modern problems of modeling from time series. In: Proceedings of the University of Sarajevo, series Physics, vol. 6, pp. 3–27. Sarajevo (2006)
3. Guo, L.Z., Billings, S.A., Coca, D.: Identification of partial differential equation models for a class of multiscale spatio-temporal dynamical systems. Int. J. Control 83(1), 40–48 (2010)
4. Matveev, M.G., Mikhailov, V.V., Sirota, E.A.: Combined prognostic model of a nonstationary multidimensional time series for constructing a spatial profile of atmospheric temperature. Inf. Technol. 22(2), 89–94 (2016)
5. Xun, X., Cao, J., Mallick, B., Carrol, R.J., Maity, A.: Parameter estimation of partial differential equation models. J. Am. Stat. Assoc. 108(503), 1–27 (2013)
6. Nosko, V.P.: Econometrica. Introduction to the regression analysis of time series. NFPK, Moscow (2002)
7. Gareth, J., Witten, D., Hastie, T., Tibshirani, R.: An Introduction to Statistical Learning. Springer, New York (2013)
8. Matveev, M.G., Kopytin, A.V., Sirota, E.A., Kopytina, E.A.: Modeling of nonstationary distributed processes on the basis of multidimensional time series. Proc. Eng. 201, 511–516 (2017)
9. Samarsky, A.A.: Theory of Difference Schemes. Nauka, Moscow (1978)
10. Matveev, M.G., Sirota, E.A.: Analysis of the properties of the OLS-estimators in case of elimination of multi-collinearity in the problems of parametric identification of distributed dynamic processes. In: Bulletin of Voronezh state University, series "System analysis and information technologies", vol. 2, pp. 15–22 (2020)
11. Matveev, M.G., Sirota, E.A.: Analysis and investigation of the conservativeness condition in the problem of parametric identification of dynamic distributed processes. J. Phys. Conf. Series 1902(1), 012079 (2021)
12. Kopytin, A.V., Kopytina, E.A., Matveev, M.G.: Application of the expanded Kalman filter for identifying the parameters of a distributed dynamical system. In: Proceedings of Voronezh State University. Series: Systems Analysis and Information Technologies vol. 3, pp. 44–50 (2018)

13. Barseghyan, A.A., Kupriyanov, M.S., Kholod, I.I., Tess, M.D., Elizarov, S.I.: Analysis of data and processes. 3rd edn. BHV-Petersburg, St. Petersburg (2009)
14. Tibshirani, R.J.: Regression shrink-age and selection via the lasso. J. R. Stat. Soc. Series B Methodol. **58**(1), 267–288 (1996)
15. Ayvazyan, S.A., Bukhstaber, V.M., Enyukov, I.S., Meshalkin, L.D.: Applied statistics. Classification and reduction of dimension. Finance and Statistics, Moscow (1989)

# Scheduling with Genetic Algorithms Based on Binary Matrix Encoding

Vladislav Korotkov$^{(\boxtimes)}$ ⓘ and Mikhail Matveev ⓘ

Voronezh State University, Voronezh, Russia

**Abstract.** Metaheuristic optimization approaches like genetic algorithms have been successfully applied to automatic scheduling in various fields. Scheduling problems are known for high computational complexity. This is especially true for problems with precedence constraints like RCPSP (resource constrained project scheduling problem). Thus, scheduling algorithms need further improvements to deal with large practical problems.

The paper proposes a novel binary matrix approach for encoding order in scheduling algorithms. Such encoding makes it simple to extract information about relative position of jobs (tasks, activities). It's shown that precedence constraints can be also encoded as binary matrix of such type. The paper also introduces mutation and crossover operators for the proposed encoding to be used in genetic algorithms. Unlike the classical crossover operators, the proposed one produces offspring individuals which preserve relative and not absolute order of some subsets of elements.

It is also proven that the proposed binary matrices can be reduced to canonical vector form. This makes it possible to apply Holland's schema theorem to justify the convergence of modified genetic algorithms based on binary matrix encoding.

The efficiency of the modified genetic algorithm in comparison with classical encodings and genetic operators has been tested on some instances of scheduling problems. It is shown that the proposed method demonstrates quite promising results on large scale problems.

**Keywords:** Scheduling · Genetic algorithms · Crossover operator · Combinatorial optimization

## 1 Introduction

Scheduling is the problem of determining the optimal plan for performing some tasks (jobs, activities). Scheduling is crucial for planning any working processes in a wide variety of areas. Optimal schedule can help one make better use of available resources and reduce execution times. But building optimal or near optimal schedule is a non-trivial task. In project management, the critical path method is usually used to build project execution schedules. It estimates the minimum time required to complete a project. However, in order to be practically feasible, the schedule must consider the available resource stocks and resource requirements. In addition, other specific requirements and

© Springer Nature Switzerland AG 2022
V. Taratukhin et al. (Eds.): ICID 2021, CCIS 1539, pp. 224–234, 2022.
https://doi.org/10.1007/978-3-030-95494-9_19

constraints may arise in various subject areas and under some conditions. This makes the problem even more complex and almost impossible to solve by humans.

An artificial intelligence approach can be used to solve this problem programmatically. Various algorithms for solving the problem have been proposed, but due to its complexity, the development of new effective approaches capable of solving large practical instances of the problem is still required.

## 2  Problem Description

Various variants of scheduling problems are known. The most general one is the so-called Resource Constrained Project Scheduling Problem (RCPSP), which arises in the field of project management. Other scheduling problems like job-shop or open-shop can be considered as its special cases [1].

RCPSP considers the project as a set of jobs (or activities). Each job requires some resources, which are available in a limited amount (e.g., the employees). The resulting schedule should be as short as possible, but precedence and resource constraints must not be violated. The problem can be formally defined as an optimization problem:

$$f_n \to \min, \tag{1}$$

subject to

$$f_i \leq f_j - d_j \quad \forall (i,j) \in A, \tag{2}$$

$$f_1 = 0, \tag{3}$$

$$\sum_{i \in s_t} r_{ik} \leq a_k \quad \forall k = 1, ..., m, \ t = 1, ..., f_n. \tag{4}$$

Here $f_i$ is job finish time, $d_i$ - its duration, $A$ - project graph that defines job precedence constraints, $r_{ik}$ - the amount of resource $k$ required for job $i$, $a_k$ - the amount of resource $k$ available.

## 3  Metaheuristic Solutions

It is known that scheduling problems are highly computationally intensive. RCPSP is proven to be strongly NP-hard [2]. Because of this, the exact methods can only solve small problems with fairly weak resource constraints. But realistic projects can have up to several hundred tasks, and solutions must often be found as quickly as possible. Consequently, metaheuristic algorithms are usually used to solve such problems. They provide the best trade-off between accuracy and computation speed.

Any metaheuristic for solving combinatorial optimization problem partially enumerates feasible solutions. The set of all feasible solutions forms the so-called search space. Heuristic search is usually performed in the space of encoded solutions which correspond to the subset of all possible schedules as the original search space is too large.

There are multiple encodings for RCPSP solution representation. The most common encoding is called "activity list" [3]. It is a vector of all jobs, arranged in an order that does not violate the precedence constraints:

$$I = \left(j_1^I, \ldots, j_n^I\right) \tag{5}$$

The order of jobs can also be implicitly specified using priority values. Such approach is called "random key" representation (Fig. 1). Other approaches to order encoding have been also presented in the literature. But comparative studies show that using activity list representation still leads to better results [3, 4].

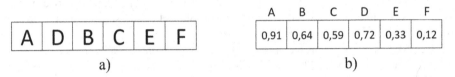

a)                                          b)

**Fig. 1.** Classical order representations: a) activity list, b) random key.

The corresponding schedule can be obtained by applying any of two decoding procedures – serial or parallel scheduling schemes. Starting with an empty schedule they sequentially construct a new partial schedule by placing the next activities to the earliest possible time so that precedence and resource constraints are satisfied [5]. The jobs are considered in the given order. The main difference between two generation schemes is that the sequential one iterates over jobs while the parallel one iterates over points in time. It is proved that optimal solution may not be present in the set of schedules obtained by the parallel scheme [6]. Therefore, serial scheduling scheme is usually used in metaheuristic algorithms to decode solutions.

The chosen representation also defines possible operators used to produce new solutions. A unary operator takes a single solution as input and returns one that is slightly different. It implements the neighborhood move in local search algorithms like tabu search and simulated annealing. A typical unary operator for activity list representation swaps two jobs if this doesn't violate precedence constraints (Fig. 2).

**Fig. 2.** Classical unary swap operator for activity list representation

A binary operator takes two parent solutions to produce one or two children, which combine features of their parents. Such operator is specific for genetic algorithms where it is used for crossover.

Genetic algorithm is a metaheuristic inspired by Charles Darwin's theory of natural evolution. It is based on three genetic operators: mutation, crossover and selection. At each iteration of the algorithm, a set of current solutions to the problem, called a population, is maintained. Each solution or individual is represented by its chromosomes, which encode the individual's features. To evaluate solutions, a fitness function is used, which measures how good any solution is in relation to the problem being solved. The best solutions have a higher chance of survival after some selection procedure. Then binary crossover operator is applied to all or some of the pairs of individuals. Mutation operator, with some low probability, makes changes in the individuals of the population, providing the required diversity. The resulting individuals form a new generation. This procedure is repeated until necessary termination conditions are reached. In the end, the best of all the solutions found is taken.

The unary swap operator for activity list representation described above is commonly used for mutation in genetic algorithms solving RCPSP. There are many possible crossover operators for activity list representation. Classical ones are: one-point crossover, two-point-crossover, uniform crossover (Fig. 3) [7]. They transfer some sequences of adjacent jobs from parent solutions to children or preserve some absolute positions unchanged.

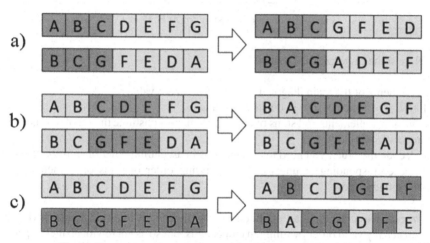

**Fig. 3.** Crossover operators: a) one-point, b) two-point, c) uniform.

Many comparative studies show that genetic algorithms outperform the majority of other metaheuristic methods in solving scheduling problems [3, 4, 8]. This allows us to think about the importance of a crossover operator. But it is worth noting that currently used crossovers do not fully take into account the specifics of the problem being solved. In particular, due to resource constraints and other indirect relations between jobs, the relative rather than absolute order of some jobs can distinguish good solutions from bad ones.

## 4    The Proposed Approach

### 4.1    Binary Matrix Encoding

An alternative approach to job order encoding is proposed. Let's define a binary $n \times n$ matrix $M$ where some arbitrary cell $M[i, j]$ contains 1 if and only if job $i$ precedes job $j$ in the encoded order (and 0 otherwise). So, ones in the $i$-th row correspond to successors (direct or indirect) of the job $i$ while ones in the $i$-th column correspond to its predecessors. An example of such matrix is shown in Fig. 4.

|   | A | B | C | D | E | F |
|---|---|---|---|---|---|---|
| A | 0 | 1 | 1 | 1 | 1 | 1 |
| B | 0 | 0 | 1 | 1 | 1 | 1 |
| C | 0 | 0 | 0 | 0 | 0 | 1 |
| D | 0 | 0 | 1 | 0 | 1 | 1 |
| E | 0 | 0 | 1 | 0 | 0 | 1 |
| F | 0 | 0 | 0 | 0 | 0 | 0 |

**Fig. 4.** Example of encoding for A, B, D, E, C, F.

Note some important properties:

- all elements of the main diagonal always have a zero value;
- if $M[i, j] = 1$ then $M[j, i] = 0$ and vice versa;
- sums of row elements and sums of column elements constitute the set of non-negative integers from 0 to $n - 1$;
- to decode the order, you need to arrange jobs in ascending order of sums of elements in the corresponding columns or in decreasing order of sums of elements in the corresponding rows.

Precedence constraints can also be encoded with partially filled binary matrix (Fig. 5). We can get any valid order by filling in empty cells, but so as not to violate the properties of the matrix.

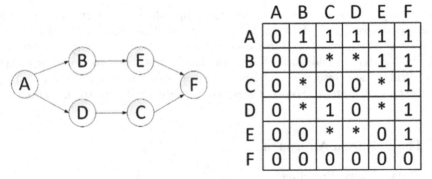

|   | A | B | C | D | E | F |
|---|---|---|---|---|---|---|
| A | 0 | 1 | 1 | 1 | 1 | 1 |
| B | 0 | 0 | * | * | 1 | 1 |
| C | 0 | * | 0 | 0 | * | 1 |
| D | 0 | * | 1 | 0 | * | 1 |
| E | 0 | 0 | * | * | 0 | 1 |
| F | 0 | 0 | 0 | 0 | 0 | 0 |

**Fig. 5.** Precedence constraints and its corresponding binary matrix.

## 4.2 Mutation Operator

Classical swap operator can still be used for the proposed encoding method. To swap two jobs we just need to swap the corresponding rows and columns of the matrix.

## 4.3 Crossover Operator

The idea of crossing over two solutions encoded by binary matrices is to preserve relative order of jobs. The main challenge here is not to violate precedence matrix properties. With this in mind, we get the following crossing over procedure.

1. Split the set of all jobs into two disjoint subsets. For this, the main parent is randomly selected and its first job is added to the first subset. Then the current subset is switched with some probability and the next job from the main parent is added to the current subset. The procedure continues until all jobs are distributed into two subsets.
2. Transfer information about the relative position of jobs of the first set from the first parent to the child.
3. Perform disjunction with precedence matrix to enforce precedence constraints.
4. Sequentially recover matrix:

   a. at the $i$-th step, we consider the columns, the sum of which equals to $i$;
   b. select column $k$ that has no blank cells or whose corresponding job is located earlier in the second parent;
   c. place the $k$-th job at the $i$-th position, filling all predecessors with ones in the $k$-th line.

5  The second child is obtained by swapping parents and repeating steps 2–4.

Stage 4 of this algorithm fills the remaining cells so as not to violate the properties of the matrix and arrange jobs from the second set in the same order as in the second parent.

Figure 6 shows an example of such crossover. The second picture shows two children after completing the first three steps of the algorithm. The precedence constraints were taken from Fig. 5. The "*" sign here denotes empty cells.

## 4.4 Genetic Algorithm

The modified genetic algorithm can be developed based on the proposed encoding method and genetic operators. It is heavily inspired by genetic algorithms proposed in [9, 10]. Its main components are described below.

**Fitness Function.**  Schedule duration is used as a fitness function. Thus, the solution must be decoded to be evaluated. Serial scheduling scheme is used for this purpose.

**Initial Population.**  The initial population is randomly generated. First, a zero matrix $M$ and a corresponding empty set of jobs $J$ are created for each individual. At each step of

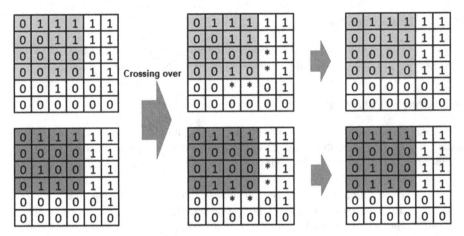

**Fig. 6.** Example of crossover

the procedure a random job *j* is selected whose predecessors are already in the set *J*. In the *j*-th column, all cells corresponding to jobs from the set *J* are filled with ones. Then *j* is added to *J*. The resulting instances obviously satisfy the precedence constraints.

**Mutation.** Mutation is applied to all individuals. For each job located at the *i*-th position in encoded order a random value $p_i \in [0, 1]$ is picked independently. If $p_i \leq P_{PM}$, where $P_{PM}$ is initially defined mutation probability, then *i*-th and *i* + 1 jobs are swapped unless this violates the precedence constraints.

**Crossover.** The entire population is split into pairs of individuals. The crossover operator described in Sect. 4.3 is applied to each pair with some predefined probability.

**Selection.** A tournament selection approach is used. Three random individuals are selected and only the best of them is added to the new generation. The procedure is repeated until the required population size is reached.

### 4.5   Justification of Convergence

The power of traditional genetic algorithms is explained by so-called schema theorem proposed by John Holland [11]. Schema is a template of binary string where values at some positions are fixed while others are arbitrary. The number of fixed positions is called the schema order. The Eq. (6) explains how low-ordered potentially good schemas increase in successive generations.

$$E(m(H, t + 1)) \geq \frac{m(H, t)f(H)}{a_t}[1 - p] \qquad (6)$$

Here $m(H, t)$ is the number of individuals that correspond to schema *H* at generation *t*, $f(H)$ is average fitness of schema *H*, $a_t$ - average fitness at generation *t*, *p* - probability of schema disruption after mutation and crossover.

Schema theorem is applicable for traditional genetic algorithms with binary string encoding. Therefore, we need to introduce the one-to-one mapping from binary precedence matrices to binary strings.

Let $Z_i$ be the number of zeroes in the $i$-th row of upper triangular part of the matrix. Then we can introduce the following conversion operator:

$$\sum_{i=1}^{n-1} (n-i)! \times Z_i \qquad (7)$$

As an example, for the precedence matrix shown in Fig. 5 we got the following conversion:

$(6-1)! \times 0 + (6-2)! \times 0 + (6-3)! \times 2 + (6-4)! \times 0 + (6-5)! \times 0 = 12 \Leftrightarrow (1\,1\,0\,0)$

Now Holland's schema theorem can be used to justify the convergence of the modified genetic algorithm too.

# 5 Computational Experiments

The proposed approach was compared with traditional ones on a subset of PSBLIB problem set [12]. Five variations of genetic algorithm were implemented with different encoding methods and crossover operators:

- activity list encoding and one-point crossover;
- activity list encoding and two-point crossover;
- activity list encoding and uniform crossover;
- activity list encoding and uniform crossover (with the method of dividing the set of jobs, which is similar to that used in the proposed crossover operator);
- binary matrix encoding and the proposed crossover operator.

All other components of genetic algorithms were the same. To eliminate the influence of random factors mutation probability was set to zero and initial population was the same for every variation of genetic algorithm. Thus, the main focus was on the ability of algorithms to identify promising combinations in fewer iterations. Population size was set to 80, generation count – 40, switching probability while jobs partitioning – 0.2.

The variation marked as "Uniform 2" uses uniform crossover with similar method of jobs division when jobs are sequentially added to the first set with a probability of switching to another set at some point. Such algorithm was included in the list to take into account the contribution of this trick into final results.

Random problem instances of three different sizes were taken from j60.sm, j90.sm and j120.sm problem sets. A total of 144 instances were taken. The results are presented in Table 1. For each problem set and each algorithm the number of cases is shown when the best result was achieved. Detailed results for j120.sm problem set are presented in Table 2. Each problem of this set is a large instance, consisting of 120 jobs.

**Table 1.** Overall benchmark results

|  | One-point | Two-point | Uniform 1 | Uniform 2 | Matrix |
|---|---|---|---|---|---|
| j60.sm<br>(48 inst.) | 31<br>(64.6%) | 34<br>(70.1%) | 30<br>(62.5%) | 31<br>(64.6%) | **46**<br>**(95.8%)** |
| j90.sm<br>(48 inst.) | 29<br>(60.4%) | 33<br>(68.8%) | 25<br>(52.1%) | 26<br>(54.2%) | **45**<br>**(93.8%)** |
| j120.sm<br>(48 inst.) | 7<br>(14.6%) | 22<br>(45.8%) | 3<br>(6.25%) | 7<br>(14.6%) | **34**<br>**(70.8%)** |
| Total<br>(144 inst.) | 67<br>(46.5%) | 89<br>(61.8%) | 58<br>(40.3%) | 64<br>(44.4%) | **125**<br>**(86.8%)** |

**Table 2.** Detailed benchmark results for j120.sm problem set

| Instance | One-point | Two-point | Uniform 1 | Uniform 2 | Matrix |
|---|---|---|---|---|---|
| j1201_1.sm | 120 | 117 | 123 | 124 | **116** |
| j1202_2.sm | 82 | **80** | 84 | 83 | 82 |
| j1203_3.sm | **100** | **100** | 103 | 102 | **100** |
| j1204_4.sm | 78 | 76 | 80 | 78 | **75** |
| j1205_5.sm | **77** | **77** | 78 | **77** | **77** |
| j1206_6.sm | 184 | 181 | 185 | 189 | **180** |
| j1207_7.sm | 141 | 138 | 142 | 141 | **136** |
| j1208_8.sm | **97** | **97** | 102 | 101 | **97** |
| j1209_9.sm | 95 | **93** | 97 | 96 | 95 |
| j12010_10.sm | 67 | 67 | 72 | 69 | **66** |
| j12011_1.sm | 203 | 197 | 201 | **195** | 197 |
| j12012_2.sm | 132 | **131** | 133 | 132 | **131** |
| j12013_3.sm | 138 | 138 | 144 | 145 | **137** |
| j12014_4.sm | 103 | **101** | 107 | 105 | **101** |
| j12015_5.sm | **87** | **87** | **87** | **87** | **87** |
| j12016_6.sm | 234 | **229** | 238 | **229** | 229 |
| j12017_7.sm | 167 | 167 | 169 | 168 | **163** |
| j12018_8.sm | 119 | **114** | 122 | 118 | **114** |
| j12019_9.sm | 96 | **94** | 99 | 99 | 96 |
| j12020_10.sm | 83 | **81** | 86 | 82 | **81** |
| j12021_1.sm | 128 | **127** | 130 | 129 | **127** |

(*continued*)

**Table 2.** (*continued*)

| Instance | One-point | Two-point | Uniform 1 | Uniform 2 | Matrix |
|---|---|---|---|---|---|
| j12022_2.sm | 119 | 119 | 118 | 119 | **114** |
| j12023_3.sm | **99** | **99** | **99** | **99** | **99** |
| j12024_4.sm | **101** | **101** | 104 | 103 | **101** |
| j12025_5.sm | **100** | **100** | **100** | **100** | **100** |
| j12026_6.sm | 222 | **208** | 219 | 215 | **208** |
| j12027_7.sm | 149 | 147 | 153 | 153 | **146** |
| j12028_8.sm | 113 | 112 | 118 | 113 | **110** |
| j12029_9.sm | 103 | 103 | 105 | 103 | **98** |
| j12030_10.sm | 90 | **88** | 91 | 90 | **88** |
| j12031_1.sm | 227 | 227 | 229 | 227 | **226** |
| j12032_2.sm | 152 | **147** | 154 | 154 | **147** |
| j12033_3.sm | 122 | **118** | 129 | 128 | **118** |
| j12034_4.sm | 111 | 110 | 114 | 115 | **106** |
| j12035_5.sm | 105 | 106 | 109 | 109 | **103** |
| j12036_6.sm | 266 | **255** | 266 | 258 | **255** |
| j12037_7.sm | 183 | **182** | 185 | **182** | **182** |
| j12038_8.sm | 141 | 138 | 144 | 140 | **137** |

The proposed algorithm outperformed classical approaches on all problem sizes. Moreover, this and two-point crossover variations showed significantly better results on larger sizes compared to other algorithms. But genetic algorithm based on matrix encoding is still the most efficient what makes it potentially usable for solving practical large-scale problems.

# 6   Conclusion

The proposed modified genetic algorithm showed better results on some test problems compared to the classical approaches. This was achieved due to the novel binary matrix encoding method and the corresponding crossover operator. The proposed approach can be used in scheduling software for scheduling projects with hundreds of jobs or solving other practical large-scale problems.

# References

1. Sprecher, A.: Resource-Constrained Project Scheduling. Exact Methods for the Multi-Mode Case. Springer, Heidelberg (1994)

2. Blazewicz, J., Lenstra, J.K., Rinnooy Kan, A.H.G.: Scheduling subject to resource constraints: classification and complexity. Discrete Appl. Math. **5**(1), 11–24 (1983)
3. Kolisch, R., Hartmann, S.: Heuristic algorithms for solving the resource-constrained project scheduling problem: classification and computational analysis. Internat. Ser. Oper. Res. Management Sci. **14**, 147–178 (1999)
4. Pellerin, R.: A survey of hybrid metaheuristics for the resource-constrained project scheduling problem. Eur. J. Oper. Res. **280**(2), 395–416 (2020)
5. Kim, J., Ellis, R.: Comparing schedule generation schemes in resource-constrained project scheduling using elitist genetic algorithm. J. Constr. Eng. Manag. **136**(2), 160–169 (2010)
6. Kolisch, R.: Serial and parallel resource-constrained project scheduling methods revisited: theory and computation. Eur. J. Oper. Res. **90**, 320–333 (1996)
7. Hartmann, S.: A competitive genetic algorithm for resource-constrained project scheduling. Nav. Res. Logist. **45**, 733–750 (1998)
8. Kolisch, R., Hartmann, S.: Experimental investigation of heuristics for resource-constrained project scheduling: an update. Eur. J. Oper. Res. **174**, 23–37 (2006)
9. Zatsarinnyi, A.A., Korotkov, V.V., Matveev, M.G.: Modeling the process of network planning of a portfolio of projects with heterogeneous resources under fuzziness. Inform. Appl. **13**(2), 92–99 (2019)
10. Korotkov, V., Matveev, M.: Individual scheduling for the multi-mode resource-constrained multi-project scheduling problem. In: Proceedings of the 1st International Conference of Information Systems and Design. http://ceur-ws.org/Vol-2570/paper5.pdf. Accessed 16 Oct 2021
11. Holland, J.: Adaptation in Natural and Artificial Systems: An Introductory Analysis with Applications to Biology, Control and Artificial Intelligence. MIT Press, Cambridge (1992)
12. Kolisch, R., Sprecher, A.: PSPLIB – a project scheduling library. Eur. J. Oper. Res. **96**, 205–216 (1996)

# Software for Researching the Processes of Curing of Polymer Compositions Using Mathematical Modeling

Sergey Tikhomirov[1]([⊠]) [iD], Olga Karmanova[2] [iD], Aleksandr Maslov[1] [iD],
and Elena Lintsova[1] [iD]

[1] Voronezh State University of Engineering Technologies, Information and Control System,
Voronezh, Russia
[2] Voronezh State University of Engineering Technologies. Technologies of Organic
Compounds, Polymer Processing and Technosphere Safety, Voronezh, Russia

**Abstract.** The work is devoted to the development of a mathematical apparatus which establishes the relationship between the consist of the elastomer composition and the projected properties of the finished rubber compound taking into account the parameters of its manufacture and processing. A software product has been developed for carrying out research-work in the study of vulcanization of rubber compounds using multi-component structuring systems. The program implements algorithms for calculating the modes of vulcanization of thick-walled reinforced composites, which allow, when designing, to assess the technological capabilities of production, to plan the choice of materials and equipment. The complex of programs has a block-modular structure, which allows its expansion without loss of functionality. With the use of the developed software product, the parameters of the isothermal vulcanization process were evaluated on a test example of the «Tire Carcass» recipe. The maximum deviation from the experimental data does not exceed 5%, which indicates the adequacy of the model.

**Keywords:** Mathematical modeling · Rubber compound · Vulcanization · Car tire · Software package

## 1 Introduction

The manufacture of automobile tires is a multi-stage process that is carried out in the factory and includes: market analysis, development of a digital model and prototype, production of a rubber compound, release of semi-finished products, assembly, vulcanization and quality control of the finished product [1]. Currently, there are no effective methods that, based on information about the components of rubber compounds, would make it possible to predict the properties of final products. The concentrations of the initial components and the parameters of the technological process have a significant effect on the formation of the vulcanization structure. In this case, the properties of the final products are determined by control samples obtained from the corresponding batch. The need to predict the properties and behavior of such products in the conditions of

© Springer Nature Switzerland AG 2022
V. Taratukhin et al. (Eds.): ICID 2021, CCIS 1539, pp. 235–253, 2022.
https://doi.org/10.1007/978-3-030-95494-9_20

their operation determines the creation of new approaches based on in-depth theoretical analysis and mathematical description [2–6].

When developing approaches to creating methods for assessing the optimal modes of vulcanization and properties of the resulting products, it should be borne in mind that vulcanization, which is a complex chemical process with a large amount of reagents, is characterized by certain regularities of flow or kinetics and consists in the formation of a single spatial network of rubber macromolecules. In the process of vulcanization, the properties of rubbers are formed, the change in which with the duration of vulcanization is not the same, i.e. kinetic curves do not coincide with each other: according to some properties they pass through a maximum, according to others - through a minimum; in most cases, the kinetic curves "property - duration of the process" are characterized by an initial (induction) period, followed by a period of predominant structuring, a vulcanization plateau (in which property levels are retained for given temperature conditions) and a final period of predominant reversion (destruction, decay of nodes, rupture of spatial grid chains) [1, 8]. Existing computational analytical and engineering methods for describing the kinetics of vulcanization are applicable mainly to particular types of kinetic curves.

When modeling the structure of elastomers, such approaches may be promising, in which the properties of each component of the composition are sequentially considered. Their practical implementation in the manufacture of products and elastomers will reduce the labor intensity of the technological process, ensure the quality of predicting the physical, mechanical and technological properties of materials, and reduce the cost of finished products [7].

In general, the methodology for optimizing tire vulcanization modes at the stages of their design and manufacture includes the following components:

- assessment of the degree of vulcanization of rubber compounds and the kinetics of vulcanization un-der actual temperature conditions by a set of properties that determine the behavior of the product in operation;
- calculation and modeling of temperature fields in vulcanizable products;
- analysis of the mechanical behavior of materials of vulcanizable products at all stages of the process from the induction period to the final post-vulcanization period;
- optimization of tire vulcanization modes, including their correction taking into account the actual process parameters.

Vulcanization is the final stage in the production of tires, at which a product made from semi-finished products consisting of reinforcing materials and rubber compounds, on vulcanization equipment acquires the required shape and new properties. In this process, heat is transferred to the tire from the mold, which is maintained at a high temperature by circulating the heat carrier. After the completion of the process, the finished tire is removed from the mold and cooled at room temperature. Table 1 shows an example of a pneumatic tire vulcanization mode. This process is energy intensive, and its optimization will allow not only to produce high quality tires, but also to increase the volume of products. To optimize the vulcanization process, an appropriate assessment of the distribution of temperature fields in different layers of the tire over time is necessary.

The main method for determining the conditions of vulcanization is the method of thermocouples, which are placed in different parts of the wet tire to measure the temperature and build profiles of temperature changes over time and determine the conditions of vulcanization [8]. However, this method is costly and very time consuming. Therefore, tire manufacturers are looking for alternative methods based on finite element process modeling to predict temperature distribution and cure rate in a product volume.

**Table 1.** Table captions should be placed above the tables.

| Process | Pressure, MPa | Time, min | Steam temperature, °C |
|---|---|---|---|
| Clamping | Segment form 150 °C/side plate 140 °C | | |
| Steam filling | 0.8–1.2 | 8 | 178 |
| First water filling | | 2 | |
| Hot water filling (cycle) | 2.4–3.0 | 77 | 165 |
| Hot water filling (no cycle) | | 70 | 165 |
| Water draining | 0,5 | 2 | |
| Water draining | 0 | 2 | |
| Vacuuming | | 2 | |
| Total cycle time | | 163 | |

The tire vulcanization process can be conventionally divided into 8 stages. It is more complex than the process of manufacturing semi-finished products. The current thermocouple method does not predict the degree of vulcanization at an arbitrary location on the tire, since temperature changes can only be obtained at the location where the thermocouple is embedded. This process is a complex procedure of thermal interaction "liquid-solid" and to obtain accurate solutions it is necessary to use the finite element method [8, 9].

In mathematical modeling of physicochemical processes of vulcanization, certain difficulties may arise in reconciling the postulated kinetic mechanism with experimental data. First, the reaction mechanism includes many stages. Second, the equations for describing the kinetics of individual stages are nonlinear due to the binary reactions and the exponential dependence of the rate on temperature. Traditional methods used to find the coefficients in the case of rigid systems of ordinary differential equations will be unstable, which will require the creation of additional algorithms for their stabilization [8, 10]. For these reasons, it is important to create simpler mathematical models that allow an adequate description of complex physicochemical reactions of vulcanization. The development of a technique for evaluating the kinetics of vulcanization of multicomponent elastomeric compositions is based on the fact that individual stages of the vulcanization process differ significantly in rate. Along with the slowly proceeding stages of the formation of the actual vulcanization agent, there are extremely rapid stages proceeding by a radical mechanism. In such cases, to simplify the equations of chemical kinetics, the conditions of quasi-equilibrium of fast stages are usually used [10].

Therefore, it is necessary to develop an alternative method of temperature control of the vulcanization process, taking into account the composition of rubber compounds. The purpose of this work is to develop a mathematical apparatus that establishes a relationship between the composition of the elastomeric composition and the predicted properties of the finished rubber compound, taking into account the parameters of its manufacture and processing.

## 2  Mathematical Software Models

For the purpose of studying the processes of vulcanization of elastomeric compositions on the basis of mathematical modeling, software has been developed. The application has several software modules (Fig. 1) for computer simulation of the vulcanization of elastomeric compositions.

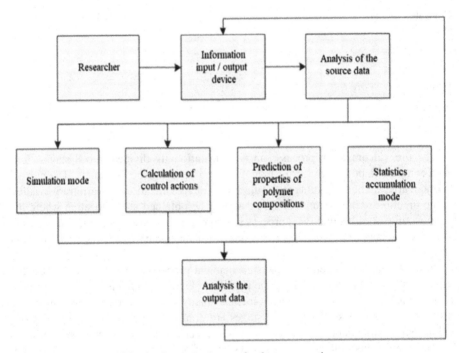

**Fig. 1.**  General scheme of software operation.

Each module implements data processing algorithms according to the selected mode: modeling heat transfer processes and the kinetics of the vulcanization process; calculation of control actions of the technological process; the mode of accumulation of statistical data. Figure 2, 3 shows the detail, sequence and logic of the user's work with software subsystems in order to obtain a set of results determined by the functions of the subsystems.

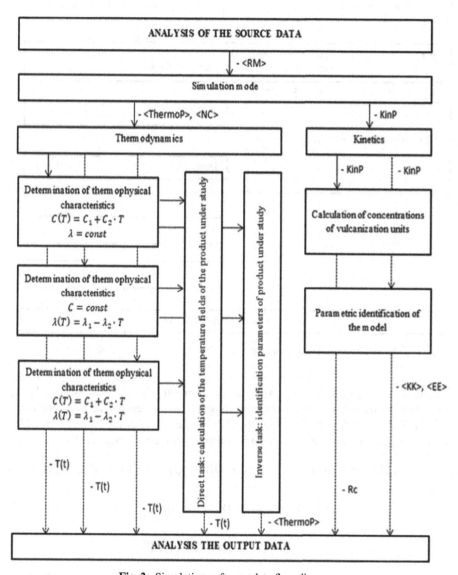

**Fig. 2.** Simulation software data flow diagrams

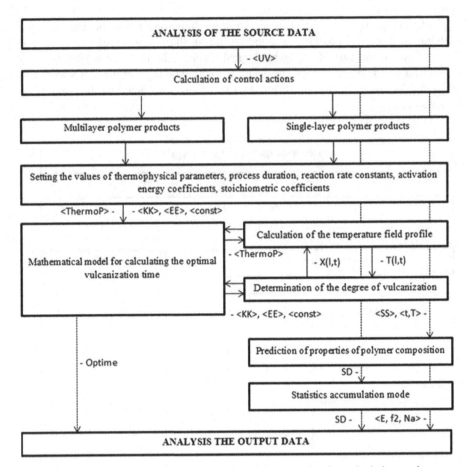

**Fig. 3.** Data flow diagram of the software of the control action calculation mode

The mode of modeling heat transfer in rubber products is intended for calculating temperature fields and evaluating the thermo-physical parameters of an elastomeric composition. The calculation methods are based on a mathematical model represented by the heat conduction equation with given initial and boundary conditions [10]:

$$C(T) \cdot \partial T(l, t)/\partial t = \partial/\partial l\{\lambda(T) \cdot \partial T(l, t)/\partial l\} + q(l, t); \; l \in [0, L], t \in [0, t\_k]. \quad (1)$$

$$T(l, 0) = T_0(l); \; l \in [0, L]. \quad (2)$$

$$T(0, t) = T_v(t), T(L, t) = T_n(t); \; t \in [0, t_k]. \quad (3)$$

where $C(T)$ – is the coefficient of volumetric heat capacity of the sample, J/(m$^3$·К); $T(l,t)$ – temperature, К; $L$ – s the sample thickness, m; $t_k$ – end point of time, sec; $T_v(t)$ and $T_n(t)$ – are the temperature values at the top and bottom of the sample, К;

$\lambda(T)$ – coefficient of thermal- conductivity of the sample, W/(m·K), $q(l,t)$ – heat release density.

The program implements algorithms for calculating the coefficients of thermal conductivity and heat capacity of elastomeric compositions depending on temperature based on experimental data (Fig. 5, 6). The researcher chooses which mathematical model to calculate the thermophysical parameters.

In particular, an initial vulcanization temperature is set; the calculated value of the heat capacity coefficient and the a priori approximation of the parameters of the approximating thermal conductivity function depending on the temperature; product thickness; end time of the vulcanization process.

The program calculates the curves of temperature fields inside the product at different levels using built-in algorithms. As a result of the calculations, an array of temperatures is formed and graphs are built. The error in relation to the experimental data is also automatically calculated. The data can be imported into the built-in database for convenient storage and further analysis of the obtained data.

It is assumed that the investigated products have the shape of a cylinder with a thermally insulated lateral surface (see Fig. 4).

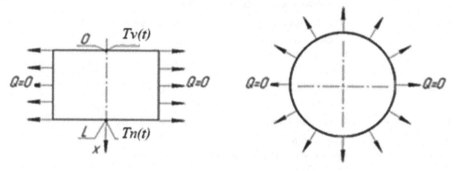

**Fig. 4.** The shape of the test articles of the elastomeric composition

Table 2 describes the information flows based on CASE technologies. The main data tuples of the information system are described.

The program implements algorithms for calculating the coefficients of thermal conductivity and heat capacity of elastomeric compositions depending on temperature based on experimental data (Fig. 5, 6). The researcher chooses which mathematical model to calculate the thermophysical parameters. In particular, an initial vulcanization temperature is set; the calculated value of the heat capacity coefficient and the a priori approximation of the parameters of the approximating thermal conductivity function depending on the temperature; product thickness; end time of the vulcanization process. The program calculates the curves of temperature fields inside the product at different levels using built-in algorithms. As a result of the calculations, an array of temperatures is formed and graphs are built. The error in relation to the experimental data is also automatically calculated. The data can be imported into the built-in database for convenient storage and further analysis of the obtained data.

**Table 2.** Information system data tuples

| Name Of The Tuple | Data Description | Tuple Content |
|---|---|---|
| Nc | Parameters for evaluating the heat capacity of the product Initial concentrations (DAV, the precursor of crosslinking, the active form of crosslinking, intramolecular-bound sulfur, sta-ble and labile nodes of the vulcanization grid, rubber and its microradial) | $<C1(0),...,C(8)>$ |
| KK | Values of pre-exponent constants of reaction rates | $<k_1,...,k_9>$ |
| EE | Activation energy values | $<E1,..,E9>$ |
| ThermoP | Parameters for evaluating the thermal conductivity of the product | $<C1,C2, \lambda1,\lambda2, Tv,Tn,Ts,L,q>$ |
| q | Heat dissipation coefficient | Q |
| SS | Mass fraction of sulfur, accelerator, activator and microradial of rubber, respectively | $<S8,Ac,Act,R>$ |
| Time | Process time simulation parameters | $<tn,tk,dt>$ |
| L | Product thickness | L |
| T | Temperatures on the upper, lower parts and in the center of the product | $<Tv,Tn,Ts>$ |
| KinP, M(t) | The value of the torque of the geometric curve | M(t), $<t,T>$, $<Nc>$, $<C0<SS>$, L |
| Const | Values of stoichiometric coefficients | $<alf,bet,gam,et,tet,dz>$ |
| Z | The specified value of the degree of vulcanization at the "cold" point | z |
| RM | Process simulation parameters | $<t,T>$, $<Nc>$, $<KK>$, $<ThermoP>$ |
| Optime | Duration and temperature of vulcanization | $<t,T>$ |

*(continued)*

**Table 2.** (*continued*)

| Name Of The Tuple | Data Description | Tuple Content |
|---|---|---|
| UV | Process simulation control parameters | \<t,T\>, \<Nc\>, \<KK\>, \<EE\>, \<Const\>, \<Time\>, \<ThermoP\>, Z |
| SD | Accumulation of data | \<ThermoP\>, \<KK\>, \<EE\>, Optime |
| Process Mp | Mechanical properties (elongation, conditional tensile strength, Shore hardness | \<E,f2,Na\> |
| X(l,t) | The degree of completion of the vulcanization process | Z |

In the simulation mode, it is possible to calculate the degree of vulcanization of multilayer products (both under isothermal and non-isothermal process conditions).

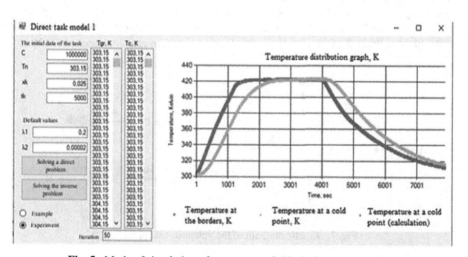

**Fig. 5.** Mode of simulation of temperature fields during vulcanization.

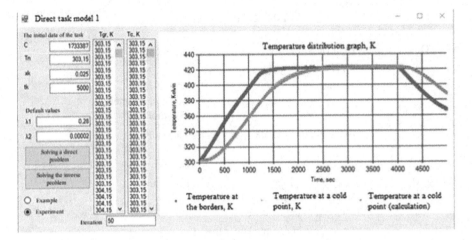

**Fig. 6.** Simulation mode. Calculation of thermophysical coefficients

The model is based on a system of differential Eqs. (4) describing the kinetic scheme (Fig. 7), where A – is the actual vulcanization agent (DAV), B–is the precursor of crosslinking, B*– is the active form of crosslinking, C – is intramolecular-bound sulfur, $Vu_{St}$, $Vu_{Lab}$–stable and labile nodes of the vulcanization network, R–rubber, R*–macro-radical of rubber formed as a result of thermal fluctuation decomposition; $\alpha$, $\beta$, $\gamma$ and $\delta$–stoichiometric coefficients, $k_1$, $k_2$,…, $k_8$, $k_9$ - are reaction rate constants [11, 12].

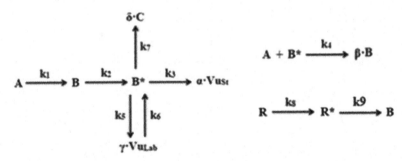

**Fig. 7.** Kinetic diagram of the vulcanization process

$$\begin{cases} \dfrac{dC_A}{dt} = -k_1 \cdot C_A - k_4 \cdot C_A \cdot C_{B^*} \\[2mm] \dfrac{dC_B}{dt} = k_1 \cdot C_A - k_2 \cdot C_B + \beta \cdot k_4 \cdot C_A \cdot C_{B^*} + k_9 \cdot C_{R^*} \\[2mm] \dfrac{dC_{B^*}}{dt} = k_2 \cdot C_B - (k_3 + k_5 + k_7) \cdot C_{B^*} + k_6 \cdot C_{Vu_{Lab}} - k_4 \cdot C_A \cdot C_{B^*} \\[2mm] \dfrac{dC_{Vu_{St}}}{dt} = \alpha \cdot k_3 \cdot C_{B^*} \\[2mm] \dfrac{dC_{Vu_{Lab}}}{dt} = \gamma \cdot k_5 \cdot C_{B^*} - k_6 \cdot C_{Vu_{Lab}} \\[2mm] \dfrac{dC_C}{dt} = \delta \cdot k_7 \cdot C_{B^*} \\[2mm] \dfrac{dC_{R^*}}{dt} = k_8 \cdot C_R - k_9 \cdot C_{R^*} \\[2mm] \dfrac{dC_R}{dt} = -k_8 \cdot C_R \end{cases} \tag{4}$$

The initial concentration of the actual vulcanization agent is determined by the expression (5):

$$[DAV] = \xi \cdot [S_8] \cdot [A_c] \cdot [Akt]^{\theta} \cdot [R]^{\eta} \tag{5}$$

where $\xi$, $\theta$, $\eta$, - some coefficients, $[S_8]$– is the initial concentration of sulfur, $[A_c]$ – is the initial concentration of the accelerator, $[Akt]$– is the initial concentration of the vulcanization activator, $[R]^{\eta}$ – is the initial concentration of the rubber macroradical.

All concentrations of reagents are in units of [mol/kg]. The dimensions of the reaction rate constants are presented in Table 3.

**Table 3.** Dimensions of reaction rate constants

| Constant | Dimension | Constant | Dimension |
|---|---|---|---|
| $k_1$ | $[s^{-1}]$ | $k_5$ | $[s^{-1}]$ |
| $k_2$ | $[s^{-1}]$ | $k_6$ | $[s^{-1}]$ |
| $k_3$ | $[s^{-1}]$ | $k_7$ | $[s^{-1}]$ |
| $k_4$ | $[kg \cdot s^{-1}/mol]$ | $k_8$ | $[s^{-1}]$ |
| | | $k_9$ | $[s^{-1}]$ |

The solution to the problem of identifying kinetic parameters is carried out by minimizing the functional (6):

$$\Phi(k_1, k_2, \ldots, k_8, k_9, \alpha, \beta, \gamma, \xi, \theta, \eta) = \int_0^{t_k} q^2(k_1, k_2, \ldots, k_8, k_9, \alpha, \beta, \gamma, \xi, \theta, \eta, t) dt \tag{6}$$

$$q(k_1, k_2, \ldots, k_8, k_9, \alpha, \beta, \gamma, \xi, \theta, \eta, t) = \frac{R(M(t) - M_{min})}{M_{max} - M_{min}} - C_{Vust} \qquad (7)$$

where $M_{max}$, $M_{min}$ – are the maximum and minimum values of the torque $M(t)$, respectively, $R$ is the scale factor.

The control action calculation mode implements the numerical simulation of the vulcanization process based on the solution of two interrelated problems: the problem of determining the temperature fields in the product and calculating the degree of completion of the process. The algorithms are implemented in such a way that a sequential calculation of the temperature fields in each section of the sample is carried out and, based on the data on them, the degree of vulcanization $X(l,t)$ is calculated. The expression for the heat release density contains the value $X(l,t)$, which characterizes the degree of completion of the vulcanization process. The degree of completion of the process is defined as the ratio of the concentration of vulcanization units to its maximum value. As a result, the module for calculating control actions makes it possible to select the temperature-time mode of the process, taking into account the recipe for the elastomeric composition and the conditions for its implementation.

The system implements a database of experiments and calculation results for various modes (Fig. 8). As an example, a database window for accumulating experimental data of rheometric curves is shown (Fig. 9).

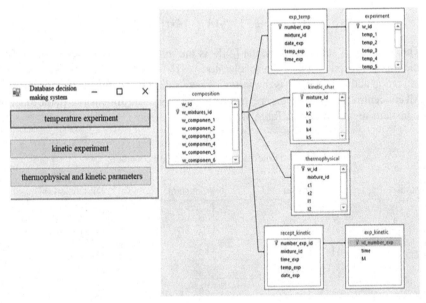

**Fig. 8.** Module for the accumulation of statistical data. The accumulation mode selection window. Scheme of interaction of database tables.

Designations: *composition* – mass fractions of the ingredients of the mixture composition; *exp_temp* – conditions and date of the experiment on the laboratory installation; *experiment* - data from thermocouples; *recept_kinetic* - conditions and date of the experiment on the rheometer; *exp_kinetic* - readings from the rheometer; *thermorhysycal* – thermophysical parameters; *kinetic_char* – kinetic characteristics.

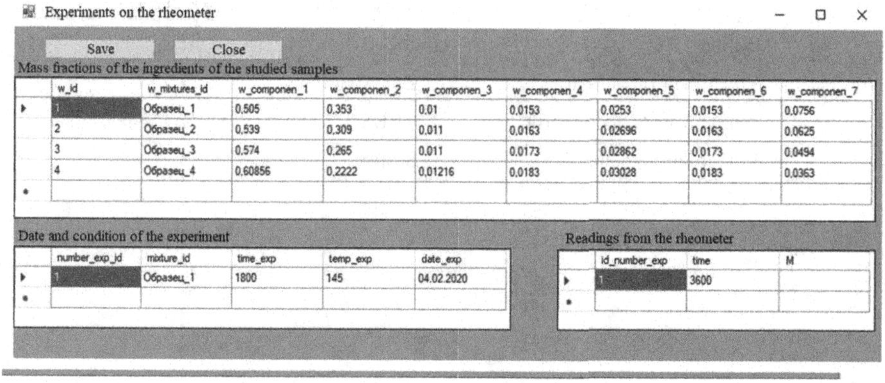

**Fig. 9.** DB window for accumulating experimental data of rheometric curves.

The software package can be used in the production of car tires at the stages of laboratory research and in production conditions for a significant increase in production volumes of products by selecting the optimal temperature and time modes of the technological process.

# 3   Experiment and Processing of Results

In order to demonstrate the operation of the software for studying vulcanization processes, an algorithm for identifying kinetic constants based on experimental data is presented. In general, the sequence of research actions can be represented as follows:

- preparation of a recipe for an elastomeric composition, production of a rubber compound and registration of the kinetic characteristics of a control sample using a rheometer;
- assessment of initial concentrations;
- processing of the rheometric curve;
- determination of the concentration of cross-links at the current time;
- identification of constants using software;
- analysis of the results.

Rheograms describe the vulcanization curve, which is measured in Nm (Fig. 10). Digitization can be done manually or using special programs (for example, Gr2Digit). Table 4 shows the basic recipe for an elastomeric composition of a control sample used in the manufacture of a car tire carcass.

The control sample was examined at a temperature of 443.15 K for 10 min. Figure 10 shows 3 plots of experimental rheometric curves. Due to the peculiarities of the study

**Table 4.** Basic recipe for the elastomeric composition of the control sample «Tire carcass».

| № | Purpose of components | «Tire carcass» | |
|---|---|---|---|
| | | Description | Weight, kg |
| 1 | Rubber | SKI-3 | 32.3 |
| 2 | Vulcanization activator | Zinc oxide | 7.75 |
| 3 | Vulcanization accelerator | TBBS | 1.15 |
| 4 | Vulcanizing agent | Sulfur | 3.2 |

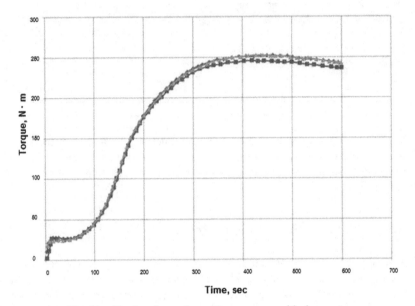

**Fig. 10.** Graphs of experimental cure curves with rheometer

on the rheometer, the control samples were processed several times in order to assess the uniformity of the process and eliminate measurement errors.

For the data obtained, mathematical processing and data normalization procedure were carried out. Initially, the torque was converted into conventional units:

$$M_{con} = \frac{M_{cur} - M_{min}}{M_{max} - M_{min}} \tag{8}$$

Then the conventional units are converted to the concentration of cross-links (mol/kg) according to the Formula (9):

$$C_{Vu_{st_i}} = R \cdot M_{ycn_i} \; ; \; i=0,\ldots,n \tag{9}$$

where $R = 5 \times 10^{-19}$, n – the number of points in the data array over time.

According to formula (5), to estimate the concentration of the actual vulcanization agent, it is necessary to know the initial concentrations of sulfur, accelerator, vulcanization activator, and rubber macroradicals. After converting them to molar concentrations, the following values were obtained: $[S_8] = 0.051$[mol/kg], $[A_c] = 0.086$ [mol/kg], $[Akt] = 2.415$[mol/kg],$[R] = 0.004$[mol/kg]. The calculation formula of the actual vulcanization agent contains the coefficients $\xi$, $\theta$, $\eta$, to be use.

The digitized and processed experimental data are entered into the program, the initial conditions, approximations, and the search range for constants are selected. The identification results for the control sample are presented in Table 5 and in Fig. 11–12.

**Table 5.** Evaluation of model constants for the control example "Tire carcass" at a vulcanization temperature of 443.15 $^0$K.

| Parameter | Meaning | | Parameter | Meaning | |
|---|---|---|---|---|---|
| | Antecedent | Calculated | | Antecedent | Calculated |
| $k_1$, [s$^{-1}$] | 0.004 | 0.016 | $\alpha$ | 1 | 1 |
| $k_2$, [s$^{-1}$] | 0.008 | 0.017 | $\beta$ | 2 | 2 |
| $k_3$, [s$^{-1}$] | 1.255e$^{-21}$ | 4.704e$^{-21}$ | $\gamma$ | 1 | 1 |
| $k_4$, [kg·s$^{-1}$/mol] | 0.008 | 0.009 | $\xi$ | 1.65 | 1.65 |
| $k_5$, [s$^{-1}$] | 4.166e$^{-22}$ | 1.102e$^{-22}$ | $\theta$ | 1.5 | 1.5 |
| $k_6$, [s$^{-1}$] | 1.666e$^{-22}$ | 1.667e$^{-22}$ | $\eta$ | 0 | 0 |
| $k_7$, [s$^{-1}$] | 0.005 | 0.017 | | | |
| $k_8$, [s$^{-1}$] | 0.021 | 0.036 | | | |
| $k_9$, [s$^{-1}$] | 0.004 | 0.026 | | | |

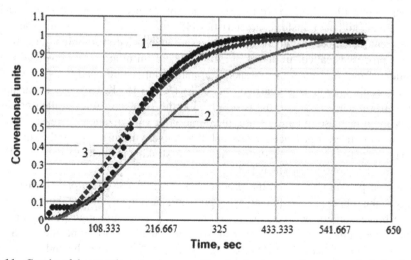

**Fig. 11.** Graphs of rheometric curves in conventional units: 1 - experiment; 2 - calculation from a priori approximations; 3 - is a graph of calculated values after identification of model parameters

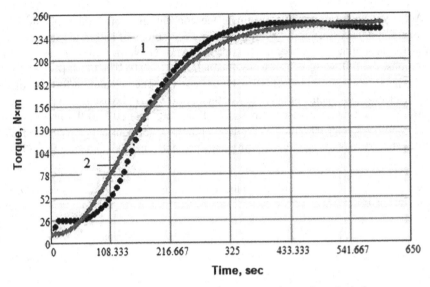

**Fig. 12.** Graphs of rheometric curves: 1 - experiment; 2 - calculation.

As a result of identifying the parameters of the model and estimating the error, it can be concluded that the pro-posed model adequately describes the kinetics of isothermal vulcanization. The average relative error for the control example was 4.29%.

Analyzing modern approaches to modeling vulcanization processes, as well as the production of products from polymer compositions, it can be concluded that the main factors affecting the formation of a spatial vulcanization structure are the conditions for carrying out spatial vulcanization (heat and energy carriers with various physical inclusions: pressure, temperature; duration of the process) and the composition of the polymer composition. In terms of thermal conditions, vulcanization occurs at variable temperatures and heat transfer characteristics [15, 16]. The most commonly used rubbers (styrene-butadiene, butadiene and natural) are vulcanized in combination with sulfur. This process is called sulfur vulcanization, where sulfur atoms are involved in the formation of intermolecular cross-links. The main temperature conditions for vulcanization of rubber products are carried out at 413.15–443.15° K. During vulcanization, a non-uniform temperature profile is observed in the volume of thick-walled products [17]. With an increase in temperature, the duration of vulcanization decreases, but for large-sized multi-layer products, the probability of over vulcanization on the surface of the final products and an uneven distribution of crosslinking in thickness increase. Thus, it degrades the quality of products [18].

External pressure acting on the product has a significant effect on increasing the solidity of the rubber [17]. With increasing pressure, the porosity of the rubber decreases and the bond strength between the layers increases. To ensure high performance of production equipment, vulcanization must be carried out in a shorter time interval. Since, due to the absorbed heat, the vulcanization process continues after the end of heating, when determining the process parameters, this circumstance must be taken into account. To ensure an increase in the homogeneity of the temperature field, the heating process can be carried out with stepwise heating or preheating the initial rubber mixture. Therefore, it is advisable not to achieve complete vulcanization of the product [18]. The time and temperature conditions for vulcanization depend on the nature of the rubber and the initial composition of the polymer composition. The vulcanization time during which the required physical and mechanical properties of the vulcanizates are achieved is called the cure optimum. The time interval during which the required or close physical and mechanical parameters are maintained is called the vulcanization plateau. When choosing the temperature-time parameters of the process, there is mainly an empirical approach that does not allow to obtain reliably reproducible results and to design modes for the manufacture of products with optimal properties at a sufficiently high scientific and technical level [19]. In industrial conditions, the vulcanization mode is selected on the basis of calculated and experimental methods that provide the choice of the program for the implementation of the process. At the same time, in order to fulfill the selected modes, it is necessary to equip the process with automation and control means, which ensure the implementation of the selected procedure.

The result of an analytical review of various approaches to describing the structuring of elastomers showed that today there is no unified approach to describing the process of non-isothermal vulcanization and a general methodology for assessing the vulcanization characteristics, properties of vulcanizates, which will allow the selection of the optimal temperature-time mode of vulcanization. [14, 20]. At present, the study and calculation of vulcanization processes is based mainly on the results of a full-scale experiment, empirical and graphic-analytical methods of calculations, while their generalized analysis is absent. The experience of research in this field of science shows the need to use calculation methods for determining the optimal vulcanization mode, as well as mathematical models and algorithms underlying them.

# 4 Conclusion

Based on the analysis of theoretical approaches to the description of vulcanization, the general kinetic scheme of the vulcanization process has been improved. Additional reactions are proposed that describe the for-mation and destruction of labile polysulfide bonds, intramolecular cyclization, and other reactions leading to the modification of the sulfur macromolecule, the formation of a macroradical and its reaction with DAB pendants. The mathematical model of structuring is supplemented with initial conditions, which are defined as functions of the initial concentrations of the components of the vulcanizing group.

A software product has been developed for carrying out research work in the study of vulcanization of rubber compounds using multicomponent structuring systems. The complex of programs has a block-modular structure, which allows its expansion without loss of functionality. The directions of its modernization are the inclusion in the mathematical description of the non-isothermal mode of vulcanization with further integration into the circuit of the process control system as an expert information and control system for issuing recommendations for managing the vulcanization process and making decisions.

With the use of the developed software product, the parameters of the isothermal vulcanization process were evaluated on a test example of the Tire Carcass recipe. The maximum deviation from the experimental data does not exceed 5%, which indicates the adequacy of the model.

# References

1. Ilyasov, R.S.: Tires. Some problems of operation and production. In: Ilyasov, R.S., Dorozhkin, V.P., Vlasov, Ya.G., et al. (eds.) Kazan: KSTU, p. 576 (2000)
2. Grishaeva, N.Y.: Thermal properties simulation of multilayer pipe. In: AIP Conference Proceedings 1623, 187 (2014); International Conference on Physical Mesomechanics of Multilevel Systems, 3–5 September 2014 - Tomsk, Russia, pp. 187–190 (2014)
3. Ghoreishy, M.H.R.: Numerical simulation of the curing process of rubber articles. In: OW, U., (ed.) Computational Materials. Nova Science Publishers, New York, pp. 445–478 (2009)
4. Vafayan, M., Ghoreishy, M.H.R., Abedini, H., Beheshty, M.H.: Development of an optimized thermal cure cycle for a complexshape composite part using a coupled finite element/genetic algorithm technique. Iran Polym. J. **24**, 459–469 (2015)
5. Khang, T., Ariff, Z.: Vulcanization kinetics study of natural rubber compounds having different formulation variables. J. Therm. Anal. Calorim. **109**, 1545–1553 (2012)
6. Wang, D.H., Dong, Q., Jia, Y.X.: Mathematical modeling and numerical simulation of the non-isothermal in-mold vulcanization of natural rubber. Chin. J. Polym. Sci. **33**, 395–403 (2015)
7. Gordeev, V.K.: Increasing the efficiency of the tire vulcanization process. In: Gordeev, V.K., Porotsky, V.G., Saveliev, V.V., Aldonina, T.N. (eds.) Questions of Practical Technology of Tire Production. N1.C, pp. 85–98 (2000)
8. Markelov, V.G.: Modeling the process of vulcanization of thick-walled rubber products. In: Markelov, V.G., Soloviev, M.E. (eds.) Modeling of the vulcanization process of thick-walled rubber products, Izvestiya vysshikh uchebnykh zavod. Series: Chemistry and Chemical Technology. T.50. no. 4. pp. 95–98 (2007)
9. Fischenkov, A.N.: The use of the finite element method in mathematical modeling of the tire vulcanization process. North Caucasian region. Technical science. no. 6(142), pp. 45–48 (2007)
10. Pyatakov, Yu.V.: Calculation of temperature fields of pneumatic tires during vulcanization. In: Pyatakov, Yu.V., Tikhomirov, S.G., Karmanova, O.V., Maslov, A.A. (eds.) Collection of Scientific Papers of the XXVII International Symposium. Problems of tires, rubber goods and elastomeric composites, pp. 405–413 (2016)
11. Tikhomirov, S.G.: A technique of calculating the kinetics of the process of nonisothermal vulcanization of large articles. In: Tikhomirov, S.G., Pyatakov, Yu.V., Karmanova, O.V., Maslov, A.A., Khaustov, I.A., Podval'nyi, S.L. (eds.) Chemical and Petroleum Engineering, vol. 53, nos. 9–10, January 2018 (Russian Original Nos. 9–10, September–October 2017)
12. Tikhomirov, S.G.: Determination of thermophysical characteristics of vulcanizable rubber products by the mathematical modeling method. In: Tikhomirov, S.G., Pyatakov, Y.V., Karmanova, O.V., Maslov, A.A. (eds.) IOP Conference Series: Journal of Physics: Conference Series 973 (2018)
13. Tikhomirov, S.G.: Software for the problem of determining the optimal time for vulcanization of rubber mixtures. In: Tikhomirov, S.G., Karmanova, O.V., Bityukov, V.K., Maslov, A.A. (eds.) Proceedings of the VSUET. no. 4, pp. 108–116 (2018)
14. Milani, G., Milani, F.: Optimization of extrusion production lines for EPDM rubber vulcanized with sulphur: a two-phase model based on finite elements and kinetic second order differential equation. Comput. Chem. Eng. **43**, 173–190 (2012)

15. Pavlov, V.V.: Development of a methodology for optimizing thermal modes of vulcanization of rubber coatings. In: Yu, R., Osipov, S., Osipov, Y. (eds.) Bulletin of the Cherepovets State University 2008, no. 3 (2008)

16. The current state of the world production of non-molded and molded rubber goods. Analytical overview information. M: NIIEMI, p. 233 (2005)

17. Lukomskaya, A.I., et al.: Rubber and Rubber. no. 6, pp. 29–31 (1972)

18. Sheludyak, Yu.: Thermophysical properties of components of combustible systems. In: Sheludyak, Yu.E., Kashporov, L.Ya., Malinin, L.A., Tsalkov, V.N. (eds.) M: NPO "Inform TEI", p. 184 (1992)

19. Dontsov, A.A.: Vulcanization of Elastomers/Translated from English. In: Dontsova, A.A., Alliger, G. (eds.) Moscow, Chemistry, p. 428 (1967)

20. Maslov, A.A.: Methodology for calculating the optimal time for vulcanization of rubber mixtures. In: Maslov, O.V., Karmanova, S.G., Tikhomirov, I.A., Khaustov (eds.) Problems of Tires, Rubber Goods and Elastomeric Composites Collection of Scientific Papers of the XXVIII International Symposium, pp. 258–264 (2018)

# An Intelligent Approach to Decentralized Control in the Agro-industrial Complex

Felix Ereshko$^{(\boxtimes)}$ ⓘ, Michail Gorelov ⓘ, and Vladimir Budzko ⓘ

Federal Research Center "Computer Science and Control" of the Russian Academy of Sciences, Vavilova 44-2, 119333 Moscow, Russian Federation

**Abstract.** The article considers the theoretical and applied aspects of the problem of decision-making in a system with a hierarchical structure under uncertainty inherent in the agro-industrial complex, due to the presence of changeable weather conditions and the freedom to make economic decisions in market conditions. It is noted that to date, market mechanisms have been implemented in the agro-industrial complex, and the role of decision makers is minimal, and this determines the feasibility of applying a decentralized scheme of the control model, which is discussed in detail in the domestic theory of hierarchical games. It is taken into account that enterprises of different forms of ownership can act in different roles, unequal relations can arise: someone shows dictates and gets the right of the first move. In these productions, along with enterprises of different forms of ownership, the participation of the coordinator of the Center, which regulates public relations arising between state authorities and subjects of the industrial sector of the economy, should be considered. Economic stimulus and regulatory measures (mandatory regulations and prohibitions) can be used as instruments of government influence on productive entities. To assess the activities of the participants, the principle of the maximal guaranteed result is used. A general scheme for obtaining correlations for the Center criterion in cases of centralization and decentralization, their comparison and a detailed conclusion for the linear case under the selected organizational conditions are presented.

**Keywords:** Agro-industrial complex · Awareness · Centralization · Decentralization · Uncertain factors · The greatest guaranteed result · The right to move · Strategies

## 1 Introduction

When developing theoretical approaches to control in the agro-industrial complex (AIC), it is very important to take into account factors that are difficult to predict. This is primarily characteristic of natural conditions, which must be taken into account when developing models for the flow of production processes. The second factor that generates uncertainty in centralized planning is freedom of choice when making decisions at the lower level of management.

These problems are investigated in this work.

© Springer Nature Switzerland AG 2022
V. Taratukhin et al. (Eds.): ICID 2021, CCIS 1539, pp. 254–264, 2022.
https://doi.org/10.1007/978-3-030-95494-9_21

Note that to date, the agro-industrial complex has implemented market mechanisms, and the role of directive decisions is minimal, which determines the feasibility of using a decentralized control model scheme, which is considered in detail in the domestic theory of hierarchical games [1–4].

It should also be borne in mind that enterprises of different forms of ownership can play different roles in mutual relations, just as unequal relations can arise in the relationship between the manufacturer and the consumer in the market: someone manifests a diktat and gets the right of the first move (Cournot, Stackelberg and Germeier models [1, 5, 6]). Finally, state-owned, joint-stock and private enterprises in some cases must follow the laws and regulations set by the state, and state-owned enterprises and private enterprises may be subject to different legal provisions, for example, different taxation schemes. In these performances, along with enterprises of various forms of ownership, it is necessary to consider the participation of the coordinator (we will use the term Center) with a description of his activities. The Center regulates public relations arising between public authorities and subjects of the industrial sector of the economy in the implementation of various instruments of state influence on the activities of companies. Measures of economic incentives and measures of state regulation (mandatory prescriptions and prohibitions) can be used as instruments of state influence on industrial activity subjects.

## 2 Decentralization of Control Mechanisms

Problems of awareness and decentralization occupy an important place in decision theory and have attracted the attention of thinkers of all eras (see, for example, https://en.wikipedia.org/wiki/Decentralization). In practice, management of rather complex organizational systems is carried out according to a hierarchical principle. Hence, we can conclude that decentralized management is more efficient than centralized. An explanation of this fact was proposed in the early 70s of the last century by Yu.B. Germeier and N.N. Moiseev: if a decision-maker delegates a part of his decision-making authority to some agents, then by joint efforts it will be possible to process large amounts of information in a timely manner and thereby manage more efficiently. These ideas formed the basis of the so-called informational theory of hierarchical systems [7–9].

For a long time, it was not possible to construct formal mathematical models to describe this effect. Various approaches to solving this problem are described in [10, 11], where the tasks of control organizational systems under external uncertainty are considered and the question of the feasibility of decentralizing control depending on the available amount of information about uncertain factors is investigated.

Of course, one would like to find the optimal organizational structure of the system that ensures the management of this or that agricultural complex. But even the formulation of such a task currently seems impossible. The first problem is already that there is no definition of "organizational structure" that could be written in the language of mathematics. The only way to solve this problem, which seems real today, is to compare the "quality" of management that control systems with a pre-fixed structure can provide. This is what is being done in this paper.

Below is a comparative analysis of the effectiveness of centralized and decentralized control methods.

## 3   Controlled System

Let's consider the simplest model of decision making under uncertainty. The system under consideration is controlled by choosing a control $w$ from a set $W$, and the control result is influenced by the value of an uncertain factor $\alpha$, the choice of which is not controlled by the decision-maker. The parameter $\alpha$ can take any value from the set $A$.

The purpose of the control is to maximize the value $g(w, \alpha)$ of the function $g$ : $W \times A \to \mathbf{R}$ (as usual, $\mathbf{R}$ is a set of real numbers).

The function $g$ will be considered continuous on the Cartesian product $W \times A$.

Let us further assume that the considered controlled system is "technologically structured": the set $W$ can be represented as a Cartesian product $W = U \times V^1 \times V^2 \times \ldots \times V^n$.

Thus, the control $w \in W$ can be written in the form $w = (u, v^1, v^2, \ldots, v^n)$, where $u \in U$, $v^i \in V^i$, $i = 1, 2, \ldots, n$.

We will assume that the choice of control $w \in W$ is made by one decision-maker (Center).

At the same time, the Center can receive reliable information about the realized value of the uncertain factor $\alpha$, but the amount of information that it is able to receive and process in a timely manner is limited.

## 4   The Amount of Information

One of the central concepts in the informational theory of hierarchical systems is the concept of information. Unfortunately, a sufficiently strict and universal definition of the concept of information does not yet exist.

Strictly speaking, the informational aspects of decision-making were already studied in the first works on game theory [12, 13]. But in this works, all information flows were set by the "rules of the game". For our purposes, we need a more detailed description, since we are interested in the limitations of the amount of information, and its content is determined by one of the players. And as a consequence, it is natural to systematically consider families of models that differ only in the amount of information processed. Until recently, no such studies have been conducted.

However, in the context of the issues discussed, the very concept of "information" is not so significant. One other concept is more important. When it comes to the time for "processing" information, the price of such "processing" requires a quantitative measure of the amount of information. There are problems here, but also, not everything is hopeless.

One of the works of A.N. Kolmogorov [14] is called, "Three approaches to the definition of the concept 'quantity of information'" (Kolmogorov 1965). The article is quite old, but has not lost its relevance. It is clear that if there are three approaches, then none of them is entirely satisfactory. Besides, these general approaches need to be somehow adapted to specific decision-making tasks.

Thus, there is a certain arbitrariness in the formalization of the concept of "amount of information". In this paper, the approach first proposed in [15] is chosen. According to Kolmogorov's classification, it should be called combinatorial.

Suppose the center is capable of processing a $l$ bits of information.

The choice of the content of this information is the right of the Center.

Hereinafter, $\Phi(X, Y)$ will denote the family of all functions that map a set $X$ to a set $Y$.

Since $l$ bits of information are available to the Center, this information can be encoded in words $s = (s_1, s_2, ..., s_l)$, where each symbol $s_i$, $i = 1, 2, ..., l$ belongs to the set $\{0, 1\}$. Thus, all messages about the value of the undefined factor that the Center can receive belong to the set $S = \{0, 1\}^l$ (Cartesian power of the set $\{0,1\}$).

## 5  Centralized System Control Model

We will assume that for each value $\alpha \in A$ the center has the right to choose a message $P(\alpha) \in S$ that corresponds to the implementation of the parameter value $\alpha$.

Thus, the Center actually chooses a function $P \in \Phi(A, S)$.

In addition, the Center has the right to choose its own control $w \in W$ for each message $s \in S$. That is, in essence, the center chooses a function $w_* \in \Phi(S, W)$.

Center's strategies are pairs $(w_*, P)$ from the set $\Phi(S, W) \times \Phi(A, S)$. If the center chooses such a strategy $(w_*, P)$ and the value $\alpha$ of the uncertain factor is realized, then the Center's payoff will be $g(w_*(P(\alpha)), \alpha)$.

If the Center does not know the value of the uncertain factor $\alpha$ in advance, then the choice of the strategy $(w_*, P)$ guarantees him a payoff equal to

$$\inf_{\alpha \in A} g(w_*(P(\alpha)), \alpha),$$

and its maximal guaranteed result will be

$$R_0 = \sup_{(w_*, P) \in \Phi(S, W) \times \Phi(A, S)} \inf_{\alpha \in A} g(w_*(P(\alpha)), \alpha).$$

Let us denote $m = 2^l$. A word $s = (s_1, s_2, ..., s_l) \in S$ can be naturally identified with a natural number from a set $\{0, 1, ..., m - 1\}$ that has a binary representation $s_1 s_2, ..., s_l$. In what follows, we will make such identification without special reservations.

Let us denote $w_s = w_*(s)$.

Then the maximal guaranteed result of the Center is equal to

$$R_0 = \max_{(w_0, w_1 ..., w_{m-1}) \in W^m} \min_{\alpha \in A} \max_{s=0, 1, ..., m-1} g(w_s, \alpha).$$

## 6  Decentralized System Control Model

Consider a different control scheme for the same system. Suppose that the Center entrusts the choice of control $v^i \in V^i$ to some agent $i$ ($i = 1, 2, ..., n$).

We will assume that the interests of agent $i$ are described by the desire to maximize the function $h^i(u, v^i, \alpha)$ (i.e., do not depend on the choices of other agents).

The Center reserves the right to choose the control $u \in U$, and it can count on receiving $l$ bits of information about an indefinite factor before the final choice of his control as in the previous model, and has the right to choose the content of this information.

Thus, the strategies of the Center will be pairs $(u_*, P)$ of functions $u_* \in \Phi(S, U)$ and $P \in \Phi(A, S)$. In this case, the payoffs of the Center and agents will be determined by the expressions $g(u_*(P(\alpha)), v^1, v^2, ..., v^n, \alpha)$ and $h^i(u_*(P(\alpha)), v^i, \alpha)$, $(i = 1, 2, ..., n)$, respectively.

We will assume that the Center has the right of the first move, i.e. he is the first to choose and communicate his strategy $(u_*, P) \in \Phi(S, U) \times \Phi(A, S)$ to the agents.

In this case, the agent $i$ knows the value of the uncertain factor $\alpha$ and the strategy $(u_*, P)$ at the moment of making a decision, and, hence, the value $u_*(P(\alpha))$, that is, the "physical" control, which the Center will have to choose. Thus, the decision-making problem turns into an optimization problem for this agent.

Let's write down an explicit formula for the maximal guaranteed result.

For any choice of strategy $(u_*, P)$, the set of values of the function $u_*(P(\alpha))$ consists of $m$ points (some of which may coincide).

Let's fix the collection $(u_0, u_1, ..., u_{m-1}) \in U^m$.

If the Center chooses control $u_s$, and the value $\alpha$ of the uncertain factor is realized, then the agent can choose any control from the set

$$E^i(u_s, \alpha) = \left\{ v^i \in V^i : h^i(u_s, v^i, \alpha) = \max_{\omega^i \in V^i} h^i(u_s, \omega^i, \alpha) \right\},$$

therefore, the Center has the right to expect to obtain a result

$$\min_{v^1 \in E^1(u_s, \alpha)} \min_{v^2 \in E^2(u_s, \alpha)} \ldots \min_{v^n \in E^n(u_s, \alpha)} g(u_s, v^1, v^2, ..., v^n, \alpha).$$

The choice of the collection $(u_0, u_1, ..., u_{m-1}) \in U^m$ is the prerogative of the Center. Thus, we get the following result.

The maximal guaranteed result of the Center in the model under consideration is

$$R_1 = \sup_{(u_0, u_1, ..., u_{m-1}) \in U^m} \min_{\alpha \in A} \max_{s=0,1,...,m-1} \min_{v^1 \in E^1(u_s, \alpha)} \min_{v^2 \in E^2(u_s, \alpha)} \ldots \min_{v^n \in E^n(u_s, \alpha)} g(u_s, v^1, v^2, ..., v^n, \alpha).$$

Hence, the question of correlating the effects of management in different ways of control (centralization or decentralization) is resolved in comparison of the maximal guaranteed results.

**Theorem** [16]. If the interests of the agents are "poorly coordinated" with the interests of the Center, then centralized control is always more advantageous. If the interests of the Center and the agents are "well-coordinated", then for large values of the amount of information available to it, centralization of control is more advantageous, and for small values, decentralized control is preferable.

In general, these conclusions are consistent with meaningful ideas. It should be emphasized that the result is valid for both interval and stochastic uncertainty.

# 7  Validity of a Qualitative Conclusion

The model discussed above, like any model, is based on certain hypotheses. Of course, it is interesting to understand how much the conclusions obtained depend on these hypotheses. The main qualitative conclusion formulated in the previous section in the form of a theorem is proved under very general assumptions. For example, assumptions about the continuity of functions or compactness of sets can hardly cause serious objections. In general, they are quite standard. Perhaps only two things can cause certain doubts.

Firstly, in the model discussed above, the attitude of the operating party to uncertainty is fixed. Namely, the operating party is assumed to be cautious, and the principle of maximal guaranteed result is used as the main one. Of course, other assumptions are possible. Some of them have already been investigated.

In the article [17] it is assumed that the external uncertainty has a stochastic character. Accordingly, it is assumed that a set of external uncertain factors is endowed with some probability measure known to the operating party. The operating party itself is assumed to be risk-neutral with respect to this uncertainty, i.e. its goal is to maximize the expected value of its payoff. If, in the case of decentralized management, the lower-level elements may have several rational responses to the well-known strategy of the center, then the operating party is still considered cautious in relation to the uncertainty of their choice.

The paper [18] considers an "intermediate" case. External uncertainty is again considered stochastic, and the principle of "Value at Risk" [19, 20] is used as the principle of optimality of operating party, i.e. it is assumed that, in general, the operating party is cautious, but is ready to exclude from consideration a certain number of "negative" events with a given total probability. In addition, another change has been made. In the decentralized management model, the assumption is made about the benevolence of the lower-level elements to the center, i.e. it is believed that from the equivalent options for him, the lower-level element will always choose the one that is preferable for the center.

Of course, many other options are possible. But from what has already been said, it is clear that these assumptions are not the main ones for the validity of the conclusion under consideration.

And secondly, as noted above, other ways of measuring the amount of information are possible.

Within the framework of the probabilistic (in terms of Kolmogorov [14]) approach, it is possible to construct at least one meaningful model. However, its investigation encounters significant difficulties. The resulting tasks are non-standard and there are no developed methods for solving them yet. However, the situation here is not completely hopeless.

Within the framework of the algorithmic approach, problems arise already at the stage of formalization of the problem. The reason is as follows. Traditionally, game-theoretic models are built on the basis of set theory. And in order to use an algorithmic approach, it is necessary to describe all the elements of such models in constructive terms. It is not yet possible to do this so that the resulting model turns out to be simple enough for analytical research.

There remains a large field for further research.

# 8  Linear Case

To illustrate the general provisions, we present the corresponding constructions for the case that is most adequate for agricultural situations, taking into account the following:

– the fundamental presence of uncertainty caused primarily by natural factors,
– the effect of decentralization, which is caused by the prevailing market relations,
– the use of specific indicators (standards), which corresponds to common economic practice.

Let there be active agents $n$. We will denote them by numbers from 1 to $n$.

Each agent can produce $m$ types of products, while spending some resources. The number of resources will be denoted by a letter $k$. For the production of a unit of a product of a type $j$, an agent $i$ spends a resource of a type $l$ in quantity $p^i_{lj}$. The agent $i$ has its own stock of a resource of the type $l$ in the amount $b^i_l$ ($l = 1, 2, ..., k$).

In addition, there are general stocks of resources in quantity $r_l$ ($l = 1, 2, ..., k$).

Products of the type $j$ can be sold on the market at a price $c_j$.

Thus, if an agent $i$ produces products of a kind $j$ in quantity $x^i_j$ ($j = 1, 2, ..., m$), then he will spend a resource of a kind $l$ in quantity $p^i_{l1}x^i_1 + p^i_{l2}x^i_2 + ... + p^i_{lm}x^i_m$, and he will be able to sell these products for an amount $c_1x^i_1 + c_2x^i_2 + ... + c_mx^i_m$.

Prices $c_j$ are positive in their meaning. We assume that the cost coefficients are non-negative $p^i_{lj}$, and for any $i$ and every $j$ at least one coefficient $p^i_{1j}, p^i_{2j}, ..., p^i_{kj}$ is positive. Stocks $b^i_l$ and $r_l$ are assumed to be non-negative.

Further, it will be convenient to use the following matrix notation. Will denote a $x^i$ column vector $(x^i_1, x^i_2, ..., x^i_m)^T$ through $x^i$ (superscript $T$, as usual, denotes transposition). Let $c = (c_1, c_2, \ldots, c_m)$, and

$$P^i = \begin{pmatrix} p^i_{11} & p^i_{12} & \cdots & p^i_{1m} \\ p^i_{21} & p^i_{22} & \cdots & p^i_{2m} \\ \cdots & \cdots & \cdots & \cdots \\ p^i_{k1} & p^i_{k2} & \cdots & p^i_{km} \end{pmatrix}.$$

Let denote $b^i = (b^i_1, b^i_2, ..., b^i_k)^T$, $r = (r_1, r_2, ..., r_k)^T$.

In this notation, the previous formulas will look like this. If the agent $i$ releases products, in quantity $x^i$, then he will be able to bail out the amount $cx^i$ for it, and at the same time resources in the quantity $y^i = P^i x^i$ will be spent, where $y^i$ is the vector column $(y^i_1, y^i_2, ..., y^i_k)^T$.

Let $Y$ be the set of all sets of vectors $y^1, y^2, ..., y^n$ satisfying the conditions

$$y^1 + y^2 + ... + y^n \le r, \ y^i \ge 0, \ i = 1, 2, ..., n$$

(hereinafter, vector inequalities must be satisfied componentwise).

In this setting, we assume that the Center does not know exactly the technological matrices $P^i$ and the agents' own reserves of resources $b^i$. It is only known that they belong to parametric families $P^i(\alpha)$ and $b^i(\alpha)$, where $\alpha$ they belong to some set $A$. For simplicity, in this model, we will assume that the stocks $r_1, r_2, ..., r_k$ are strictly positive.

Agents, on the other hand, know for sure their own technologies and capabilities. We will assume that the Center disposes of the division of "general" resources, but in this case it allocates resources for specific production programs. Agents have the right to choose these programs on their own. Of course, the center is not obliged to ensure the feasibility of any program proposed by the agent. On the contrary, the agent is forced to choose his program based on the resources allocated to him. In addition, agents can transmit information to the Center about the realized value of the undefined factor. These reports are not necessarily reliable and the Center is aware of this.

These considerations are formalized as follows.

The center chooses a collection $y_* = (y_*^1, y_*^2, ..., y_*^n)$ of functions $y_*^i : \mathbf{R}_+^m \times A \to \mathbf{R}_+^k$ such that

$$y_*^1(x^1, \beta^1) + y_*^2(x^2, \beta^2) + ... + y_*^n(x^n, \beta^n) \le r$$

for any plans $x^1, x^2, ..., x^n$ from $\mathbf{R}_+^m$ and any messages $\beta^1, \beta^2, ..., \beta^n$ from $A$. The set of all such sets will be denoted by $Y_*$.

After that, the $i$-th agent ($i = 1, 2, ..., n$) selects a vector $x^i$ and a message $\beta^i$ from the set

$$X^i(y_*^i, \alpha) = \left\{ \left(z^i, \beta^i\right) \in \mathbf{R}_+^m \times A : P^i(\alpha)z^i \le b^i(\alpha) + y_*^i(z^i, P^i(\alpha)z^i) \right\}.$$

After these choices have been made, the Center receives a payoff

$$g(x^1, x^2, ..., x^n) = cx^1 + cx^2 + ... + cx^n,$$

and the $i$-th agent receives a payoff $h^i(x^i) = c^i x^i$.

Thus, we get a game with forbidden situations $\langle N, Y_*, X^1, X^2, ..., X^n, g, h^1, h^2, ..., h^n, A \rangle$, where $N = \{C, 1, 2, ..., n\}$ is the set of players (the symbol $C$ is reserved for the Center, and the agents, as before, are numbered from 1 to $n$). Let's use the notation $\Gamma_*$ for this game.

To close the model, it is necessary to describe the relationship of the Center to uncertainty. As before, we will assume that the Center has the right of the first move; it considers all agents to be rational and is careful with respect to the remaining uncertainty.

We formalize what has been said as follows.

Let $H^i(y_*^i, \alpha)$ be the least upper bound for the values of the function $h^i(x^i) = c^i x^i$ on the set $(x^i, \beta^i)$ of pairs satisfying the conditions

$$P^i(\alpha)x^i \le b^i(\alpha) + y_*^i\left(x^i, \beta^i\right), \ x^i \ge 0.$$

If this upper bound is attained, then we define the set $BR^i(y_*^i, \alpha)$ as the set of pairs $(x^i, \beta^i)$ satisfying the conditions

$$c^i x^i = H^i(y_*^i, \alpha), \ P^i(\alpha)x^i \le b^i(\alpha) + y_*^i(x^i, \beta^i), \ x^i \ge 0.$$

Otherwise, we define the set $BR^i(y_*^i, \alpha)$ by the conditions

$$c^i x^i \ge H^i(y_*^i, \alpha) - \kappa, \ P^i(\alpha)x^i \le b^i(\alpha) + y_*^i(x^i, \beta^i), \ x^i \ge 0,$$

where $\kappa$ is a given positive number.

Then the maximal guaranteed result of the Center $R(\Gamma_*)$ is equal to

$$\sup_{(y_*^1,y_*^2,...,y_*^n)\in Y_*} \min_{\alpha\in A} \min_{((x^1,\beta^1),(x^2,\beta^2),...,(x^n,\beta^n))\in BR^1(y_*^1,\alpha)\times BR^2(y_*^2,\alpha)\times...\times BR^n(y_*^n,\alpha)} \left(cx^1 + cx^2 + ... + cx^n\right).$$

The resulting expression reflects the fact that the Center, choosing its program of action, focuses on some guaranteed payoff, the uncertainty of which in the initial expression is determined by the multiple choice of many participants and the uncertain external situation.

Direct calculation of the quantity $R(\Gamma_*)$ is quite difficult, but by now a set of effective tools for solving problems of this class has been developed, either by reducing them to optimization problems or by using the scenario approach and simulation experiments.

# 9 Conclusion

The paper describes a general approach to decision-making by the Center under the conditions of the receipt of data, the values of which are undefined in advance. The situation is absolutely typical for the agro-industrial complex, when the grain yield can differ from year to year by 30%. In addition, in market conditions, the lower-level systems (individual farms and holdings) have the freedom to choose their decisions, and the Center, having, for example, the goal of providing the population with food according to medical standards, is faced with the need to take into account the shortage of products, due to the disadvantage of its production by enterprises on the market prices.

The applicability of the general approach is illustrated by the example of linear models, when generally accepted standards and specific indicators are used for estimated economic calculations.

The concrete result of this work is the formal relationship for assessing the guaranteed results of the Center, which performs coordinating functions.

The general scheme of rational reasoning is presented for the conclusion of the necessary recommendations and methods.

General control theory currently has powerful formal tools: model tools, mathematical apparatus, computational tools, its own methodology, which include decision theory, operations research, control of dynamic systems, game theory, and systems analysis.

However, this methodology in the context of digitalization requires significant adaptation efforts in practical applications.

One of the most important areas of further research for the final practical application of the results obtained should be the development of an adequate formalized understanding of the essence of governance and power in a digital format.

As we mentioned, and again in conclusion note, with the emergence of the concept of "amount of information", a new definition of "optimal" information exchange can be given. If the amount of processed information is not limited, then there is a simple optimal solution - you need to process all available information. If it becomes possible to estimate the amount of information received by the player, then the costs of obtaining and processing the information can be associated with it. And it is natural to take these costs into account in the player's objective function. And then the answer to the question about the optimal way to exchange information becomes more meaningful. This kind of model can be constructed and investigated.

Until recently, the main focus was on models in which the top-level player receives only reliable information. As we found out, this is no coincidence. It can be shown that in games without external uncertainty, the appearance of the possibility of obtaining unreliable information does not increase the maximum guaranteed result of the top-level player.

The situation when the reliability of the information received by the player of the upper level is guaranteed corresponds to the case when he himself "extracts" information about the actions of the partner. In practice, it is much more common for lower-level players to submit reports on their activities upstairs, which may contain incorrect information.

First, information can be distorted during transmission (undetected integrity violation). And if only "part" of the transmitted information is distorted, but the player who receives it does not know which part turned out to be distorted, then when using models without restrictions on the amount of information transmitted, it is not possible to obtain a meaningful formal description of such a situation.

In the case of games with limited volumes of transmitted information, it is possible to implement both interval and stochastic versions of the model. In both versions, the problem of calculating the maximum guaranteed result of the operating side can be solved. And the results obtained have a fairly reasonable interpretation.

Secondly, a lower-level player may deliberately distort information about his actions. The case where the top-level player cannot control the accuracy of this information is not of interest. When the amount of information to be transmitted is limited, the top-level player can "selectively check" the validity of the transmitted messages. Here again we get a whole range of meaningful problem statements that can be effectively solved. In this context, it is even possible to study some corruption schemes, which is very important today.

**Acknowledgement.** This work was supported by a grant from the Ministry of Science and Higher Education of the Russian Federation, internal number 00600/2020/51896, agreement No. 075-15-2020-914.

# References

1. Germeier, Y.: Nonantagonistic Games. D. Reidel Publishing Co., Dordrecht (1986)
2. Vatel, I., Ereshko, F.: Mathematics of Conflict and Cooperation. Knowledge, Moscow (1973)
3. Vatel, I., Ereshko, F.: Games with a hierarchical structure, In: Mathematical Encyclopedia, pp. 478–482. Soviet Encyclopedia, Moscow (1979)
4. Ereshko, F.: The theory of hierarchical games in the application to lawmaking in the digital society bus law. Comput. Nanotechnol. **2**, 52–58 (2017)
5. Tirole, J.: The Theory of Industrial Organization, MIT Press, Cambridge (1988)
6. Von Stackelberg, H.: Market Structure and Equilibrium, 1st edn. Translation into English, Bazin. Springer, Urch & Hill (2011)
7. Germeier, Yu. B., Moiseev, N.N.: On some problems in the theory of hierarchical systems. In: Problems of Applied Mathematics and Mechanics, pp. 30–43. Nauka, Moscow (1971)

8. Moiseev, N.N.: Elements of the Theory of Optimal Ssystems. Nauka, Moscow (1975)
9. Moiseev, N.N.: Mathematical Problems of System Analysis. Nauka, Moscow (1981)
10. Ereshko, F., Turko, N., Tsvirkun, A., Chursin, A.: Design of organizational structures in large-scale projects of digital economy. Autom. Remote. Control. **79**(10), 1836–1853 (2018)
11. Gorelov, M., Ereshko, F.: On models of centralization and decentralization of control in a digital society. In: Ivanov, V.V., Malinetsky, G.G., Sirenko, S.N. (eds.) The Contours of Digital Reality: The Humanitarian And Technological Revolution and the Choice of the Future, pp. 187–202. Lenand, Moscow (2018)
12. Von Neumann, J., Morgenstern, O.: Theory of Games and Economic Behavior. Princeton University Press, Princeton (1953)
13. Kuhn, H.W.: Extensive games and the problem of information. In: Contributions to the Theory of Games, vol. II, pp. 193–216. Princeton University Press, Princeton (1953)
14. Kolmogorov, A.N.: Three approaches to the definition of the concept of the quantity of information. Probl. Peredachi Inform. **1**(1), 3–11 (1965)
15. Gorelov, M.: Maximal guaranteed result for limited volume of transmitted information. Autom. Remote. Control. **72**(3), 580–599 (2011)
16. Gorelov, M., Ereshko, F.: Awareness and control decentralization. Autom. Remote. Control. **80**(6), 1063–1076 (2019)
17. Gorelov, M., Ereshko, F.: Awareness and control decentralization: stochastic case. Autom. Remote. Control. **81**(1), 41–52 (2020)
18. Ereshko, F., Gorelov, M.: Information and hierarchy. In: Recent Advances of the Russian operations research society, pp. 2–28. Cambridge Scholars Publishing, Cambridge (2020)
19. Dempster, M.A.H. (ed.): Risk Management. Value at Risk and Beyond. Cambridge University Press, Cambridge (2002)
20. Gorelov, M.: Risk management in hierarchical games with random factors. Autom. Remote. Control. **80**(7), 1221–1234 (2019)

# Methodical Approach for Impact Assessment of Energy Facilities on Environment

Vladimir R. Kuzmin$^{(\boxtimes)}$ ⓘ and Liudmila V. Massel ⓘ

ESI SB RAS, Lermontova Street, 130, 664033 Irkutsk, Russia
`rulisp@vigo.su`, `massel@isem.irk.ru`

**Abstract.** The issues of influence and assessment of pollutant emissions from industrial facilities, that also include energy facilities, are attracting more attention in the world. However, the process of conducting such studies is complicated by the fact that: it is required to attract experts from the field of energy and ecology, the availability of a large amount of data on the technical characteristics of energy facilities and burned fuels, it is necessary to consider various factors such as weather, relief, and many others. The article discusses the pro-posed methodological approach to assessing the impact of power facilities on the environment, technology and algorithms for supporting such studies, tools that allows to perform such research. The article also presents the results of computational experiments carried out as a result of approbation of technology and tools.

**Keywords:** Impact assessment · Energy objects · Energy · Environment

## 1 Introduction

The issues of influence and assessment of pollutant emissions from energy facilities, are attracting more and more attention in the world. According to EU directive 2016/2284, by 2030 EU countries must reduce emissions of harmful substances into the atmosphere, including sulfur oxides, by an average of 70%, nitrogen oxide emissions - by 60% compared to 2005 [1, 2]. In the fall of 2021, the World Health Organization (WHO) published new guidelines for air quality [3] based on the analysis of more than 500 scientific studies on the effects of air pollution on human health. This analysis showed that harm is caused at lower concentrations of pollutants. Compared with the recommendations of 2005, [4] the maximum allowed concentrations were reduced:

- four times - for nitrogen dioxide ($NO_2$),
- two times - for PM2.5,
- four times - for PM10.

In Russia, according to the passport of the national project Ecology, which has been in effect since 2018, it is envisaged to eliminate the most dangerous objects of accumulated environmental harm and ecological restoration and rehabilitation of water bodies (which includes the lake Baikal) and a twofold reduction in emissions of hazardous pollutants

© Springer Nature Switzerland AG 2022
V. Taratukhin et al. (Eds.): ICID 2021, CCIS 1539, pp. 265–276, 2022.
https://doi.org/10.1007/978-3-030-95494-9_22

that have the greatest negative impact on the environment and human health [5, 6] until 2024. It is also worth noting that, according to the Energy Strategy of the Russian Federation for the period up to 2035 [7], ones of the priorities of the strategy are:

- transition to environmentally friendly and resource-saving energy
- rational use of natural resources and energy efficiency

The primary sources of information for environmental assessments of the activities of energy facilities are measurements and monitoring of emissions of pollutants into the environment, State reports, and reports of enterprises. In cases when such information is absent, then an assessment of the impact of energy facilities can be carried out using the results of a computational experiment utilizing approved regulatory methods. However, the existing methods used to assess the impact of energy facilities are applied separately, and it complicates the conduct of such studies. During the analysis of the sources, the authors were unable to find information on attempts to integrate these methods.

In this article, we will discuss the proposed methodical approach for impact assessment of energy facilities on the environment, technology, and tools for conducting complex studies to assess the impact of energy facilities on the environment and the results of a computational experiment obtained as a part of approbation.

## 2  State of the Research Area

In Russia, several approved regulatory methods are used to determine emissions of pollutants into the atmosphere during fuel combustion in power plant boilers, e.g., "Methodology for determining gross emissions of pollutants into the atmosphere from boiler plants" and "Methodology for determining emissions of pollutants into the atmosphere when burning fuel in boilers with a capacity less than 30 tons of steam per hour or less than 20 Gcal/hour". The scope of their application includes:

- establishment of maximum and temporarily allowed concentrations of emissions
- planning of measures to reduce emissions
- control of emissions of harmful substances into the atmospheric air

These methods allow to determine the volumes of emissions of gaseous harmful substances (such as carbon oxides, sulfur oxides and nitrogen oxides) according to the data of instrumental measurements and the results of calculations, the volumes of emissions of particulate matter. These methods are universal and can be used both to assess the impact of energy facilities [8, 9] and other industrial facilities [10].

To assess the spreading of pollutants in the atmospheric air, the "Methods for calculating the spreading of emissions of harmful (polluting) substances in the atmospheric air" (MRR-2017) [11] and [12]. MRR-2017 is used for:

- development of standards for emissions of pollutants into the air and measures to protect the environment, as well as measures that affect the level of air pollution, including the assessment of their results.

- justification of the boundaries of the sanitary protection zones
- assessment of the impact of the proposed activity on the quality of atmospheric air, as well as assessment of short-term and long-term levels of air pollution and corresponding concentrations of air pollutants

However, MPR-2017 has a drawback associated with a small range of calculations - 100 km. Such distance may be sufficient for small energy facilities, however for large power plants, the detection range of emissions of harmful substances can reach 200–250 km.

The methods described by Berland in [12] are also applied. He describes various methods and information for conducting research related to the assessment of the spreading of pollutants. Criteria for the hazard of atmospheric pollution and how they can be used in the forecast are given. The physical foundations of forecasting air pollution are described - prognostic equations, concentration averaging, accounting of fog and smog, as well as the effect of the relief. The described methods for forecasting air pollution can be divided into two groups: numerical and statistical. Numerical methods help to estimate the maximum one-time concentrations of pollutants from individual sources and integral from areal and aggregate sources, as well as smog forecasting. Statistical methods allow to estimate the annual and daily air pollution, estimate the concentration of pollutants, and calculate the integral indicators of air pollution in the city. The described methods also allow to predict adverse meteorological conditions and regulate pollutants emissions into the atmosphere. The application of these methods for assessing the dispersion of pollutants was described in [13–16].

In the world, for such studies the HYSPLIT model (Hybrid Single-Particle Lagrangian Integrated Trajectory) [17] and the AERMOD [18] model are used. HYSPLIT allows to compute air parcel trajectories, chemical transformations, and simulation of the sedimentation of substances on the surface. One of its main areas of application is back trajectory analysis to determine the source of certain air masses and to establish "source-recipient" relationships [19–21]. But, HYSPLIT has its own drawbacks such as it is unable to account secondary chemical reactions and is highly dependent on meteorological data - frequency of obtaining these data and their geo-referencing. U.S. Environmental Protection Agency (US EPA) suggests using AERMOD that consists of:

- stationary model of short-term spreading of pollutants emissions into the atmospheric air
- meteorological data preprocessor AERMET
- AERMAP landscape preprocessor

Currently, AERMOD is used in various studies related to the assessment of spreading and identification of emission sources [22–24].

As mentioned above, the authors were unable to find information on attempts to integrate various regulatory methods for carrying out calculations.

# 3 Description of Methodical Approach

To assess the impact of energy systems and complexes on the environment, we propose the following methodological approach, which includes:

- principles of integration of methods for calculating emissions and pollutants spreading
- algorithms for calculating emissions and pollutants spreading
- ontological engineering to build a system of ontologies of subject areas - energy and ecology
- Information models and principles of database development
- Principles of construction and architecture of intelligent DSS (IDSS) based on the agent-service approach

Next, we will consider in more detail some of the components of the proposed approach.

Principles of integration of methods for calculating emissions and pollutants spreading are:

- Applicability of integrated techniques to a territory of research
- Consistency and comprehensiveness of the approach to assessing the impact of an energy facility on the environment.
- Results of the calculation using one method must be usable (comparable) for the second method - same units of measurements, types of emissions.
- Confirmation of the recommendations developed at the qualitative level using a computational experiment

The impact assessment of the of energy facilities on the environment can be attributed to a poorly structured research area. First, this is because both energy and ecology are large research areas with a significant number of concepts and relationships between them. Secondly, the process of impact assessment of energy facilities on the environment requires a large amount of different information: information about the technical characteristics of energy facilities, volume and type of fuel consumed (including its characteristics), meteorological data, knowledge about the emitted harmful (polluting) substances and the processes of their interaction both with each other and with elements of the environment (water, air). To model the subject area, we used ontological modeling, which is one of the tools of semantic modeling Semantic description allows one to create a representation of knowledge, with which one can identify the basic concepts and their relationships. As a result of the ontological engineering of subject areas, we, together with Vorozhtsova T.N., Ivanova I.Yu. and Maisyuk E.P., built a system of ontologies. Figure 1 and Fig. 2 show some of the ontologies.

The developed system of ontologies was also used to design a database for storing the results of calculations, this was described in more detail in the article [25].

To carry out complex studies for impact assessment of energy systems and complexes on the environment, we developed the architecture of a typal IDSS (Fig. 3). Usage of agent-service approach for development of IDSS was described in [26].

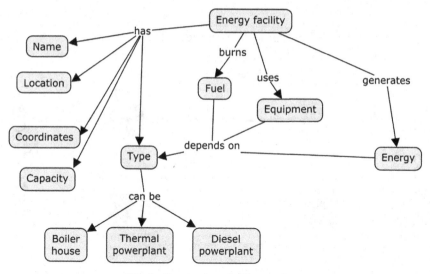

**Fig. 1.** Typal energy facility ontology

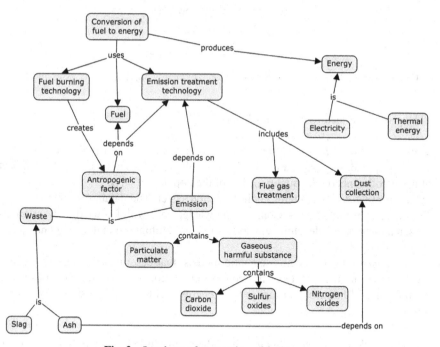

**Fig. 2.** Ontology of conversion of fuel to energy

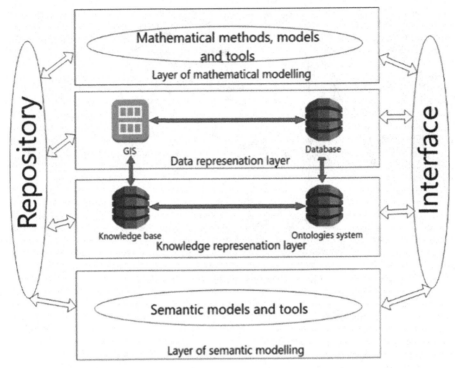

**Fig. 3.** Architecture of typal IDSS for research of energy objects' impact assessment

The given architecture includes 4 levels:

- *Mathematical modeling* – contains programs for calculation of pollutants emission volumes, spreading of pollutants into the atmospheric air, and assessment of the impact of these pollutants on the quality of life of the population
- *Data representation* – contains geoinformation system and databases
- *Knowledge representation* – Contains knowledge base with knowledge descriptions for semantic models' development and system of ontologies that describe knowledge about subject areas
- *Semantic modeling* – contains semantic models, that describe connections between factors, affecting the quality of life considering antropotechnogenic factors such as the influence of harmful substances from energy facilities and the supply of energy resources

Principles of integration of methods for calculating emissions and pollutants spreading and development of the algorithms for calculating emissions and pollutants spreading will be discussed in future publications.

# 4   Technology and Tools

As a part of the proposed methodical approach, we developed the technology for impact assessment of energy systems and complexes on the environment and support of decision-making to reduce their harmful effects, which includes the following stages:

- Stage of quantitative calculation of emissions based on methods for determining emissions of pollutants into the atmosphere
- Stage of calculation of emissions spreading
- Storage and work with the results of the analysis of snow samples containing quantitative data of pollutants found at the points of measurement

To support the technology, we developed a scientific prototype of the IDSS "WIAIS" (Web-oriented Impact Assessment Information System). The system is based on the proposed typical architecture. For impact assessment of energy systems and complexes on the environment, we also developed the following set of basic components:

- Component that handles quantitative pollutants emission calculation
- Component that handles pollutants spreading calculation
- Component for work with the results of the analysis of snow samples
- Visualization component
- Service component that handles work with the databases

The technology can be used to conduct environmental assessments of both existing energy facilities and when planning the construction of new energy facilities. Also, the technology can be utilized for the development of recommendations to reduce emissions of pollutants from energy enterprises.

# 5   Computational Experiment and Approbation

To test the proposed technology and the developed IDSS, a computational experiment was carried out. As a part of the experiment, an assessment of the impact of emissions from boiler houses located in the Baikal natural territory (BNT) and working on coal was made. The approbation was carried out based on information about 48 boiler houses using various equipment, having different installed capacities, and located in different places within the BNT.

As a result, we obtained information on the quantitative indicators of emissions from the selected energy facilities, according to which:

- The total volume of pollutants emitted is 18.33 thous.t/year. 11.27 thous.t/year are emitted by facilities located in the Republic of Buryatia, and 7.06 thous.t/year are emitted by facilities located in the Irkutsk region
- The main contribution to the emissions of pollutants emitted into the atmosphere is made by dust emissions - 14.3 thous.t/year or 78% of the total volume
- The main sources of emissions of pollutants into the atmosphere are facilities operating in Slyudyanka, Elantsy, Severobaikalsk and Nizhneangarsk

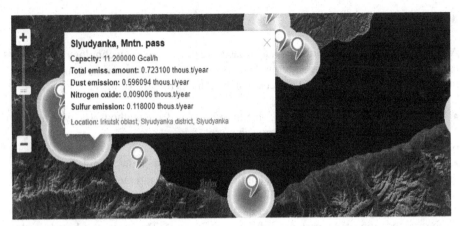

**Fig. 4.** Geovisualization of the calculation results of the emissions' quantitative indicators

Figure 4 shows the geovisualization of the calculation results of the emissions' quantitative indicators using tools of the geovisualization and Table 1 shows results for a Elantsy central boiler house.

**Table 1.** Table representation of emissions of Elantsy central boiler house

| Facility | Fuel | Total emission, tons/year | Dust emission, tons/year | $SO_4$ emission, tons/year | $NO_X$ emission, tons/year | Dust removal, % |
|---|---|---|---|---|---|---|
| *Elansty central boiler house* | Cheremkhovsky coal | 0,364 | 0,252 | 0,112 | 0,0001 | |
| Boiler 1 | | 0,052 | 0,036 | 0,016 | 0,00002 | 0 |
| Boiler 2 | | 0,052 | 0,036 | 0,016 | 0,00002 | 0 |
| Boiler 3 | | 0,052 | 0,036 | 0,016 | 0,00002 | 0 |
| Boiler 4 | | 0,052 | 0,036 | 0,016 | 0,00002 | 0 |
| Boiler 5 | | 0,052 | 0,036 | 0,016 | 0,00002 | 0 |
| Boiler 6 | | 0,052 | 0,036 | 0,016 | 0,00002 | 0 |
| Boiler 7 | | 0,052 | 0,036 | 0,016 | 0,00002 | 0 |

As the result of a computational experiment of calculation of pollutants spreading in the atmospheric air, the following information was obtained:

- Main directions of the pollutants spreading have been identified
- The main part of pollutants emitted into the atmosphere settles mainly within a radius of 150–250 m from the emission source. This is due, firstly, to the low height of the

chimneys of the considered facilities, and secondly, to the low temperature of the outgoing gases.

Figure 5 shows geovisualization of the calculation results of emission spreading, energy facilities (mainly boiler houses) are marked with blue markers. The results of the spreading calculation are preliminary since the relief is not taken into account at this stage – Lake Baikal is surrounded by hills and mountains, and therefore, emissions of pollutants will mainly accumulate near energy facilities.

**Fig. 5.** Geovisualization of the calculation results of emission spreading

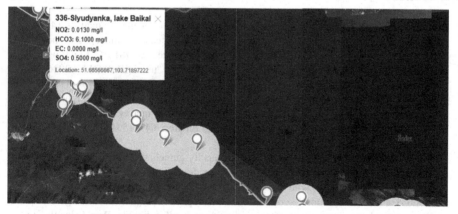

**Fig. 6.** Geovisualization of snow samples analysis results

We also developed a component for work with the results of the analysis of pollutants in snow. These results contain data on pollutants emission, such as $SO_4$, $NO_x$, $NH_4$, ions ($Ca^{2+}$, $Mg^{2+}$) and others, obtained from a set of measurement points. The results of the analysis of snow samples can be used for long-term monitoring of pollutant emissions, since the sampling is carried out in the same places over the years. Also, the results can be used to check the correctness of calculations of emissions and their dispersion in the air. The component provides storage and visualization (in the form of graphs or geovisualization) of the results obtained, as well as export of the results for the required point in the form of an Excel file. Figure 6 shows an example of geovisualization of results.

# 6  Conclusion

The article considered the proposed methodological approach for assessing the impact of energy systems and complexes, and its main components were briefly described. The technology, developed on the basis of proposed approach, was presented along with the tools for technology's support. The results of approbation (such as quantitative indicators of the emissions and the results of calculation of emission spreading) of technology and tools for its support were also shown. In the future, we plan to add relief accounting to the pollutants spreading calculation component, and accounting of the maximum allowed concentrations and notification if they will be exceeded according to the results of the calculation.

The research was carried out under State Assignment Project (№ FWEU-2021-0007 AAAA-A21-121012090007-7) using the resources of the High-Temperature Circuit Multi-Access Research Center (Ministry of Science and Higher Education of the Russian Federation, project no 13.CKP.21.0038), some aspects were made as part of following project, supported by RFBR grant №19-07-00351.

# References

1. Directive (EU) 2016/2284 of the European Parliament and of the Council of 14 December 2016 on the reduction of national emissions of certain atmospheric pollutants, amending Directive 2003/35/EC and repealing Directive 2001/81/EC (Text with EEA relevance). https://eur-lex.europa.eu/legal-content/EN/TXT?uri=uriserv:OJ.L_.2016.344.01.0001.01.ENG&toc=OJ:L:2016:344:TOC. Accessed 15 Sep 2021
2. Serrano, H.C., Oliveira, M.A., Barros, C., Augusto, A.S., Pereira, M.J., Pinho, P., et al.: Measuring and mapping the effectiveness of the European air quality directive in reducing N and S deposition at the ecosystem level. Sci. Total Environ. **647**, 1531–1538 (2019). https://doi.org/10.1016/j.scitotenv.2018.08.059
3. World Health Organization. WHO global air quality guidelines: particulate matter (PM2.5 and PM10), ozone, nitrogen dioxide, sulfur dioxide and carbon monoxide. World Health Organization6 https://apps.who.int/iris/handle/10665/345329. Accessed 10 Oct 2021
4. World Health Organization. Regional Office for Europe. Air quality guidelines: global update 2005: particulate matter, ozone, nitrogen dioxide and sulfur dioxide. World Health Organization. Regional Office for Europe. https://apps.who.int/iris/handle/10665/107823. Accessed 10 Oct 2021
5. Semenova, G.: Global environmental problems in Russia. In: Key Trends in Transportation Innovation (KTTI-2019), vol. 157, 8 p. E3S Web of Conferences (2020). https://doi.org/10.1051/e3sconf/202015702023

6. Egorchenkov, A., Egorchenkov, D.: Social aspects of environmental issues in the context of the national project "Ecology". In: IOP Conference Series: Earth and Environmental Science, vol. 579, 5 p. (2020). https://doi.org/10.1088/1755-1315/579/1/012099

7. Alekseev, A.N., Bogoviz, A.V., Goncharenko, L.P., Sybachin, S.A.: A critical review of Russia's energy strategy in the period until 2035. Int. J. Energy Econ. Policy **9**(6), 95–102 (2019). https://doi.org/10.32479/ijeep.8263

8. Detsuk, V.S.: Methods for reducing emissions of pollutants during the combustion of organic fuels in boilers of thermal power plants. Vestnik Belorusskogo gosudarstvennogo universiteta transporta: Nauka i transport **2**(35), 27–29 (2017). (in Russian)

9. Sozaeva, L.T., Shungarov, A.G., Khegai, A.G.: Air pollution by heat supply facilities of the city of Nalchik. Proc. Voeikov Main Geophys. Observatory **950**, 190–198 (2018). (in Russian)

10. Berezutskii, A., Katin, V.D.: Analysis of emissions of harmful substances by oil refineries and the mechanisms of their formation during fuel combustion. Sci. Tech. Econ. Cooperation Asia Pac. Countries XXI Century **2**, 315–322 (2012). (in Russian)

11. Decree of Ministry of Natural Resources and Environment of the Russian Federation no. 273, dated 06.06.2017 "On Approval of Methods for Calculation of Emissions of Harmful (Polluting) Substances in the Atmospheric Air". http://publication.pravo.gov.ru/Document/View/0001201708110012. Accessed 20 Sep 2021. (in Russian)

12. Berlyand, M.E.: Prediction and Regulation of Air Pollution. Gidrometeoizdat, Leningrad (1985). (in Russian)

13. Kormina, L.A., Sukach, O.O.: Introduction of resource-saving technologies in the energy sector. Chemistry. Ecology. Urbanism 2020-16, pp. 120–123 (2020). (in Russian)

14. Semakina, A.V., Platunova, G.R., Mansurov, A.R.: Atmospheric air condition in the territory of the republic of Bashkortostan. Bull. Udmurt Univ. Ser. Biol. Earth Sci. **30**(3), 278–284 (2020). https://doi.org/10.35634/2412-9518-2020-30-3-278-284 (in Russian)

15. Litvinova, N.A., Azarov, V.N.: On the model of the vertical distribution of pollutant concentrations along the height of buildings, taking into account the type of local development. Bull. Volgograd State Univ. Arch. Civil Eng. Ser. Build. Arch. **3**(84), 108–121 (2021). (in Russian)

16. Biliaiev, M.M., Slavinska, O.S., Kyrychenko, R.V.: Numerical prediction models for air pollution by motor vehicle emissions. Sci. Transp. Prog. Bull. Dnipropetrovsk Nat. Univ. Railway Transp. **6**(66), 25–32 (2016). https://doi.org/10.15802/stp2016/90457. (in Russian)

17. Stein, A.F., Draxler, R.R., Rolph, G.D., Stunder, B.J.B., Cohen, M.D., Ngan, F.: NOAA's HYSPLIT atmospheric transport and dispersion modeling system. Bull. Am. Meteor. Soc. **96**(12), 2059–2077 (2015). https://doi.org/10.1175/BAMS-D-14-00110.1

18. Aermod: Description of Model Formulation. https://nepis.epa.gov/Exe/ZyPDF.cgi/P10 09OXW.PDF?Dockey=P1009OXW.PDF. Accessed 12 Sep 2021

19. Fleming, Z.L., Monks, P.S., Manning, A.J.: Review: untangling the influence of air-mass history in interpreting observed atmospheric composition. Atmos. Res. **104–105**, 1–39 (2012). https://doi.org/10.1016/j.atmosres.2011.09.009

20. Ma, Y., Wang, M., Wang, S., Wang, Y., Feng, L., Wu, K.: Air pollutant emission characteristics and HYSPLIT model analysis during heating period in Shenyang, China. Environ. Monit. Assess. **193**(1), 1–14 (2020). https://doi.org/10.1007/s10661-020-08767-4

21. Chang, L., Jiang, N., Watt, S., Azzi, M., Riley, M., Barthelemy, X.: The use of 'HYSPLIT in NSW' in air quality management and forecasting. In: CASANZ 2021: the 25th International Clean Air and Environment Conference, 6 p. (2021)

22. dos Santos Cerqueira, J., de Albuquerque, H.N., de Assis Salviano de Sousa, F.: Atmospheric pollutants: modeling with Aermod software. Air Qual. Atmos. Health **12**(1), 21–32 (2018). https://doi.org/10.1007/s11869-018-0626-9

23. Gopi, R., Saravanakumar, R., Elango, K.S., Chandrasekar, A., Navaneethan, K.S., Gopal, N.: Construction emission management using wind rose plot and AERMOD application. In: IOP Conference Series: Materials Science and Engineering, vol. 1145, 5 p. (2021). https://doi.org/ 10.1088/1757-899X/1145/1/012106
24. Pandey, G., Sharan, M.: Application of AERMOD for the identification of a point-source release in the FFT-07 experiment. Air Qual. Atmos. Health **14**(5), 679–690 (2021). https:// doi.org/10.1007/s11869-020-00971-y
25. Kuzmin, V.R.: Development of information subsystem for calculation and visualization of harmful emissions from energy objects. Inf. Math. Technol. Sci. Manage. **1**(17), 142–155 (2020). (in Russian)
26. Massel, A.G., Galperov, V.I., Kuzmin, V.R.: Agent-service approach for development of intelligent decision-making support systems. In: Proceedings of the 6th International Workshop "Critical Infrastructures in the Digital World" (IWCI-2019), pp. 211–215, Atlantis Press (2019). https://doi.org/10.2991/iwci-19.2019.37

# Young Scientists Research in the Areas of Enterprise Digitalization

# Disadvantages of Relational DBMS for Big Data Processing

Svetlana Borisova(✉) 🄳 and Ali Zein 🄳

National Research University "Moscow Power Engineering Institute", Krasnokazarmennaya 17, Moscow 111250, Russia

zeynaln@mpei.ru

**Abstract.** In this paper, we are analyzing relational DBMS while working with big data. All experiments and performance tests are implemented in MS SQL Server 2019 Developer Edition and MongoDB. In the beginning, we start by creating a database that contains non-linked tables (flat tables) with maximum memory consummation. The performance test starts with filling (Insert) the database with millions of rows. Then we continue with data modification operations (Update). Finally, we launch the data cleaning (Delete) test. When all tests pass by and execution time is picked up for further analysis, we continue our experiment with relational tables and NoSQL collections. That's why in the second part of this paper we take the same database structure where we add primary and foreign key constraints with cascade update and delete options. When repeating the performance test, we notice that the table size and execution time have significantly increased.

**Keywords:** Relational DBMS · Performance test · Big data · Link constraints · DML commands · MongoDB · Data collection · JSON

## 1 Database Creation

### 1.1 Flat Database Creation

In [1] we see an actual problem related to unstructured data. In this paper, we are solving a reverse problem related to relational (structured) data. In the past, it has been common practice to store all data in relational databases (MS SQL, MySQL, Oracle, PostgresSQL). At the same time, it was not so important whether relational databases are suitable for storing this type of data or not.

For the first part experiment tests, we have created a database that contains three flat tables (Fig. 1).

Each table contains 6 columns with maximum memory consummation: we have used long value data types as varchar (max) and bigint. Below we present the SQL-code for creating a table, which will be used in our experiments. The other two tables have the same structure.

© Springer Nature Switzerland AG 2022
V. Taratukhin et al. (Eds.): ICID 2021, CCIS 1539, pp. 279–290, 2022.
https://doi.org/10.1007/978-3-030-95494-9_23

| table1_OO_A_11M_20 | table2_OO_A_11M_20 | table3_OO_A_11M_20 |
|---|---|---|
| col1_OO_A_11M_20<br>col2_OO_A_11M_20<br>col3_OO_A_11M_20<br>col4_OO_A_11M_20<br>col5_OO_A_11M_20<br>col6_OO_A_11M_20 | col1_OO_A_11M_20<br>col2_OO_A_11M_20<br>col3_OO_A_11M_20<br>col4_OO_A_11M_20<br>col5_OO_A_11M_20<br>col6_OO_A_11M_20 | col1_OO_A_11M_20<br>col2_OO_A_11M_20<br>col3_OO_A_11M_20<br>col4_OO_A_11M_20<br>col5_OO_A_11M_20<br>col6_OO_A_11M_20 |

**Fig. 1.** Three flat tables representing a non-relational database structure.

```
CREATE TABLE table1_OO_A_11M_20(
col1_OO_A_11M_20 bigint,
col2_OO_A_11M_20 varchar(max),
col3_OO_A_11M_20 varchar(max),
col4_OO_A_11M_20 varchar(max),
col5_OO_A_11M_20 varchar(max),
col6_OO_A_11M_20 bigint)
```

For the flat table creation, we do not use primary and foreign keys. Indexes and constraints are eliminated.

### 1.2 Relational Database Creation

In [2] we have used a relational database with a tree structure. That's why, for the second part of experiment tests, we have taken the same database structure as in 1.1 where we have made relational links: we added primary and foreign key constraints with cascade update and delete options.

We all know Dr. Edgar F. Codd's 12 rules, which can be applied to any database system that manages stored data using only its relational capabilities. This is a foundation rule, which acts as a base for all the other rules.

As a result, we have a relational database with relational restrictions and rules (Fig. 2). It is well known that relational databases ensure data integrity. So, it is interesting to understand how much this integrity affects performance.

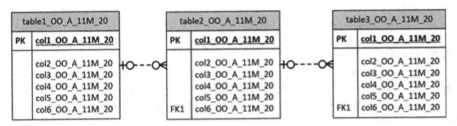

**Fig. 2.** Relational database structure.

We have decided to create three tables with six columns each. In addition, we added keys. Below we present the SQL-code of creating a relational database structure, which is used in our experiments.

```
CREATE TABLE table1_OO_A_11M_20(
    col1_OO_A_11M_20 bigint Primary key,
    col2_OO_A_11M_20 varchar(max),
    col3_OO_A_11M_20 varchar(max),
    col4_OO_A_11M_20 varchar(max),
    col5_OO_A_11M_20 varchar(max),
    col6_OO_A_11M_20 bigint);
CREATE TABLE table2_OO_A_11M_20(
    col1_OO_A_11M_20 bigint Primary key,
    col2_OO_A_11M_20 varchar(max),
    col3_OO_A_11M_20 varchar(max),
    col4_OO_A_11M_20 varchar(max),
    col5_OO_A_11M_20 varchar(max),
    col6_OO_A_11M_20 bigint
foreign key references
table1_OO_A_11M_20(col1_OO_A_11M_20));
CREATE TABLE table3_OO_A_11M_20(
    col1_OO_A_11M_20 bigint Primary key,
    col2_OO_A_11M_20 varchar(max),
    col3_OO_A_11M_20 varchar(max),
    col4_OO_A_11M_20 varchar(max),
    col5_OO_A_11M_20 varchar(max),
    col6_OO_A_11M_20 bigint
foreign key references
table2_OO_A_11M_20(col1_OO_A_11M_20));
```

By default with relational database creation, we get indexes and constraints. If we compare ER-diagrams in Figs. 1 and 2, we understand that we have the same data structure, where we can load, unload and process data.

### 1.3 NoSQL Database Creation in MongoDB

MongoDB takes a new approach to building databases without tables, schemas, SQL queries, foreign keys, and many other things that are inherent in object-relational databases.

Unlike relational databases, MongoDB offers a document-oriented data model, which makes MongoDB faster, more scalable, and easier to use. However, even taking into account all the disadvantages of traditional databases and the advantages of MongoDB, it is important to understand that tasks are different and methods of solving them are different. In some situation, MongoDB will really improve the performance of your application, for example, if you need to store data that is complex in structure.

Otherwise, it might be better to use traditional relational databases. Alternatively, you can use a mixed approach: store one datatype in MongoDB and another datatype in traditional databases. An entire MongoDB system can represent more than just one database residing on one physical server. MongoDB functionality allows multiple databases to be located on multiple physical servers, and these databases can easily exchange data and maintain integrity.

When working with MongoDB we should mention the JSON format that is used to work with data sets and data collections. First, we start by creating a collection:

```
db.createCollection("coll_OO_A_11M_20")
```

Then, we continue by filling these collections using the function "insertOne()":

```
db.coll_OO_A_11M_20.insertOne(
{
  coll_OO_A_11M_20: 1,
  col2_OO_A_11M_20: <long string>,
  col3_OO_A_11M_20: <long string>,
  col4_OO_A_11M_20: <long string>,
  col5_OO_A_11M_20: <long string>,
  col6_OO_A_11M_20: 1
} );
```

The other two collections (col2_OO_A_11M_20, col3_OO_A_11M_20) are created in the same way. Using a loop, we can fill up all the three collections with the needed number of rows for further analysis.

## 2   Commands for Performance Test

In order to make performance tests, we have decided to implement the main DML operators (insert, update and delete). In our performance tests, we are going to check the execution time of these operators for different amounts of data rows: 1, 10, 100, 1 000, 10 000, 50 000, 100 000, 200 000.

### 2.1   DML Insert Command

In order to start filling tables with data, we have decided to use char() and rand() functions as well "while" loop structure.

```
declare @i int = 1;
/* Initializing strings */
declare @string1 varchar(1000) = '', @string2 var-
char(1000) = '', @string3 varchar(1000) = '', @string4
varchar(1000) = '';
while @i<1001
BEGIN
  /*Appending random chars into every string
    rand()*25+65 returns random number from 65 to 90
    char(N) returns char from ASCII table with code N
    char(65) = A
    char(66) = B
    ..
    char(89) = Y
    char(90) = Z*/
  set @string1 += char(rand()*25+65)
  set @string2 += char(rand()*25+65)
  set @string3 += char(rand()*25+65)
  set @string4 += char(rand()*25+65)
  set @i = @i+1;
END
set @i = 1;
  /*Saving the start time to calculate the table filling
time*/
declare @start datetime = getdate(), @end datetime;

/*table filling loop*/
while @i<1000001
BEGIN
  insert into table1_OO_A_11M_20
    (col1_OO_A_11M_20,
     col2_OO_A_11M_20,
     col3_OO_A_11M_20,
     col4_OO_A_11M_20,
     col5_OO_A_11M_20,
     col6_OO_A_11M_20)
   values
    ( @i, @string1, @string2, @string3, @string4, @i)
  set @i = @i+1
END
set @end = getdate();
select datediff(SS, @start, @end) as duration_in_seconds
```

## 2.2 DML Update Command

We execute the UPDATE command on a various number of records. Column col2_OO_A_11M_20 will is replaced with the same data. The number of entries is be governed by the "where" condition.

```
declare @start datetime = getdate(), @end datetime;
update table1_OO_A_11M_20
  set col2_OO_A_11M_20 = col2_OO_A_11M_20
  where col1_OO_A_11M_20<=5000
/*records count condition */
set @end = getdate();
select datediff(SS, @start, @end)as duration_in_seconds
```

## 2.3 DML Delete Command

For deleting table rows, we use DELETE TOP(N) statement and a few more code lines to calculate the execution time.

```
/*Saving the start time to calculate the deleting time*/
declare @start datetime = getdate();
declare @end datetime;
delete top(1) from table1_OO_A_11M_20
set @end = getdate();
select datediff(SS, @start, @end)as duration_in_seconds
```

## 2.4 MongoDB Insert Loop

In order to start filling MongoDB collections with data, we have decided to use a loop with "insert()" function. In this case, for the three collections that we have created before, we are going to use this program code:

```
for (var i = 1; i <= 1000000; i++) {
    db.table1_OO_A_11M_20.insert( {
    col1_OO_A_11M_20: 1,
    col2_OO_A_11M_20: <long string_i>,
    col3_OO_A_11M_20: <long string_i>,
    col4_OO_A_11M_20: <long string_i>,
    col5_OO_A_11M_20: <long string_i>,
    col6_OO_A_11M_20: 1
} );
}
```

## 2.5 MongoDB Update Loop

To start updating data in MongoDB collections, we have decided to use "foreach()" loop with row limitation. In this case, for the three collections that we have created before, we are going to use this program code:

```
db.table1_OO_A_11M_20.find().limit(<number to update>)fo-
rEach(function(doc){
  var updated = doc.col2_OO_A_11M_20 + "qwerty";
  db.table1_OO_A_11M_20.update({col1_OO_A_11M_20:
  doc.col1_OO_A_11M_20},{$set:{"col2_OO_A_11M_20":
updated}});
})
```

## 2.6 MongoDB Delete Loop

To complete our experiment, we have also prepared a "foreach()" loop limiting the number of affected rows during data deletion. For the three collections that we have created before, the following program code is used:

```
db.table1_OO_A_11M_20.find()limit
(<number to delete>).forEach(function(doc){
    db.table1_OO_A_11M_20.remove({
    col1_OO_A_11M_20: doc.col1_OO_A_11M_20});
})
```

# 3  Performance Test Results

All commands are executed for flat tables (non-relational database structure) and relational database. We run the same commands for the same data. The keys of the table take up space, so the size of each table has increased by 50%. Tables with foreign keys take up a little more space than without them. There was also a time difference. It was 52% for the third table while filling, updating, or deleting rows. From this, we can conclude that tables with foreign keys are filled slower than tables without them. Additional resources are spent on filling tables for foreign keys. However, the time to fill the tables also depends very much on the availability of free RAM at the moment, the fullness of the hard disk (SSD) cache. Therefore, it was possible to notice that when filling in the third table, there was almost no free RAM left, and the write speed on the SSD dropped to 19 MB/s.

The results of the update test are presented below (see Fig. 3).

Based on the data obtained, it can be concluded that the existence of key columns in the table slightly speeds up the work with non-key columns, but significantly slows down the update of data in PK columns.

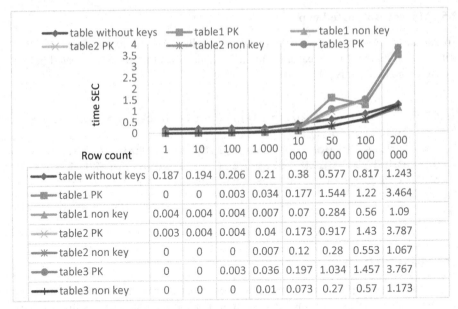

**Fig. 3.** Performance test on update operand in MS SQL Server.

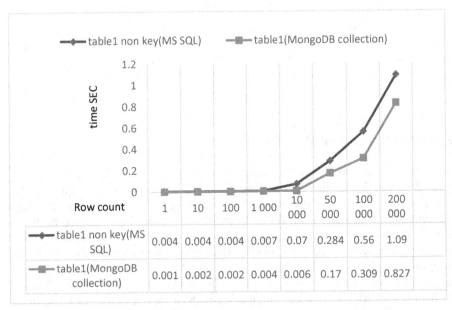

**Fig. 4.** Performance test on update operand in MongoDB.

If we start comparing relational and non-relational databases, we can conclude that non-relational databases (e.g. MongoDB) are even faster than flat tables in relational databases. MongoDB is working faster on significant data volumes, on Fig. 4 we can see that MongoDB is faster than MS SQL by 24% when modifying more than 200 000 rows.

While deleting rows in the first table due to "ON DELETE CASCADE", the associated data in the second and the third tables are deleted. The operations were performed in the following order: Non-key Time table 3, Non-key time table 2, Non-key Time table1, PK Time Table1, PK Time Table2. The difference between the time of deleting one record and 10 can be explained by caching (Fig. 5). On the first query, SQL Server caches the table, so deleting 10 or more rows was faster than deleting a single row.

We can see on the graph (Fig. 5), that the first operation DELETE ROWS (table without any keys) was very fast. Operations with non-key columns were slower. And the slowest operations were "with PK"-columns. As in the case of UPDATE, key columns slow down execution DELETE operand.

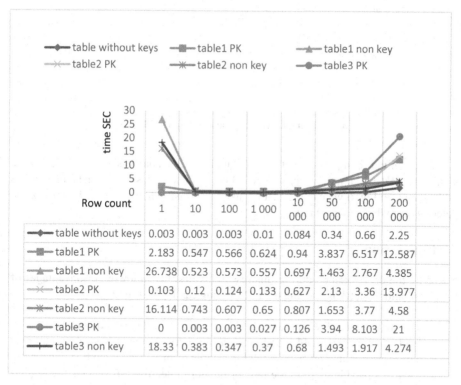

| Row count | 1 | 10 | 100 | 1 000 | 10 000 | 50 000 | 100 000 | 200 000 |
|---|---|---|---|---|---|---|---|---|
| ◆ table without keys | 0.003 | 0.003 | 0.003 | 0.01 | 0.084 | 0.34 | 0.66 | 2.25 |
| ■ table1 PK | 2.183 | 0.547 | 0.566 | 0.624 | 0.94 | 3.837 | 6.517 | 12.587 |
| ▲ table1 non key | 26.738 | 0.523 | 0.573 | 0.557 | 0.697 | 1.463 | 2.767 | 4.385 |
| ✕ table2 PK | 0.103 | 0.12 | 0.124 | 0.133 | 0.627 | 2.13 | 3.36 | 13.977 |
| ✳ table2 non key | 16.114 | 0.743 | 0.607 | 0.65 | 0.807 | 1.653 | 3.77 | 4.58 |
| ● table3 PK | 0 | 0.003 | 0.003 | 0.027 | 0.126 | 3.94 | 8.103 | 21 |
| ┼ table3 non key | 18.33 | 0.383 | 0.347 | 0.37 | 0.68 | 1.493 | 1.917 | 4.274 |

**Fig. 5.** Performance test on delete operand in MS SQL Server.

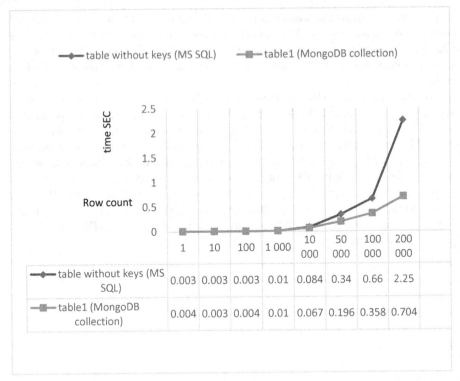

**Fig. 6.** Performance test on delete operand in MongoDB.

If we continue comparing relational and non-relational databases, the result is more significant. We can conclude that MongoDB is significantly faster than flat tables in MS SQL Server. In Fig. 6 we can see that MongoDB is faster than MS SQL by 69% when deleting more than 200 000 rows.

## 4   Conclusion

Based on the data obtained during this research work, the following disadvantages of using a relational database for big data processing can be identified:

1. As the size of the tables increases, the access time to the records increases. And with a large number of tables and relationships between them, huge computational resources are required even to search for a single record. In case you find one record, you will not be able to go to the next or previous one. You must repeat the request and change its conditions to get access to other data.
2. After adding key columns and related tables, the data access time (SELECT operand) will decrease, but the execution time for other operations (DELETE, UPDATE) will increase.
3. For transactions to work correctly, their size and execution time must be minimal, since access to data is blocked during their execution.

4. In multi-user mode, some transactions can be canceled by the server during deadlocks, which will cause problems on the client's side.

After comparing relational with non-relation databases (e.g. MS SQL Server and MongoDB) we paid attention to the following moments:

1. Small amounts of data modification is not perceptible: MongoDB works with a slight advantage.
2. The real benefit of the usage of non-relational database can be revealed with significant data modification/deletion.

**Aknowledgments.** The investigation was carried out within the framework of the project "Development of Analytical Information System for Storage and Intellectual Processing of the Results of Experimental and Numerical Studies of Physical Processes Following in the Elements of Energy Processing" with the support of a grant from NRU "MPEI" for implementation of scientific research programs "Energy", "Electronics, Radio Engineering and IT", and "Industry 4.0, Technologies for Industry and Robotics in 2020–2022".

# References

1. Zein, A.N., Borisova, S.V.: The distribution problem of unstructured data when solving data mining tasks on computer clusters. Paper Presented at the CEUR Workshop Proceedings, p. 2570 (2020)
2. Komarov, I., Vegera, A., Zein, A., Borisova, S., Blazhenova, S., Gavrilov A.: Design of tree-like database structure for solving test modeling tasks of energy equipment. In: Proceedings of the 2020 5th International Conference on Information Technologies in Engineering Education, pp. 14–17. Russian Federation, Inforino, Moscow, April 2020
3. Faroult, S., Robson, P.: The Art of SQL. p. 370. O'Reilly Media, Sebastopol (2006)
4. Perkins, B., Hammer, J.V., Reid, J.D.: C# 7 Programming with Visual Studio 2017, p. 884. Wrox, Indianapolis (2018)
5. Plugge, E., Membrey, P., Hawkins, T.: The Definitive Guide to MongoDB: The NoSQL Database for Cloud and Desktop Computing, p. 327. Apress (2010). ISBN:1-4302-3051-7
6. Chodorow, K.: MongoDB: The Definitive Guide, 2nd edn. p. 432. O'Reilly, Sebastopol (2013)
7. Hows, D., Membrey, P., Plugge, E., Hawkins. T.: The Definitive Guide to MongoDB: A Complete Guide to Dealing with Big Data Using MongoDB, 3rd edn. p. 376c. Apress (2015). ISBN: 978-1-842-1183-0
8. L'Heureux, A., Grolinger, K., Elyamany, H.F., Capretz, M.A.M.: Machine learning with big data: challenges and approaches. IEEE Access. 5, 7776–7797 (2017). https://doi.org/10.1109/ACCESS.2017.2696365.ISSN2169-3536
9. Kitchin, R., McArdle, G.: What makes big data, big data? Exploring the ontological characteristics of 26 datasets. Big Data Soc. 3(1), 205395171663113 (2016). https://doi.org/10.1177/2053951716631130. ISSN: 2053-9517
10. Dedić, N., Stanier, C.: Towards differentiating business intelligence, big data, data analytics and knowledge discovery. In: Piazolo, F., Geist, V., Brehm, L., Schmidt, R. (eds.) ERP Future 2016. LNBIP, vol. 285, pp. 114–122. Springer, Cham (2017). https://doi.org/10.1007/978-3-319-58801-8_10

11. Alekperov, R.: Development and Implementation of an Algorithm for Processing Fuzzy Queries to Relational Databases, vol. 307. SCOPUS (2022). www.scopus.com. https://doi.org/10.1007/978-3-030-85626-7_104
12. Baumann, P., Misev, D., Merticariu, V., Huu, B.P.: Array databases: concepts, standards, implementations. J. Big Data 8(1), 1–61 (2021). https://doi.org/10.1186/s40537-020-00399-2
13. Vershinin, I.S., Mustafina, A.R.: Performance Analysis of PostgreSQL, MySQL, Microsoft SQL Server Systems Based on TPC-H Tests. SCOPUS (2021). www.scopus.com. https://doi.org/10.1109/RusAutoCon52004.2021.9537400
14. Gomes, A.: et al.: An Empirical Performance Comparison between MySQL and MongoDB on Analytical Queries in the COMEX Database. SCOPUS (2021) www.scopus.com. https://doi.org/10.23919/CISTI52073.2021.9476623

# Development of an Application for Data Structuring with the Possibility of Displaying 3D Models

Alena Kurushkina$^{(\boxtimes)}$ ⓘ and Ali Zein ⓘ

National Research University "Moscow Power Engineering Institute", Krasnokazarmennaya 17, Moscow 111250, Russia

{KurushkinaAA,zeynaln}@mpei.ru

**Abstract.** In the modern world in many industries, including energy, 3D modeling is the most visual means for demonstrating a part or assembly unit. Therefore, when creating a digital catalog for the presentation of new developments, along with the display of text information and images, it is important to provide the possibility of demonstrating 3D models. Two approaches are considered for displaying 3D objects: the first one - from the thread of the main application, we launch a 3D model as an external process; the second is the import of information about the 3D scene and its construction, then the configuration of the components required to display the scene is carried out.

**Keywords:** CAD systems · Power equipment design · 3D modeling · 3D files · 3D model display · Win32 window embedding · HwndHost · HelixToolkit · WinAPI · Digital catalog

## 1 Introduction

In line with the digitalization trends various digital design tools have been developed [1–3]. One of these tools is CAD systems that allow you to develop 3D models of equipment. They improve the product description process, shorten the initial stages of the life cycle and reduce the cost of design and construction of equipment [4].

All CAD systems are divided into several levels. Lower-level systems are graphics editors for creating various drawings (older versions of AutoCAD). Middle-level systems allow you to develop 3D models and assembly units (SolidWorks) [5], and higher-level systems enable you to specify requirements to drawings as well as to the models, which allows you to turn to the so-called "digital twins of products" (Siemens NX) [6].

The research was carried out within the framework of the project "DEVELOPMENT OF ANA-LYTICAL INFORMATION SYSTEM FOR STORAGE AND INTELLECTUAL PROCESSING OF THE RESULTS OF EXPERIMENTAL AND NUMERICAL STUDIES OF PHYSICAL PRO-CESSES FLOWING IN THE ELEMENTS OF ENERGY PROCESSING" with the support of a grant from NRU "MPEI" for implementation of scientific research programs "Energy", "Elec-tronics, Radio Engineering and IT", and "Industry 4.0, Technologies for Industry and Robotics in 2020–2022".

While creating 3D models in CAD systems of the middle or higher level, geometric models can take different file extensions. There is wireframe, surface, volumetric (solid) geometric models. The wireframe model represents the shape of a part as a finite set of lines lying on the surfaces of the part. For each line, the coordinates of the endpoints are known and their incidence to edges or surfaces is indicated. A surface model displays the shape of a part by defining its bounding surfaces, e.g., a collection of face, edge, and vertex data. Models of parts with complex shape surfaces, the so-called sculptural surfaces, have a special place. They include the hulls of vehicles (e.g., ships and cars), parts streamlined by flows of liquids and gases (turbine blades, aircraft wings), etc.

One of the problems of modern engineering corporations and machine-building enterprises is that 3D models of equipment created in CAD systems are stored on local computers of employees or on enterprise servers, which does not provide quick access to the digital model of equipment outside the enterprise. This problem is especially relevant when promoting products on the market. The solution to this problem is the creation of a digital catalog of the company's products, which will present both 3D models of products and a list of its technical and economical parameters.

The possibility of demonstrating 3D models in the digital catalog of the company's developments provides several benefits:

- Visibility - The three-dimensional display method is the most realistic and descriptive. It allows you to fully present the product;
- Integrity of the description - The 3D model will be presented in conjunction with a list of technical and economical parameters of the product, which will describe it more comprehensively and integrally;
- No restrictions when viewing – Without any doubt, you can use third-party programs to view 3D models, but not all such tools can provide the necessary functionality to demonstrate the model. Moreover, many users do not have the necessary software products at all.

The main task in the development of a digital catalog is the creation of a way to display 3D models inside the client application.

The algorithms described in this article are developed in the C# programming language for use in WPF, a subsystem for building graphical interfaces of the Windows OS.

It's worth noting that WPF applications use DirectX. This means that regardless of the complexity of the graphics, all drawing work goes through the DirectX pipeline. As a result, even the most ordinary business applications can take advantage of rich effects like transparency and anti-aliasing. We also benefit from hardware acceleration because DirectX delegates as much work as possible to the graphics processing unit (GPU), which is a separate processor on the graphics card.

## 2   Novelty

There are a lot of software products and firms that deal with information modeling, particularly, with BIM (Building Information Modeling) [7] and 3D infrastructure modeling

[8]. The 3D models in such technologies contain not only graphic information but also different properties (material type, cost, performance data, etc.) attached to construction components. These 3D models help to make planning, designing, managing, and operating of units and systems more efficient. The creation of such models is justified for large companies and enterprises for building and engineering complicated and complex systems as it requires a lot of financial resources and specialists [9].

In addition, there are various 3D viewers/players (for example, GLC_Player) that can display 3D models, but they don't provide the functionality of showing pictures, texts, and videos.

Anyway, for demonstrating new developments in exhibitions and presentations it would be quite appropriate to have 3D models in conjunction with description and technical parameters. But usage of described technologies isn't expedient. In this regard, the mentioned digital catalog will provide an accessible way of demonstrating new developments. It has an intuitive interface and combines opportunities for displaying pictures, texts, videos, as well as 3D models. Tree view navigation provides a hierarchy of the catalog's sections that can be changed by an administrator. Thus, the catalog helps to collect all data about the developments in one place and does not require special knowledge for filling, maintaining, and usage.

In future, we can use new popular technologies as virtual reality (VR), augmented reality (AR), mixed reality (MR) or even extended reality (XR).

VR is simulations created using virtual reality headsets. The difference from AR is that real objects completely disappear from the field of view; you only see the virtual environment. The technology of VR is as follows: there are two small monitors in front of your eyes; the picture on them reacts to head turns and/or movements in space as if you were seeing real objects. Due to this, the user is more deeply involved in what is happening in the virtual space, can look in all directions and in some cases even interact with virtual objects.

AR is a simulation when we do not look at the world directly, but through some kind of "filter" that embeds virtual objects into the real world as if they were really there. Unlike VR, the real world does not leave the field of vision. Most often, a smartphone or tablet is used as a "filter" for AR. AR can be used in education, games, navigation, art, advertising, museums, events, retail. A less common method is AR on large screens: usually used in shopping malls, at stops as part of advertising campaigns, etc. The screen becomes either a "window" in which, in addition to the backside, additional objects are shown, or a TV set showing spectators and virtual objects nearby.

MR without software, equipment is useless. To immerse yourself in VR or AR, you need someone to create this reality first. Different helmets work on different software systems (GearVR, SteamVR, Oculus). Microsoft is developing one of these systems. Perhaps to stand out, perhaps for some other reason, they named their platform Windows Mixed Reality. PC headsets from manufacturers such as Samsung, Lenovo, HP and others work on this platform. Most often, the abbreviation MR is used in this very meaning, in the form of WMR. People who are far from the details of technology sometimes put MR in a row with VR and AR because of the similarities.

It remains to tell about the last abbreviation - XR. Everything is simple here - this term is sometimes used to generalize modern immersive technologies - VR and AR. Basically, XR = VR + AR, without any additional meaning.

## 3   Designing of a Digital Catalog

We have created a relational database to structure, store and modify data (Fig. 1). The main Tree table contains information about the tree list [10] of developments: product descriptions are grouped into subsections of the lowest levels, and those form sections of the highest levels. The rest of the tables store file paths for 3D models, images, videos, etc. The client application provides a convenient user interface for displaying and editing the tree view and the data of its sections, as well as for interacting with the 3D model. When you select a section from the list, the database is accessed, information about the files' location related to the selected section is retrieved, and these files are displayed.

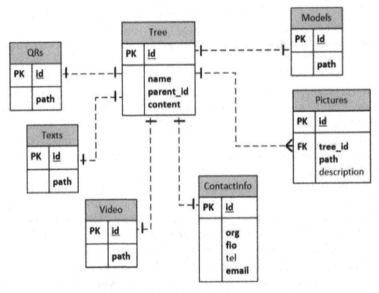

**Fig. 1.** ER-database diagram

3D models contained in this catalog often represent a part or assembly unit of power equipment, e.g., a gas turbine blade or a steam turbine rotor.

In addition to the ability to display files of standard 3D extensions, it is also important to enable the user to view standalone applications with the 3D model developed in the Unity environment.

# 4  Embedding a Standalone Application Inside a User Interface Control for Displaying 3D Models

Sometimes you may need to embed a standalone application inside a user interface control of the main application, for example, when it is necessary to facilitate access to the additional application when working with the main one.

In our case, it is necessary to provide the ability to demonstrate 3D models, which are executable files created in the Unity software.

Unity allows building standalone applications for Windows that store information about 3D models, and have advanced display options. For example, such an application might contain multiple 3D scenes and an extended user interface. Thus, it has a number of advantages over 3D files of 3DS, OBJ, FBX formats, etc.

## 4.1  Embedding Algorithm

In order to display the 3D model stored in the executable file, this file is launched from the main application thread as an external process [11], and the window of the running process is embedded in the user control of the main application. The algorithm is in Fig. 2.

To embed the application window, the HwndHost class from the System.Windows. Interop namespace is used [12], which hosts a Win32 window as an element within WPF content, e.g., a Border control. The embedded window must be initialized as a child of the window that will host it.

Thus, it is necessary to create a class that derives from HwndHost and overrides the BuildWindowCore and DestroyWindowCore methods inside it. The BuildWindowCore method takes a handle of the parent window as an input parameter, the body of the method implements the creation of a child window, the output parameter of the method is a handle of the child window. The creation of a window and interaction with it becomes possible only with the help of system functions. Platform Invocation Services (PInvoke) provide access to WinAPI tools. E.g., to send a message to a window, you must declare the following method:

```
internal static extern IntPtr SendMessage
(IntPtr hWnd, UInt32 Msg, IntPtr wParam, IntPtr lParam);
[DllImport("user32.dll", CharSet = CharSet.Auto)]
```

To start the process we use an instance of the System.Diagnostics.Process class. The following parameters are specified for process initialization [13]: path to the executable file; command line argument; a value that specifies the necessity to use the operating system shell to start the process, and a value that indicates no need to start the process in a new window. The string "-parentHWND  <parent window handle>" is specified as a command line argument to make the window child when starting the application. Below is an example of setting a process [11].

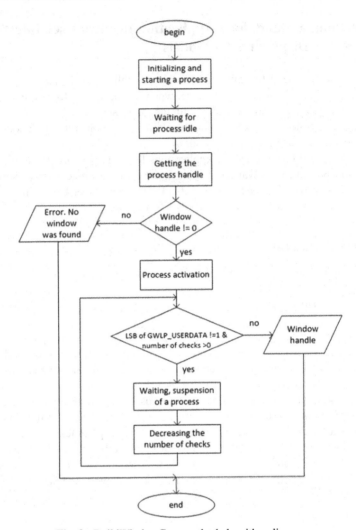

**Fig. 2.** BuildWindowCore method algorithm diagram

```
process = new Process();
process.StartInfo = new ProcessStartInfo
{
        FileName = unityExeName,
        Arguments = "-parentHWND " + hwndParent.Handle,
        UseShellExecute = true,
        CreateNoWindow = true
};
process.Start();
```

We use WaitForInputIdle (<timeout >) command to wait for the associated process to enter an idle state. If we have used the Unity environment for embedded applications, then before providing the user with the functionality of resizing an embedded window, we need to make sure that the window's GUI is initialized. To do this the attributes of the embedded window, particularly, the value GWLP_USERDATA [11], must be checked using the GetWindowLongPtr method [13]. If the value's least significant bit is set to one, then the GUI is loaded and the window can be scaled. To make the child window user interface available send WM_ACTIVATE message to the window. To close the application window, we must send in the body of the DestroyWindowCore method the CloseMainWindow method of the child process.

In Fig. 3 you can see the window of the application displaying the 3D model in the format of the executable file. The demonstrated standalone application retains all of its functionality. The user can click on the buttons, rotate the 3D model, and interact with it in the ways that were provided by the model developer.

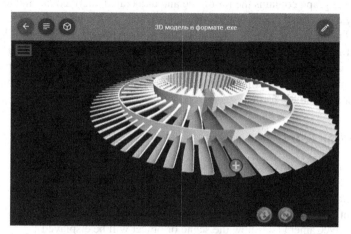

**Fig. 3.** The application window displaying 3D model as an executable file

When hosted by another application, a Unity application does not control the runtime lifecycle, so it may not work in some scenarios. It isn't possible to run multiple instances of a Unity app at the same time.

## 5   Using the HelixToolkit Library to Display 3D Models

HelixToolkit is an open-source 3D library that is under the MIT license (X11). The library used in the WPF platform makes working with 3D models much easier and provides capabilities that are unavailable in standard WPF tools. It can be used to import most 3D files of common formats such as STL, OBJ, FBX, 3DS, X3D, etc.…

This method uses the HelixToolkit.Wpf.SharpDX package based on a managed SharpDX.NET framework that is used for creating applications using DirectX 11 graphics technology.

## 5.1  Importing a 3D Model File

To load a 3D file, the Importer class from the HelixToolkit.Wpf.SharpDX.Assimp namespace is used. This class is based on the Assimp library, which allows you to import most of the widely used 3D file formats.

We should create an instance of the AssimpImporter class. This class loads the model file and extracts the model data into a format that we can access. Before loading the model file, using ImportFile(), it is better to check if the model file is of a format supported by Assimp. If so, we continue on to import the model file. Assimp provides a number of optional post-processing steps that can be performed on the imported model, namely, we can ensure that the imported model will have normal and tangent vectors by using the GenerateSmoothNormals and CalculateTangentSpace flags. If the mesh does not have normals or tangents, Assimp will calculate them from the mesh geometry. This function will return an Assimp.Scene object if it was able to successfully import the model, which contains all of the mesh information.

The scene graph contains the root node and associated intermediate and leaf nodes. In the body of the Load method of the mentioned class, information about the 3D scene is imported using the ImportFile method of the Assimp library. The ImportFile method extracts information about a 3D scene (meshes, materials, animations, etc.) from a file and places it in managed memory. Building a HelixToolkit scene includes processing the loaded information, building a scene graph on its basis, and loading the animation if it is provided by the file format.

## 5.2  Setting up a 3D Scene

Several components are required to display a 3D scene:

1. Viewport3DX;
2. Element3DPresenter (the container for the model);
3. PerspectiveCamera (sets how the scene or object will be displayed);
4. DirectionalLight3D (illuminates a 3D scene).

Viewport3DX is a class from the HelixToolkit.Wpf.SharpDX namespace that is a container for placing 3D content. In general, the viewport can be represented as a surface onto which a 3D scene is projected. This class contains properties such as Camera, CameraMode, and CameraRotationMode. Inside the Viewport3DX container, there must be an Element3DPresenter, which attaches the loaded model to the viewport. The camera determines the way the 3D scene is projected onto the 2D surface of the viewport. The Camera property can be set to PerspectiveCamera, which provides a perspective projection of the 3D scene, or OrthographicCamera, which does not apply a perspective viewpoint. The first value was chosen, therefore the objects that are farther away seem to be smaller, which is more familiar to the user.

It is also necessary to position the camera at some point in space and orient it in the desired direction. The point $(1,-1,1)$ was selected as the value for the property Position. It should be mentioned that the loaded models can be of different sizes. So, the ZoomExtends method is applied to the viewport to automatically scale the scene if the camera is initially placed too close to the model. The LookDirection property defines the orientation vector of the camera (the direction the camera is looking at). The UpDirection property determines the tilt of the camera. Usually, this property has the value (0,1,0), which means there is no tilt.

The CameraMode property determines the camera movement mode and can take the following values:

- Inspect - The camera rotates around the observation point (the position of the object is fixed, when zooming in the camera approaches the object);
- WalkAround - When the position rotation of the camera is fixed, the camera moves along its orientation vector when zooming;
- FixedPosition - The camera's observation point is fixed, so the scene size changes when zooming.

The first value was chosen as the most convenient for use.

The values of the CameraRotationMode property set different axes around which the model can rotate.

First, it should be mentioned that the HelixToolkit library uses a right-handed coordinate system by default: the origin is in the center of the rendering area, the Y-axis is upward, the Z-axis is toward the viewer, and the X-axis is to the right, as shown in Fig. 4.

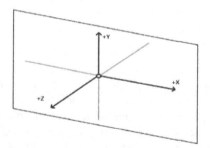

**Fig. 4.** HelixToolkit coordinate system

The CameraRotationMode property has the following possible values:

- Turntable - Object (model) can rotate around two axes (Y and X axis of the object);
- Turnball - Rotation can occur around three axes (the axis of the object coinciding with the vector of the camera's orientation, the X and Y axes of the object);
- Trackball - Rotation using a virtual trackball.

In order to simplify the use of this functionality the "Turntable" mode was selected.

To illuminate the scene, it was decided to use DirectionalLight value, which fills the scene with parallel light rays going in the direction of the camera orientation vector.

In Fig. 5 we can see the application window displaying the model in the FBX format. Using the HelixToolkit library, the following methods of interacting with the model are available: rotating and scaling the model, displacing it relative to the viewport, highlighting parts of the model, selecting and playing model animation.

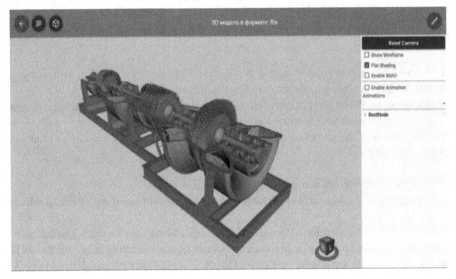

**Fig. 5.** The application window displaying FBX 3D model

## 6   Conclusion

The article describes two ways of displaying 3D models inside the digital catalog. The first method allows you to embed a standalone application containing 3D scenes, the second - to import a 3D file. As for the embedding method, you can develop and embed unity applications with extended functionality that is not available in the method that uses the HelixToolkit library. At the same time, the 3D model as an application takes up more memory than the same model in 3D formats that are supported in another method. Finally, you cannot integrate more than one Unity runtime. Therefore, your choice of the method will depend on the 3D model format.

A prospect for the development of this project may be the introduction of clickable models that give out information and technical characteristics of the design unit of the model on which the user clicked. This will allow integrating the description of the model so as not to separate it from its three-dimensional image. The HelixToolkit library allows you to realize a response on a part of the model click. This response will be to display information about the part.

Considering the trend towards paperless production and, as a consequence, the increased demand for three-dimensional models, the described digital catalog, and its methods of displaying a 3D model are very relevant and affordable means for demonstrating the benefits of new developments. 3D models can be used for exhibitions and presentations in order to convey the main concept in a clear and structured form that is understandable to the majority of listeners.

# References

1. Abramovici, M.: Future trends in product lifecycle management (PLM). In: Krause, F.L. (ed.) The Future of Product Development. pp. 665–674. Springer, Heidelberg (2007). https://doi.org/10.1007/978-3-540-69820-3_64
2. Terzi, S., Bouras, A., Dutta, D., Garetti, M., Kiritsis, D.: Product lifecycle management–from its history to its new role. Int. J. Prod. Lifecycle Manag. **4**(4), 360–389 (2010)
3. Stark, J.: Product Lifecycle Management, vol. 2, pp. 1–35. Springer, Cham (2016). https://doi.org/10.1007/978-3-319-17440-2
4. Sarcar, M.M.M., Rao, K.M., Narayan, K.L.: Computer Aided Design and Manufacturing. PHI Learning Pvt. Ltd., Delhi (2008)
5. Lombard, M.: SolidWorks 2007 Bible, vol. 529. John Wiley & Sons, Hoboken (2008)
6. Tornincasa, S., Di Monaco, F.: The future and the evolution of CAD. In: Proceedings of the 14th International Research/Expert Conference: Trends in the Development of Machinery and Associated Technology, 2010, vol. 1, pp. 11–18 (2010)
7. Saka, A.B., Chan, D.W.M.: Adoption and implementation of building information modelling (BIM) in small and medium-sized enterprises (SMEs): a review and conceptualization. Eng. Constr. Archit. Manag. **28**(7), 1829–1862 (2021)
8. Zhang, Y., Abourizk, S., Han, S.: Benefiting from three-dimensional infrastructure design and construction management. In: 11th International Conference on Construction Applications of Virtual Reality, Germany (2011)
9. Azhar, S.: Building information modeling (BIM): trends, benefits, risks, and challenges for the AEC industry. Leadersh. Manag. Eng. **11** (2011)
10. Komarov, I., Vegera, A., Zein, S., Borisova, S., Blazhenova, A., Gavrilov, A.: Design of tree-like database structure for solving test modeling tasks of energy equipment. In: Proceedings 2020 5th International Conference on Information Technologies in Engineering Education, Inforino, Moscow, Russian Federation, pp. 14–17, April 2020
11. Integrating Unity into Windows and UWP applications. Unity Documentation. https://docs.unity3d.com/2019.4/Documentation/Manual/UnityasaLibrary-Windows.html. Accessed 24 Dec 2020
12. Nathan, A.: WPF 4. Unleashed. Person Education, London (2010)
13. Troelsen, A., Japikse, P.: Pro C# 7 With .NET and .NET Core. 8th edn. APress, Berkley (2017)

# Key Concepts for Experience-Driven Innovation

I. A. Firsova[1]([⊠]) [iD] and V. V. Taratukhin[2] [iD]

[1] Faculty of Computer Sciences, Voronezh State University, Voronezh, Russia
[2] HSE University, Cand. Sc. (Technology), Moscow, Russia
victor.taratoukhine@sap.com

**Abstract.** Companies that seek to evoke a certain experience from their customers should not only change their design process, but also reorganize their innovation processes. Companies can innovate in their business in an effort to provide a certain consumer experience, and they can hire designers to create such an experience for their customers. Design thinking is the driving force of business, which corresponds to its status as a popular subject in leading international universities. Using design thinking, teams can generate great ideas and apply practical methods to find innovative answers. The analysis of the scientific publication in this article helped to identify and reveal the essence of innovations based on experience, their impact on three levels of the organization: the company, the brand and the level of the product/service. The identified creative process at each level can be characterized by four stages: the description of the context, the formulation of the vision, the generation of the concept and the final result. The article considered organizational measures and tools to support the design process.

**Keywords:** User experience · Radical innovation · Design-driven innovation · Experience-driven design · Design thinking

## 1 Introduction

A lot of everyday human experiences in industrialized societies are triggered by products. In the markets, consumers are no longer choosing products for their functional roles, but for superior usability features or engaging experiences. As a result, use and purchase of a product are increasingly dependent on whether the product or service is capable of eliciting a distinctive, relevant experience and the company's profitability may depend on the empirical qualities of the company's image and brand.

As experiential qualities become more and more important to market success, companies become more and more interested in creating certain experiences. Companies can innovate their businesses to provide a specific customer experience and they can hire designers to create that experience for their clients. But how does a company need to adapt its innovation process to provide the experience-driven design? This is the central issue for discussion in the article.

Desmet and Schifferstein described an experiential design approach that was developed at the Delft University of Technology and submitted 35 graduation projects, each

© Springer Nature Switzerland AG 2022
V. Taratukhin et al. (Eds.): ICID 2021, CCIS 1539, pp. 302–312, 2022.
https://doi.org/10.1007/978-3-030-95494-9_25

of them focused on creating a specific user experience [4]. Design projects in industry tend to involve multiple sides.

Several departments may be involved in a project, such as engineering, marketing, design and finance. Furthermore, the design project has to fit with the brand positioning and the company image.

In this paper, we would like to focus on how experiential design fits into the business context. Experience design is a holistic approach: the design team tries to make different design elements (e.g. function, look, grip, sounds, communication) support the customer experience [6].

## 2  Communication Usability and User Experimentation

Developers are working on creating systems that are simple and understandable for people. Terms such as user-friendly and easy to use often point to these characteristics, but the general technical term for them is usability. The GOST R ISO 9241-210-2012 standard on the ergonomics of human interaction with the system defines usability as:

A property of a system, product or service, in the presence of which an established user can apply products under certain conditions of use to achieve established goals with the necessary effectiveness, efficiency and satisfaction.

Efficiency is defined as the accuracy and completeness with which users achieve certain goals; efficiency is defined as the resources expended in relation to the accuracy and completeness of achieving the goals. satisfaction is defined as "freedom from discomfort and a positive attitude towards the use of a product [system, service or environment].

Also, many companies have been considering the following aspects of the usability part for a long time:

flexibility: the extent to which the system can adapt to changes desired by the user, in addition to those previously specified.;

learnability: the time and effort required to achieve a given level of system efficiency (also known as ease of learning);

memorability: the time and effort required to return to a given level of usage performance after a given period away from the system;

security: aspects of the system related to protecting the user from dangerous conditions and undesirable situations.

ISO standards on software quality refer to this broad view of usability as the quality of use, as it is the general user perception of the quality of the product.

We see that usability is not an absolute definition, but refers to users, goals, and usage contexts corresponding to a specific set of circumstances.

User Experience (UX) is the newest term in the set of criteria by which a system should be evaluated. It arose from the realization that as systems become more and more ubiquitous in all aspects of life, users are looking for and expecting more than an easy-to-use system. Usability emphasizes the feasibility of achieving specific tasks in specific usage contexts, but with new technologies such as the Internet and portable media players such as iPods, users are not necessarily eager to complete the task, but also entertain yourself. Therefore, the term user experience, originally popularized by Norman (1998), it appeared to encompass components of user interaction with systems and reactions to them that go beyond efficiency, effectiveness and traditional interpretations of satisfaction.

Different authors have emphasized different aspects of UX.

Dillon sharing the opinion that design needs to go beyond usability, and the evaluation of systems suggests that an emphasis is also needed on three key issues of user interaction with systems:

* Process: what the user does, for example, navigating a website, using certain functions, helping, etc. This allows you to develop an understanding of the movements, attention and difficulties of users;
* Results: what the user achieves, for example, what constitutes the purpose of the interaction. This allows you to understand what it means for the user to feel the shutdown.
* Affect: What the user feels; this includes the concept of job satisfaction, the definition of usability, but goes beyond that to include all the emotional reactions of users that can be amplified, irritated, enriched or confident. This allows us to develop an understanding of the emotional interaction of users with systems.

Bevan suggests that the definition of usability can be extended to encompass user experience, interpreting satisfaction as including:

* Likability: the degree to which the user is satisfied with their perceived achievement of pragmatic goals, including acceptable perceived results of use and consequences of use;
* Pleasure: the degree to which the user is satisfied with the perceived achievement of goals and the evocation of related emotional reactions.
* Comfort: The degree to which the user is satisfied with physical comfort.
* Trust: The degree to which the user is satisfied that the product will behave as intended.

## 3    What is Experience-Driven Innovation?

Experience-Driven Innovation (ExpDI) is characterized by an innovation process in which design plays a central role, combined with a design process in which the creation of a specific user experience is the starting point for the design [14]. Using experience as a driving force for innovation not only implies a change in the design process but also has implications for the entire organization.

A lot of companies create user experiences for customers at different levels. For example, a company can communicate with customers through a corporate image or through a series of brand images. All of these images can evoke certain emotions. In addition, the company strives for keep in touch with consumers through the products they manufacture and the services they provide.

It is impossible to fully predict the experience a customer will have, as that experience depends not only on design, but also on context-related factors such as the user's instantaneous mood, the situation in which the product is encountered, people who are involved, and so on. However, Expedia's goal is to create optimal conditions for users to have a specific experience.

Although innovation as a subject for IT companies and scientists has recently become one of the main sources of competitive advantages, the relationship between consumer experience and innovation is not something new.

Experience-driven innovations focus not only on the question of "how" (service), but also "what" (product) - an assessment of the entire customer experience in order to find new ways to introduce innovations. We need to focus on user values when introducing a new product.

Thus, most of the academic literature on innovations based on customer experience, was largely focused on the importance of common ground.

Firstly, each point of contact offers a unique chance for innovation - to distinguish the quality of customer service from competitors. This means that every time a customer interacts with a product, the organization has a chance to stand out from the competition. Thus, it is extremely important to get the right experience.

Secondly, innovation through customer experience can be achieved by defining different points of contact to identify different potential business opportunities for innovation. Once the points of contact are identified, it is necessary to determine the potential of each point of contact to "add value" to the consumer.

Finally, since the value is perceived individually, this makes the client a co-author of his own experience. Organizations can consider its role as intermediaries, creating the potential for value creation, however, the creation of the unique is in the hands of the consumer.

# 4   The Experience-Driven Innovation Model

Experience-Driven Innovation (ExpDI) can have implications for a company at least on three different hierarchical levels: at the company level, at the brand level within a company, and at the level of individual product or service offerings.

1. **Company level.** Several authors have suggested that companies and organizations can benefit from a more prominent role for designers and design thinking. This would mean that the organization has a fairly holistic and long-term strategic view of innovation. Rozendaal describes the relationship between company, designers and end users in experiential design as democratic because the relationship between different parties becomes quite equal: mutual trust and understanding are very important. [10].
2. **Brand level.** According to Roscam Abbing, brands can be a solid foundation for innovation when viewed as the relationship between the vision, culture, resources and capabilities of an organization on the one hand, and the needs, wants dreams, and aspirations of the user on the other. Building a brand-based innovation strategy requires the use of a brand that functions as a lens to view and interpret external influences, and as a projector to focus and filter internal influences. [eight]. Consequently, a brand can act as an intermediary in communication between a company and its customers.
3. **Product/service level.** Individual product or service offerings provide the most specific and direct opportunities to interact with potential users. Here, all touchpoints and personal interactions directly affect the customer experience.

The creative process at each of these levels can be described in four different general stages. These stages are called here the description of the context, the formulation of the vision, the generation of the concept and the final result. Since the creation of a certain experience is central to the creative process, experience-based design involves at least two important tasks: the first is to determine which experience to strive for, and the second is to create a design that is expected to evoke this experience.

1. **Context description** includes a description of the context for which the innovation is being developed: it gives an overview of the current situation. It involves the analysis of external states and events. In addition, it assesses the internal qualities of the company. He assesses how the company and its environment will change over time, as innovation will only come to fruition soon.
2. **Vision formulation** includes the development of the company's vision, brand and new proposals. Based on the outline of the future context, the company must develop ideas about what they would like to offer to future clients. This is where experience comes in too: In ExpDI, visions typically describe what the company would like its customers to experience: how will they feel, what they will think, how they will feel?
3. **Concept generation:** Once the company has decided what they would like to offer, they can start generating ideas for how that promise can be fulfilled, and they can develop concepts to match the proposed proposal, according to the current set of constraints. The following questions should be asked here: Who will be the users of the design? When will they use it? Under what circumstances? What will the design do? How does he work?
4. **End result:** In the final stage, the concept is transformed into reality by finding materials, products and people who will make it possible.

Ultimately, the results obtained in various creative processes will contribute to the future context for subsequent projects. Thus, the circle has closed, and new innovation processes can begin again. This leads to the following outline of the innovation model (Fig. 1):

The sketch of the innovation model shows how the company (the core of this model) relates to customers (consumers, users) in the outside world. The experience that a company's offerings evoke in its customers will affect how the outside world perceives the company.

The ExpDI model explicitly recognizes that innovation occurs at the company, brand, and product level in an organization simultaneously and interdependently. Many innovation models are presented linearly, from the strategic decision to initiate innovation to the launch of a product or campaign (for example, Ulrich and Eppinger [13]). In contrast, the ExpDI model follows the circular shape suggested by Buijs, indicating that each innovation adds to the existing world and thus to the context for innovation processes [3].

## 5   Organizational Measures

Companies can take some organizational measures to support experiential innovation. We have collected some of them below. These indicators were mainly derived from the

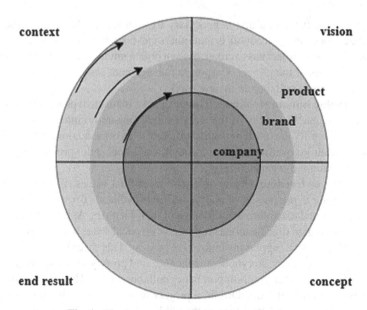

**Fig. 1.** The innovation model based on experience

innovation management literature, which emphasizes the centrality of the design process (e.g., Beverland and Farrelly [1]; Brown [2]; Dougherty [5]; Verganti [16]).

- **Develop overarching company mission.** The company we are looking for has an overarching goal to provide a sense of direction for individual projects. The company should be able to indicate what and how they would like to improve the lives of consumers.
- **Integral project management over time.** For experiential projects to deliver superior results, project management must ensure that original ideas, including design richness, are not lost in the future. Teams must avoid distraction due to restrictions (Verganti [16]). Experience design processes are generally quite unpredictable, especially in the early stages (Rozendaal [10]), and therefore it is important to share the initial points of view among all stakeholders.
- **Strategic road mapping.** Roadmap enables an organization to make strategic decisions about the optimal time frame for specific activities, such as the sequential introduction of new technologies, customer experiences, or product options in the market.
- **Interdisciplinary teams.** Departments within a company and people in different disciplines tend to use their mental models and interpretation schemes. Each discipline has the knowledge that contributes to the final design, and therefore the formation of interdisciplinary teams in which all departments are represented is essential for successful innovation (Dougherty [5]).
- **Break through organizational routines.** Organizational routines tend to reinforce the separation of thought worlds by limiting the interactions between people from different departments. Existing procedures typically promote within-department efficiency and

inhibit the collective action that is necessary for innovation. Another problem with standardized tools and procedures is that others can copy them. Therefore, in order to be innovative, each project has to create its own organizational structure, depending on the specific needs of the project. Corporate management should establish incentives for business units to collaborate in new ways.

- **Use and develop human resources.** The success of innovative processes in products largely depends on the quality of the work of the company's employees. Companies must learn to identify the most innovative thinkers in an organization. In addition, they can develop their innovative potential through workshops, pilot projects, and sharing inspiration within the organization.

- **Create employee freedom.** Create an environment that values risk-taking and constant questioning in the pursuit of continuous improvement (Brown [2]). Employees should be permitted to explore the full range of their faculties. Any experiment should have a chance to gain organizational support. A physical space where a group of people can be creative will support the innovative process. This space should allow project materials, such as photos, storyboards, concepts, and prototypes to be available all of the time. The project space will support better collaboration and better communication with partners and clients. In addition, a project website or wiki helps to keep team members in touch when they are out in the field. Flexibility is a key element of design thinking. Therefore, the designated physical and virtual space, and also the budget should be adaptable.

- **Build external relationships.** Many innovation projects benefit from external input. It is important for the company to set up a privileged network of relationships with external parties. Relational knowledge is tacitly preserved and nurtured by the people in the organization. As it involves their personal relationships to others, this cannot be easily copied by others (Verganti [16]). The nature of these assets is cumulative: the more privileged relationships you have, the more you are likely to use these contacts to develop breakthrough innovations and develop seductive visions. You can also hire external experts on a project basis: The active participation of partners will yield more ideas and creates a web of loyalty that will be hard for your competitors to erode. Alternatively, you could set op open innovation projects, or involve external designers by initiating design competitions.

- **Internal technological development.** You must master unique technologies to be an interesting partner for collaboration or discussion. Invest resources in research and development that provide new technologies to develop new solutions. Develop technologies that you or others would like to use (Verganti [16]).

- **Build knowledge on latent user needs.** To create a successful new product, you would like to know what your future customers are doing, thinking and feeling. Therefore, you should try to uncover hidden needs that users may not even be aware of to meet their future needs. Typically, this requires unconventional marketing research methods, including observational work, ethnographic research, and in-depth interviews. It is important to focus on the quality of the information received, not the quantity. Try to connect with the people you observe by entering their role: experience the design from the client's point of view.

# 6   Tools Supporting the Design Process

Companies can use several tools to support the experiential design process, some of which we have collected below. Again, our list is intended as a source of inspiration and is unlikely to be complete.

- **Develop open context vision.** When designing the context for the creative process, it may be important to create as much design freedom as possible and thereby keep the actual constraints to a minimum (Hekkert and van Dijk [7]). Determining the context implies defining the area for which the design is being created.
- **Develop experience vision statement.** The company develops a vision on what type of experience you would like to evoke among future customers. This goal can be formulated in terms of an intended experiential user effect or 'target experience'. It is also a statement on what you would like to contribute to the future world. It is important for the statement to carry authenticity: Ideally, the statement should address a fundamental customer need.
- **Pay attention to multiple layers in the user interface.** The perception of design can include different levels: materiality, function, interaction, sensory perception, aesthetics, meaning, emotional reactions, and so on. If different layers are considered during the creative process, it will increase depth and richness in the design process. For a design to evoke a certain experience, the client will have to interact with it in a certain way, and the product needs qualities that support that interaction. Therefore, it can be useful to try to translate the characteristics of the target experience into a set of interaction qualities, which, in turn, can be translated into a set of desirable product qualities (Hekkert and van Dijk [7]). The qualities of interaction must support the authenticity of the target experience.
- **Include the time dimension of user experience.** The user experience usually evolves as the user experience evolves. Storytelling helps connect ideas and create custom scenarios. A story can also be an open narrative that engages and motivates people to move it forward and fill in their conclusions. Writers may be involved from the start of the design process to help move the story along in real time. A scenario may grow and develop into a description of a consumer journey, including all the events a consumer experiences while engaging with a brand, product, or service. The story could also be an open-ended narrative that engages people and stimulates them to carry it forward and fill in their own conclusions (Brown [2]).
- **Engage in multiple design disciplines.** To bring proposals to life, multiple design disciplines can work together to deliver optimal impact. You can include product design (engineering, design design, packaging design), communication design (graphic design, advertising, digital media, corporate identity, signage), interaction design (physical interaction with buttons, controls and levers, interface design), design POS environments (architecture, interior design, exhibition design), service design (guarantee forms, personal interactions, complaint forms, call center procedures) and so on (Roscam Abbing [8]).
- **Empathy tools.** Using empathy tools helps the development team stay connected to the targeted user experience. In addition to the vision statement, the team can create a mood board that communicates the emotional impact of the target outcome.

Alternatively, a persona can be created that provides a vivid representation of a fictional end-user, designated by a realistic name, whose life is described and clearly expressed through several images (e.g. Sleeswijk Visser [12]).

- **Formalize brainstorming routines.** Brainstorming sessions are valuable in the creative process, not only for the ideas that pop up during the session, but also for the concepts and solutions that occur to people later, at home, due to the seeds planted in their minds. Some companies have formalized the routine procedure used during brainstorming to optimize the output of these sessions. For instance, IDEO uses the following rules: Defer judgment, build on the ideas of others, hold one conversation at a time, stay focused on the topic, and encourage wild ideas (Brown [2]).
- **Dirty and quick prototyping.** Encourage fast, cheap, and dirty prototyping as part of the creative process, not just as a way of testing out ideas. Use visual and physical tools, not just abstract words, to generate ideas. By creating material, physical objects, you are using different abilities and different types of knowledge compared to when you just talk and think. In addition, you can act out situations to learn how prototypes work and what is needed to improve them (Brown [2]).
- **Create and present conceptual prototypes.** Concept prototypes, such as the concept cars shown at auto fairs, embody the company's vision for the future. They make ideas physical and tangible, and their creation helps explore the feasibility of ideas. They can be an important tool within a company to optimize the flow of creative ideas and show the direction in which the company is heading. In addition, by presenting a new product at a trade show or by making a book for researchers, colleagues will learn about the product, they will tell others about it and can describe it in journals to the general public. Ideally, a new product should generate buzz. Thus, books, exhibitions, cultural events, concept presentations at fairs, magazine articles, websites, and design contests all serve to create free advertising for the new design (Roskam Abbing [8]; Verganti [16]).
- **Co-create with end users.** End users may be involved actively in the design of the new brand, product, associated services (e.g., websites, interactions with sales staff, social networks), and retail experiences. For instance, designers may develop tools that enable consumers to create their own products. Researchers can study consumers while they are using such tools. Also, designers can create experiences, by involving consumers actively in the store (e.g. cooking in a supermarket). Involving consumers in creating and using products may increase the chance of success in the market.
- **Formalize choice among ideas.** In order to make the right choices when you evaluate ideas, develop a good procedure and involve the right people. Criteria that are often used to evaluate ideas include the expected functionality of the end product and people's expected motivations for buying the product. Furthermore, the available production facilities and the expected costs are likely to affect decision making. Another criterion may be the product's potential for a long life cycle: does it satisfy deeper needs? Is it clearly linked with the brand identity? Furthermore, ideas that create a buzz should be favored. Measurement of impact helps to make the business case and ensures that resources are appropriately allocated. If each team member chooses the three best ideas by putting a mark on them, you can select the ideas with the largest number of votes (Brown [2]).

# 7 Conclusion

This work is devoted to the analysis of an innovative model based on experience (Experience-Driven Innovation). Its influence on the development of technology and the derivation of its key levels and stages. A sketch of the innovation model was presented, we provided a list of organizational measures that support innovation processes, and a list of tools that can be useful in the creative design process.

This data will be used in further work in the direct design of interfaces to improve the user experience.

In the future, I would like to introduce such an innovation as "Design Thinking" into the work of my organization. The organizational measures and tools given in the article that support the design process will help me in implementing the innovation process.

I will face the following tasks:

Build an "AS IS" model to identify the weakest and most vulnerable areas of the company's activities.

Conduct a web analysis to understand what users usually pay attention to, what nuances and shortcomings there are in the work of the team, and what needs to be corrected or supplemented.

To build a "TO BE" model to reflect the idea of new principles of the software testing stage within the framework of the design thinking methodology.

Test the model and apply it in a working project.

# References

1. Beverland, M., Farrelly, F.J.: What does it mean to be design-led? Des. Manag. Rev. **18**(4), 10–17 (2007)
2. Brown, T.: Change by Design: How Design Thinking Transforms Organizations and Inspires Innovation. Harper Collins Publishers, New York (2009)
3. Buijs, J.: Modeling product innovation processes, from linear logic to circular chaos. Creat. Innov. Manag. **12**(2), 76–93 (2003)
4. Desmet, P., Schifferstein, R.: From Floating Wheelchairs to Mobile Car Parks. A Collection of 35 Experience-driven Design Projects. Eleven International, The Hague (2011)
5. Dougherty, D.: Interpretive barriers to successful product innovation in large firms. Organ. Sci. **3**, 179–202 (1992)
6. Ceccacci, S., Giraldi, L., Mengoni. M.: From customer experience to product design: reasons to introduce a holistic design approach. In: 21st International Conference on Engineering Design (ICED 2017) (2017)
7. Hekkert, P., van Dijk, M.: Vision in Design: A Guidebook for Innovators. BIS, Amsterdam (2011)
8. Roscam Abbing, E.: Brand-driven Innovation. Strategies for Development and Design. AVA, Lausanne (2010)
9. Roozenburg, N.F.M., Eekels, J.: Product Design: Fundamentals and Methods. Wiley, Chichester (1995)
10. Rozendaal, M.C. Investigating experience design in industry. In: Sato, K., Desmet, P.M.A., Hekkert, P., Ludden, G., Mathew, A. (eds.) Proceedings of the 7th International Design & Emotion Conference, 4–7 October 2010, Chicago, USA (2010)

11. Schifferstein, H.N.J. Multi-sensory design. In: Hooper, C.J., Martens, J.B., Markopoulos, P. (eds.) ACM Proceedings of the DESIRE 2011 Conference - Creativity and Innovation in Design, 19–21 October 2011, Technical University, Eindhoven, the Netherlands, pp. 361–362 (2011)
12. Sleeswijk Visser, F. Bringing the everyday life of people into design. Ph.D. dissertation, TU Delft (2009)
13. Ulrich, K.T., Eppinger, S.D.: Product Design and Development. Mc Graw-Hill, New York (1995)
14. Luenendonk, M.: The Innovation Process: Definition, Models, Tips
15. Van Erp, J.: Don't mind the gap. In: Desmet. P., Schifferstein, R. (eds.) From Floating Wheelchairs to Mobile Car Parks: A Collection of 35 Experience-Driven Design Projects,' pp. 225–232. Eleven, The Hague (2015)
16. Verganti, R.: Design-Driven Innovation: Changing the Rules of Competition by Radically Innovating What Things Mean. Harvard Business Press, Boston (2009)

# Development of a Robot for Building a Map of the Area

A. A. Papina$^{(\boxtimes)}$ 🆔

Voronezh State University, Universitetskaya Square 1, Voronezh, Russia

**Abstract.** The active development of the field of robotics makes it necessary to study the possibilities of new approaches. The purpose of this study is to identify the operability of the Kalman filter as a tool for calculating the SLAM (Simultaneous Localization And Mapping) method for the orientation of a two-wheeled mobile robot. The robot itself is a moving platform on which a servo drive with an ultra-sonic sensor is installed to scan the front hemisphere. Data about the surrounding area is transmitted from the robot to the computer as the robot moves. For data transmission, a wireless Wi-Fi network is used for autonomous operation of the robot or an RS-232 interface for debugging. The program filters out the received data using the Kalman filter and builds a map. The result of the experiment is to obtain an adequate map of the room, which confirms the efficiency of the pro-posed models.

**Keywords:** SLAM · Kalman filter · Robot · Arduino · Mapping

## 1   SLAM Method

In order for the robot to build a map of the area and at the same time determine its position in this area, the SLAM method was used, which means Sim-ultaneous Local-ization And Mapping [1].

Currently, there are many implementations of the SLAM method, differing in complexity and various platform capabilities.

To display the state of the environment, modern systems use probabilistic methods and benchmarks. During the localization of the robot, its state $x\_t$, for example, the position of the robot, should be evaluated by all previous readings of the sensor $z(1, \ldots, t-1)$ and previous actions of the robot $u(1, \ldots, t-1)$. Localization of robots is often solved using the Markov property, that is, it is assumed that the world is static, noise is independent and there is no approximation error during simulation. Thus, due to the Markov property, the localization problem is reduced to the Bayes network shown in Fig. 1.

In addition to the state of the robot, the map m must be evaluated, as shown in Fig. 2. Usually the map consists of n landmarks, so the position of these landmarks must be estimated to build the map.

© Springer Nature Switzerland AG 2022
V. Taratukhin et al. (Eds.): ICID 2021, CCIS 1539, pp. 313–322, 2022.
https://doi.org/10.1007/978-3-030-95494-9_26

314    A. A. Papina

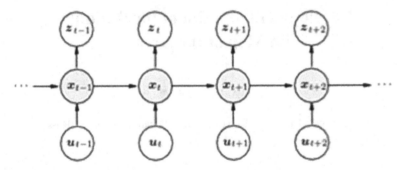

**Fig. 1.** Bayesian network for robot localization.

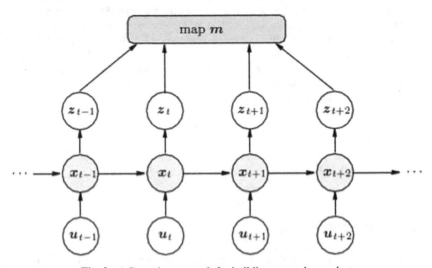

**Fig. 2.** A Bayesian network for building a map by a robot.

When the robot moves in a known area, that is, it is given a map, the uncertainty of the robot's position is minimized. However, if the robot is moving in an area un-known to it, the uncertainties of the robot's positions can be large, since odometry errors are summed up. Therefore, it is necessary to observe the landmarks in order to accurately assess the position of the robot up to the uncertainty of the position of the landmark and observation. Sensors operating in the global coordinate system, such as GPS/GLONASS, are a workaround using odometry, but they are not always avail-able, for example, inside buildings.

Closed loops play an important role in SLAM algorithms. If the robot determines the position in which it was previously and correctly compares the landmarks, then the accumulation of error is limited. The probabilistic method used in this experiment is the Kalman filter.

## 2 Kalman Filter

The Kalman filter is a sequential recursive algorithm that evaluates the state vec-tor of a dynamic system using a series of incomplete and noisy measurements.

The algorithm consists of two repeating phases: prediction and correction. At the first stage, the prediction of the state at the next moment of time is calculated (taking into account the inaccuracy of their measurement). In the second, the new information from the sensor corrects the predicted value (also taking into account the inaccuracy and noise of this information): The equations in Fig. 3 are presented in matrix form. In the case of one variable, the matrices degenerate into scalar values [2].

The subscript denotes a moment in time: k is the current one, (k-1) is the previous one, the minus sign in the superscript indicates that this is the predicted intermediate value.

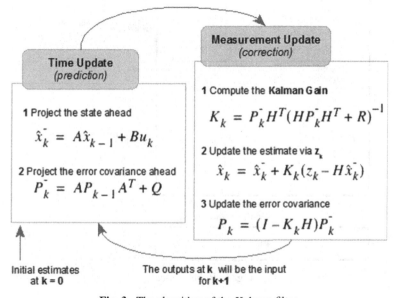

**Fig. 3.** The algorithm of the Kalman filter.

The Kalman filter in the SLAM method combines all measurements of landmarks in the covariance matrix of the positions of the landmarks and the positions of the robot. For linear measurement models, this corresponds to the maximum likelihood method, and for nonlinear measurement models, this will be true only if the measurement function can be linearized. Linearization is performed during measurements, at the linearization point it may be erroneous if the robot moves through an unknown area of the map. This error is critical, since trigonometric functions have a small quasi-linear interval. The result of erroneous linearization is that closed loops are too large compared to the basic truth [3].

# 3  Robot Device

## 3.1  Scheme

All elements of the robot are mounted on a moving platform. The platform is driven by two electric motors. The platform's autonomy is supported by two batteries. The ultrasonic sensor is mounted on a servo, thanks to which it is possible to scan the front hemisphere. The Arduino Uno training microcontroller board was chosen to control the robot platform. An ultrasonic sensor was chosen as a sensor for assessing the surrounding space. Communication with the computer is carried out using a COM port and a Wi-Fi module. The diagram is shown in Fig. 4.

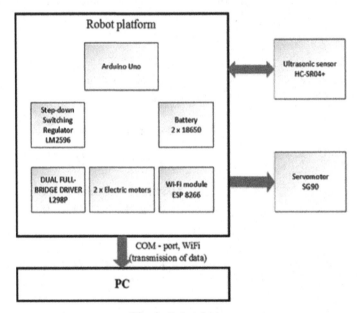

**Fig. 4.** Robot device.

## 3.2  Algorithm of the Robot's Actions

At the first step, the coordinates of the robot are transmitted via the COM port, at the initial moment of time they are zero.

At the second step, the measurement takes place up to obstacles in the front hemisphere, due to the rotation of the ultrasonic rangefinder by a servo drive with a step of 1 degree, from 0 to 180. First, the robot makes a measurement to the obstacle, then the coordinates of these points are calculated using the formulas:

$$x = \cos(pos * 3.1416/180) * dist);$$

$$y = \cos(pos * 3.1416/180) * dist);$$

where dist is the distance to the obstacle, pos is the angle of rotation of the ultrasonic sensor.

The obtained x and y values are processed using the Kalman filter, and those values that are not within the sensor's operating range (from 2 to 400 cm) are filtered out, since they are erroneous. Errors occur due to the reflection of the ultrasonic wave from the walls of obstacles [4].

The correct values are transmitted at each step from the Arduino via the COM port to the computer for further work with them.

At the third step, the robot moves, which is shown in Fig. 5, upon completion of which new coordinates of the robot will be received and transmitted via the COM port to the computer.

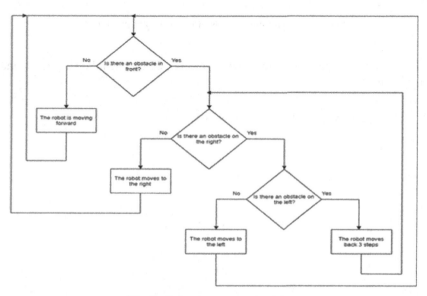

**Fig. 5.** Robot movement algorithm.

## 4 Building a Map of the Area

According to the received data, a map is being built in real time in the MapBuilder program, a large red dot indicates the position of the robot, and blue dots indicate obstacles when the sensor moves clockwise, black dots indicate counterclockwise. The resulting map is saved in the map.png file in the root folder with the program. An example of the program's working window is shown in Fig. 6 [5].

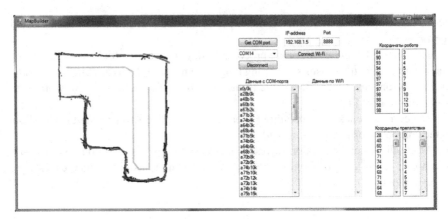

**Fig. 6.** The working window of the computer program.

To illustrate the operation of the robot algorithm, we use the following terrain map shown in Fig. 7.

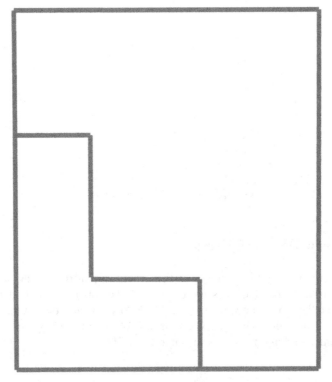

**Fig. 7.** Map of the area.

To begin with, let's consider the construction of a terrain map at a single step, based on "raw" data obtained from an ultrasonic sensor. When passing clockwise, we get the following image, Fig. 8.

**Fig. 8.** Map of the area based on "raw" clockwise data.

Superimpose the data from the ultrasonic sensor on the resulting map when moving counterclockwise, Fig. 9.

**Fig. 9.** Map of the area based on "raw" counterclockwise data.

Thus, from these figures it can be concluded that the "raw" data do not give a real idea of the terrain.

Now consider the data that comes from ultrasonic sensors and processed by the Kalman filter. Figure 10 shows the data obtained by rotating the sensor clockwise.

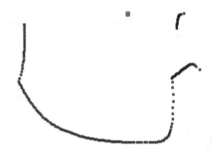

**Fig. 10.** Map of the area based on filtered clockwise data.

We will apply filtered data from the ultrasonic sensor to the resulting map when moving counterclockwise, Fig. 11.

**Fig. 11.** Map of the area based on filtered counterclockwise data.

The received card will also not be fully valid. Therefore, in the next step, it is necessary to remove inaccuracies in the construction, namely, it is required to leave points that intersect or are at close distance from each other when the sensor rotates clockwise and counterclockwise. The result is shown in Fig. 12.

**Fig. 12.** Map of the area based on real data.

These actions are repeated at each step of the robot (the robot's step in operation is 10 cm), resulting in an approximate map of the area. The resulting map is shown in Fig. 13.

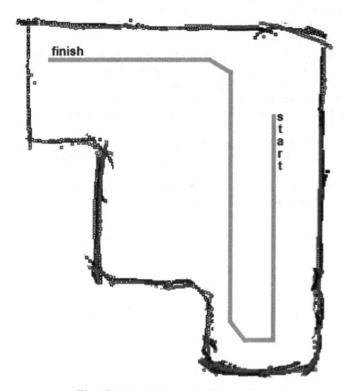

**Fig. 13.** Map of the area built by a robot.

## 5  Result of the Experiment

The result of the experiment is a constructed map of the room. This map corresponds to the actual positions of the obstacles, which confirms the correctness of the algorithm. The presence of minor noises in the resulting result is caused by errors in the measurements of ultrasonic sensors.

LiDAR or Kinect sensors can be used to improve the final result. To obtain a circular terrain map, the sensors will need to be rotated in the XY plane. It should also be noted that these sensors will be able to work to build a three-dimensional map of the area. Some LiDAR models will get a circular picture of the terrain, which allows you to reduce the time taken to build at a single step and the time to build a general map.

## References

1. Zunino, G.: Simultaneous Localization and Mapping for Navigation in Realistic Environments (2002)

2. Kalman filter. https://habr.com/ru/post/166693, Accessed 25 July 2020
3. Castellanos, J.A.: Mobile robot localization and map building: a multisensor fusion approach (2012)
4. Leonard, J.J.: Directed sonar sensing for mobile robot navigation (2012)
5. Troelsen, A., Japikse, P.: Pro C# 7. Apress (2017)
6. Welch, G.: An Introduction to the Kalman Filter. Department of Computer Science University of North Carolina (2006)
7. Valencia, R., Andrade-Cetto, J.: Mapping, Planning and Exploration with Pose SLAM (2018)
8. Jan, G.E., Chang, K.Y., Parberry, I.: Optimal path planning for mobile robot navigation. IEEE/ASME Trans. Mechatron. 13, 451–460 (2008)
9. Guibas, L., Motwani, R., Raghavan, P.: The robot localizationproblem. In: Goldberg, K., Halperin, D., Latombe, J.C., Wilson, R. (eds.) Algorithmic Foundations of Robotics, pp. 269–282. A. K. Peters (1995)
10. Mata, M., Armingol, J.M., Escalera, A., Salics, M.A.: A visual landmark recognition system for topological navigation of mobile robot. In: International Conference on Robotics and Automation (2001)
11. Warren, J.-D., Adams, J., Molle, H.: Arduino Robotics. Apress, Berkeley (2011). https://doi.org/10.1007/978-1-4302-3184-4

# Towards to Extended Empathy Methods in Design Thinking

Valeriia Skibina[1]([⊠]) [iD] and Victor Taratukhin[2] [iD]

[1] Lomonosov Moscow State University, Moscow 119991, Russia
skibina.vm19@physics.msu.ru
[2] Higher School of Economics, Nizhny Novgorod 603000, Russia
vtaratoukhine@hse.ru
https://istina.msu.ru/profile/foxoonake,
https://www.hse.ru/staff/taratukhin

**Abstract.** The paper analyzes existing empathy methods and offers new approaches so far as of empathy being an important element for design thinking (a Stanford methodology for creative problem-solving) and companies post-COVID era redesign. It comprises an overview of such methods as: VR Technology, VR Empathy Mirror, Dare Dream World Modeling, Fingers Magic, Cosplay, Scenes, and Role games. The paper takes into account different views on the origin of consciousness: molecular biology, neuroscience, pharmacology, quantum information theory, quantum gravity and etc. Many different hypotheses on the related topics and existing approaches are discussed. The article pays attention to issues related to both the opportunities of practical implementation and the moral aspects, possible consequences of the inventions in empathy like a glance on perspective. One's Experimental basis for each of the approaches is provided. After all, understanding of human thinking/feeling is placed at the heart of the global understanding of all vital processes that surround us.

**Keywords:** Design thinking · Virtual reality · Empathy · Cosplay · Role game · Consciousness · VRE-IP experiment · Dream world modeling · Fingers magic · Scenes · Companies redesign · Empathy · COVID-19 · Coronavirus

## 1 Introduction. Where Consciousness Lies?

### 1.1 A Subsection Sample

While everybody thinks that **_consciousness_** comes from our brain, Orchestrated Objective Reduction postulates that it actually originates at the quantum level among the el-ementary particles through statistically random events being acquired [9, 10]. That controversial hypothesis has absorbed several absolutely different areas of knowledge: molecular biology, neuroscience, pharmacology, quantum information theory, quantum gravity - on the border between which innovations are more likely to lie.

So far - consciousness is based on non-computable quantum processing performed by qubits (the basic units of quantum information). That might just mean that the human mind should be assumed as a result of quantum random statistical distribution process.

© Springer Nature Switzerland AG 2022
V. Taratukhin et al. (Eds.): ICID 2021, CCIS 1539, pp. 323–335, 2022.
https://doi.org/10.1007/978-3-030-95494-9_27

However, much of criticism of the Orchestrated Objective Reduction hypothesis is based exactly on the strong statement of the human brain's unsuitability.

Anyway, whatever processes are, some of the eternal questions remain:

- *How to read human minds?* Can we transfer so-called brain neuron impulses through space? Can the human mind be controlled? (By the way, never believe people who say that they can read minds by EEG, cause if they actually can read some kind of texts generated inside your brain (which is doubtful) - that is not even close to being the word of "thoughts" meaning. Cause thoughts have never been just texts. It is way far more than just that. It is qubits - units of quantum information, neuron impulses - whatever, but not just simple "littera" (from Latin). The simplest interpretation of what is to be said here is the thoughts picturing. Do you think when you imagine something? Or do you want to argue with the idea that at this moment your brain stops working? So thoughts are rather of being physical impulses of information transmission that does not have to be represented only by letters)
- *What is empathy* in the language of neuron processes and human interaction? Can it be controlled or upgraded?
- While avatar's project is under construction, can a small step be done - just on the way to understand whether it is possible *to put one person in the shoes of another*? Would it mean that we set the sustainable neuron connections between two persons and whether it is dangerous or beneficial?
- *How do new technologies help us understand each other* and why role games are more likely to sound like an attempt to discover the world inside out?
- And so on...

In order to find a way to answer the questions listed above and moreover - current world situation has been analyzed, new ideas have been generated and questioned from the different sides of view.

*The current world situation* may be presented as a puzzle of many interrelated circumstances pieces.

Coronavirus has changed the world around us, so it is never going to be the same any-more. It has become obvious at least while observing perception and transmission of information through devices with limited technical characteristics.

This year, full of coronavirus town lock-downs, people have learned distance empathy including microphones control, chat emoji usage (raised hand, applause), screen demonstrations, waiting hall, network connection problems, and so on. Many nuances affected the perception of communication (Fig. 1).

In instance, zoom killed the fine art of interrupting. Despite the fact that there already has been a lot of cultural differences with respect to how much overlap can be considered cooperative, distance discussions made interrupting less intensive and less productive in its most creative and productive form (ideas generation). On the other hand, a good opportunity of commenting during conversation appeared to be possible for shy students. In the real world, they are unlikely to get that chance to be heard.

To be highlighted - disputes about the relevance of personal meetings and their practical necessity. Moreover, there is no obvious tendency about whether it is useful to go offline or online. People are inclined to think rather contrasting. While some of them

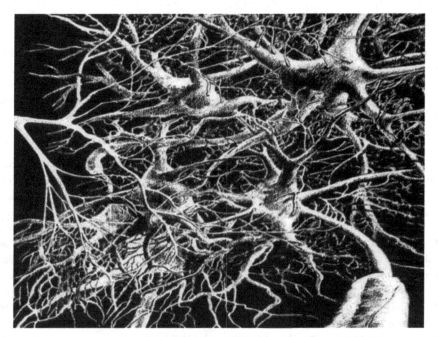

**Fig. 1.** Our brain contains over 100 billion neurons each that may have the ability to connect with 5.000 to 20.000 other neurons. The possibilities are endless. ©Pinterest

got used to setting comfortable meetings offline, other people prefer online meetings and find them more autonomous nowadays. While some of them see online meetings as benefits in new technical tools usage, other people point that technics sets its specific limits on your original creative abilities, when you feel like inside a box being unable to realize all of your potentials.

One thing is certain - *existing companies are inclined to be redesigned* in order to adapt to the changing world and to take a leading place in the new market, full of innovative approaches - to be able to solve problems and to function successful remotely if necessarily [2, 4].

New-modern empathy and interaction methods might be extraordinarily useful in that case [1].

## 2  Design Thinking

Design thinking is a Stanford methodology for creative problem-solving. The results of the design thinking can be a framework for real-world projects [3, 4].

Empathy is an important element in Design Thinking and Human-Centred Design. Empathy helps design thinkers create solutions that work and, conversely, how a lack of empathy can result in product failure [17].

## 3  VR Technology

Commercially profitable *virtual reality* equipment uses a headset with sounds, multi-projected images, and at least one joystick to be able to move around and to make choices in virtual space.

Interesting thing is that in virtual reality person *does not feel his actual physical body* the way he got used to. He distracts from it, forgets about it, focuses on the space around him.

Amazing words had been written by *Jaron Lanier in his book "You're not a gadget"*:

The body and the rest of reality no longer have a prescribed boundary. So what are you at this point? You're floating in there, as a center of the experience. You notice you exist, because what else could be going on? I think of VR as a consciousness-noticing machine.

From the personal experience, it can be said, that the appearance in virtual space of breezy unexpected objects (aquarium, Pikachu, and other), new possibilities (such as sculpting), fundamentally original environment, and principally other bases of interaction *develops creativity* and contributes to the generation of ideas, despite its unusual-ness [5]. This was shown by SAP University Alliances *Virtual Reality Experiment - Innovation Project (VRE-IP)* carried out in online form between people all over the 5 world to prove the effectiveness creativity of using VR - technologies for teamwork innovation, to explore advanced Design Thinking approach [3] (Fig. 2).

**Fig. 2.** Screenshot of virtual reality room where VRE - IP experimental seminar was held. There are participants sharing their emotions using virtual emoji-es there.

## 4  VR Empathy Mirror: Its Benefits and Dire Consequences

A kind of mind-reading is possible through empathy - to be able to become anybody and anything and understand the system from the inside. Such everyday reading of minds is the basis of productive interaction (in fact, it becomes possible to extract benefits purposefully: connect and further read the emotions and directions of another organism).

Thus, VR technologies open up the *following opportunities* for us on the event horizon:

- *To extract the experience* of people being far away from us (mentally/physically), to understand each other better
- *Force someone to forcibly plunge into the world of another person.* That to be considered as one of the most crucial weapons, machine of suffering itself and not recommended for further usage (Fig. 3).

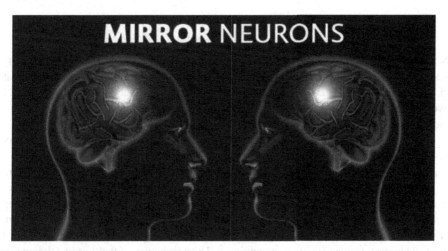

**Fig. 3.** A mirror neuron is a neuron that fires both when an animal acts and when the animal observes the same action performed by another. Thus, the neuron "mirrors" the behavior of the other, as though the observer were itself acting. Such neurons have been directly observed in human and primate species, and birds" [6–8]. It is believed that Mirror Neurons give us empathy. ©Google images

Each of the options can be called an *"empathy mirror"*. And as mentioned above, it can bring the world both good and destructive consequences. Moreover, it might be found quite similar to far-known *psychokinesis* which allows a person to influence a physical system of another organism without any physical interaction. Talking about a physical system here means an actual potential effect on the person's life (physical as well, because all the things are related).

Here are some *live examples* of Empathy Mirror creating attempts.

*Rapid Design Pivot* is a project that began with funding from Skoltech in Moscow, Russia. The project encompassed research from all three facets of Entrepreneurship, Technological Innovation and Contemporary Art. Rapid Design Pivot aimed to partner inventors and artists to produce art exhibitions and technology prototypes using state-of-the-art scientific discoveries and technological advances as a creative medium. Researchers approve that introducing and establishing a Rapid Design Pivot model early in the process of development will allow developers to discover diverse and multiple uses and markets more efficiently [13].

After the end of Skoltech funding, Rapid Design Pivot continued with additional funding from UT Austin, and the art installation related to the project was called *Omnibus Filing*. In each artwork, there was an intersection of research and art—exploring themes of empathy through multiple lenses such as interspecies architecture and artificial intelligence.

Another attempt of creating an actual Empathy Mirror [15] should be mentioned. The team explored such questions as "What are the ways that technology can/does promote empathy?" The members of the project pointed themselves as of being developing an art, science, and technology installation for soon release.

## 5    Dare Dream World Modeling

Nowadays there is no secret to anyone that virtual reality can be used for a better understanding of others. Another question that arises and becomes incredibly relevant is that of **dream modeling**. It is not a surprise if your friend says that he might have control over his own dream while sleeping. It happens, right? Now think of you, being able to model somebody's dream.

In fact, dream control leads to the direct ownership of the world. Because in this way you can stimulate faith, acceptance, feelings, and so on.

Of course, not everyone takes their dreams seriously. Moreover, many perceive them as computer games. But it is more like that no one will argue that those feelings that we experience in a dream completely *affect our productivity* during the day, and even our choices being made, our feelings. It all depends on the correct setting of a dream. A person may wake up in a cold sweat from a terrible nightmare or with a feeling of absolute harmony, peace, and quiet happiness. The correct use of feelings in a dream can be a powerful tool to build faith, inclinations, principles, and to influence completely (or almost completely) *the formation of the concrete personality*.

Moreover, *extra hours of "active life"* in astral would have delighted those workaholics who always need more hours per day to realize all of their projects. Talking not about workaholics, haven't you been delighted when your dream was a lovely movie to distract from reality and rest in peace some?

It should be noted that dreams modeling can also *go beyond* the possible, reasonable, imaginable. This is the kind of extra upgraded design thinking that we deserved. In a dream, it is even possible to accept an additional limb as your own (and this, by the way, applies to VR technology as well).

Another question remains, whether it would be possible *to unite dreams* of different persons, so they would be able to arrange a meeting while sleeping, to discuss a project or just to chill together.

So dreams modeling is way far more than just to be recommended for further *elaboration.*

Interesting thing is that in virtual reality person **does not feel its actual physical body** the way he got used to. He distracts from it, forgets about it, focuses on the space around him.

That is to say, what is the body for a person if he so easily abstracts from it in virtual reality?

The issue that needs to be discussed in this paragraph is the appropriateness of using the body within the framework of human existence. It becomes obvious that our body is nothing more than *a vessel for our presence in the very reality* in which we were born, while it is not quite necessary in virtual reality.

One must admit that in the very reality of the real world, our body is an indisputable *key* to discovering all the sides of the world, to feeling the world, the key to mental, moral, spiritual, and mental development.

In Japan, children are good taught to develop fine motor skills - the fact is that this contributes to the development of intelligence.

So-called "fingerplays" help children with fine motor development, pre-writing skills, language development [12], and way much more, according to many people reviews, including their teachers and parents. Many unique *types of fingerplays* can be used [11]:

– as attention getters
– to extend learning/understanding of a concrete topic
– to increase creativity, body awareness, etc.
– to promote language skills and vocabulary
– to help with listening skills

The first three points listed above are extremely close to the principles of the *design thinking* itself, aren't they? That confirms the fact that our body is useful precisely in the reality in which humanity exists at the exact moment. The body and all kinds of interactions in a certain way have an impact on our consciousness, affecting even design thinking processes themselves (people think they might have control over it). At the same time, it becomes apparent that *in other realities* such attachment to the body may be *potentially irrelevant.*

## 6   Life to Play - Cosplay

With attention to empathy creation, so-called *"cosplay" opportunities* should be discussed.

Cosplay means *costume play*. It unites people from completely different areas, creating empathy between people, being maybe way too far away from each other in real life (in matters related to emotions, everyday necessities).

People imagine themselves to be someone else, they try on emotions, characters, hobbies that are not peculiar to them, due to which a *"virtual" favorable environment* is built for new unexpected acquaintances and opportunities. As a great person said once upon a time:

> *"It makes no sense to continue to do the same and wait for other results."* (Albert Einstein).

In ordinary life, it might be, that these people would have never even spoken to each other. The point is, that in that costume play they are **_connected by an imaginary imitation_** of some characters, thereby discovering up opportunities of a new level of mutual understanding, distracted from every day worries.

Cosplay is perspective in both empathy way and experience.

## 7  Scenes - Role Game

"Scenes" uses pre-defined building blocks to create **_a story_** (characters, buildings, devices, backgrounds, office furniture, arrows, transportation elements signs - that can be manually divided into two groups: movable and stable objects).

"Scenes" is a unique tool widely used in companies to create **_a good team environment_** and innovations as a result. "Scenes" shows the value of ideas and product visions in a concrete appropriate well comprehended (for all of the participants) context. There are **_two main directions of "scenes"_** that should be appointed (two main flows of infor-mation, interaction):

1. **IN.** That is the direct way participants get information and experience from the storyboard. Usually, business leaders and industry professionals share their ideas and scenarios in the form of illustrative stories. Information and experience got from the "scenes" are way more memorable and understandable.
2. **OUT.** But there is another option to be highlighted as well. It happens quite often when the narrator itself gets suddenly new information, ideas, and knowledge from the story processing. That new kind of cognition may lead to improved understanding and further quality project development. Such cases take place because of "scenes" clarity, obvious interpretation, as well as interaction with real people, other participants in the discussion, exchange of opinions. Errors and inaccuracies in the story processing are easy to be timely revealed, detected, and neutralized. It is more like a safe practice of a futuristic project when you lose nothing but acquire usefully. Moreover, established team discussion is an interaction that leads to a total/teilweise co-understanding which is always beneficial (Fig. 4).

**Fig. 4.** SAP "Scenes" for the design thinking processes [16]. ©SAP SE

To put it in a nutshell, "scenes" precisely is an ultimate empathy and insights strong modeling method.

On the other hand, role games to be mentioned, that is not used to extract or share exact concrete information, but for a *powerful interaction and empathy building* for sure. If to think about it carefully, similarities between "scenes" and role games become obvious. From both of them, an exceptional experience can be got that would otherwise have been impossible.

The significant difference is that in the role-playing game there is a more *direct* interaction between all of the participants themselves, and different people act as generators of the possible outcome. Thus, the outcome of such a process is not obvious and usually happens to be more creative than any profit of the "scenes". A *random* like that expands the horizons of all participants (because it is way closer to real life, full of unpredictable events), develops strong creative thinking.

To be noted, some *companies* like Yandex use the role game method in their online courses for students. For instance, even the simplest programmed role-playing game, used in Python training courses from Yandex, can immerse the trainee in a high-quality atmosphere of the company without involving the company's human resources themselves. The process becomes fundamentally interesting, the trainee feels his involvement in the activities of the company while being in a kind of "sandbox" that does not require from the company some extra money, time, and human resources. In such a "sandbox" a new employee can make decisions that will not negatively affect the company's processes. And this is only the simplest role game model that has been described above. Only good consequences would be followed if we have upgraded the model some. So why wouldn't we try so?

# 8  Research

It should be highlighted that, as a *practical part of this work*, a series of experiments have been conducted, revealing some facts about the methods suggested by the paper.

Talking about *THE VR TECHNOLOGY*, already mentioned above SAP University Alliances Virtual Reality Experiment - Innovation Project (VRE-IP 24–25th of April 2021) was carried out in online form between people all over the world to be considered. It had shown an actual opportunity for people to work together in the *virtual reality* as a part of the {online design thinking process}: to empathize, to define, to ideate, to collaborate, to present. It is useful to note *sculpturing* as an effective method to imagine in 3D virtual reality, to get inspiration from it, to establish sustained empathy. VR experiment participants had a unique chance to present their design thinking results in virtual reality, to discuss work being done, and so on. Nevertheless, there had still been things to work on: connection, hardware, and software problems. Still, that experiment remains to be extra useful in the global understanding of far way going perspectives of distant collaboration.

Talking about *THE EMPATHY MIRROR* itself, one can say for sure that VR technologies may allow a direct transfer of pictures from somebody's point of view if required. A simple experiment of *such can easily be carried out*. Through VR we can share our knowledge, experience, feelings - like through the prism of our own consciousness.

Talking about *THE DARE DREAM WORLD MODELING*, many factors had been analyzed, thus, the reasoning is given that just may lead to an innovative technology in that sphere:

– *Falling asleep phase prediction - comparison of breathing during passive wakefulness and during sleep.*

The way human breaths during the sleep phase differs from the way he does while awake. That strictly leads us to keep on looking for some answers in bio-engineering, neuroscience.

– *Predicting the presence of dreaming during sleep.*

An accurate experiment was conducted that showed that the exact time of the dream phase can be successfully (more or less) predicted from some statistical/mathematical/sound analysis. "Sleep Monitor" app [14] was used, which predicted the MAJOR percentage of sleep phase acts - being based only on collecting such data as time/duration, body movements, and sound. The app provided deep insights into the quality of sleep by the following metrics: Sleep Score, Sleep Cycle Graphic, Sleep Statistics, Sleep Noise. The more movements or the more sleep noises - the lighter sleep is. REM phase where dreaming occurs is finally a result of an estimation being provided). That high-lights some great work being done by sleep specialists and appoints significance of the mathematical approach to that field of knowledge as well.

Insofar as of our consciousness being laid in/from some quantum process - that approach of being required as well

– *Researching the possibility of influencing the content of the exact dream from outside.*

That aspect remains to be a tricky and complicated question. It just might be assumed that the answer to the question (as well as to the possibility of creating a REAL Artificial Intelligence*) lies at a fundamental understanding in the intersection of the following sciences: biological approach (bio-engineering, neuroscience, etc.), quantum approach and simply some sort of an elementary mathematical approach.

*[P. S.]* - Here REAL *Artificial Intelligence** means an actual intelligence that can choose what to learn and pick/design the method/model of its learning processes itself; bind events, the binding of which is not written in the program code. To the point, it seems that Neural Nets themselves will not be able to succeed in carrying out that task and will finally face a dead-end associated with the need to continue the search for an ultimate solution in biological/quantum structures.

Talking about how *"THE FINGERS MAGIC"* can affect *human brain capabilities* in real life, an actual experiment had been conducted, revealing an exact correlation between brain activity capability and the degree of development of fingers tactile sensations.

With the framework of the International IT-Ideathon from SAP University Alliances in Divnomorskoe, Krasnodar Krai (3–5th of September 2021) participants (which are both students and young employees) were derived into three groups of people [5]. The first group started solving a special test without any preparation at all. The second group was asked to spend about 5 min making an "elementary level" origami just before solving the test. And the third group spent about 10 min on the so-called "hard level" origami just before the special test. It was not necessary to complete the origami being given as a task before the test. The experiment time management was highlighted as an important issue - that was exactly *to control the time being provided to have the tactile sensations, tangible experience.*

A special test was made of some basic questions, the answers to which are placed some-where at the bottom of each person's consciousness. Those were some well-known facts, in order to remember which person was turned to use the full potential of his/her brain activity. Completing the entire test for a full score in a given time was not seemed to be possible. Ideally, the test should contain an *infinite number of equivalent (?) questions.*

It is to be mentioned that the quality of the test (which is combined with the structure, content, some direct methodologies and etc.) should be improved based on further-going edge-cutting researches in the coming great future.

The results of the experiment revealed that the group of people who activated their tangible experience right before solving the test - *achieved higher scores* than those who skipped the origami handwork. It is to be featured that origami was used just as a simple well-known live example of tactile sensations. It might be replaced by an improved "fingers magic" method in the future.

One can be said, in the real world, human needs its body. It is not just the bearer of soul and consciousness. Undoubtedly, it is exactly the *key* to consciousness, cognition, and interaction in the *realities of this world*.

Talking about *THE COSPLAY*, one can say for sure - that it actually helps the acquisition of non-standard experiences and relationships. So as not to be unfounded, the description of a definite experiment and its results are described below.

Person-X had never ever been in a nightclub. So Person-X had had no background right then. When he decided to go there, Person-X absolutely had changed demeanor - be-cause of being in an absolutely unfamiliar situation. He imagined as of being another person and had done some of self-opening to the world. Person-X knew that he'd not probably meet the people around again and stopped any worries about his behavior. It is a sort of cosplay-tendency.

During his office work, Person-X never revealed that he had really liked dancing. No-body had seen him in that condition. In the nightclub, Person-X met people from different areas of activity, who also adored dancing. It happened that one of Person-X's new friends was fluent in PR management. Person-X had got interested in his new friend's PR project and suggested a partnership.

From this partnership both of the persons (Person-X and his friend) had got a unique experience and relationship, that would not have been possible if they had not met on the dance pole.

That story experiment has *numerous variations and sequels* in real life and is only a faint introduction to cosplay. Nevertheless, this experiment allows us to make sure that the cosplay method is effective, takes place in real life, and can be further used both in the framework of design thinking and in the companies redesign.

Talking about *THE ROLE GAME* potential, an emphasis can be put on the already used and mentioned above Yandex *role game scheme* in its training Python courses for trainees. The Yandex role game scheme is more like "scenes" being realized in the real-life - attempt to interact with one actual real person - many outputs (users) through one input (program). No doubt, similar techniques will still show their productivity and feasibility in the future.

# 9 Conclusion

To sum it all up, nowadays humanity is placed at the beginning of many *new approaches to discovery* combining bio-engineering, psychology, neuroscience, quantum physics, mathematical analysis, and others - to understand the mental structure of a person to build a better society and to create breakthrough innovations. It concerns revolutionary design thinking methods in empathy and companies redesign.

Technologies and methods described above (VR, cosplay, "scenes", role game) might help both design thinking to create innovations, and companies redesign, taking into consideration *the current situation* in the world due to the coronavirus.

*The list* below barely reveals just *a couple of ideas* (which we hope to be found useful) on how to construct these empathy methods in the design thinking/companies redesign; the list can certainly be expanded and modified for the great purposes of the common good and innovations.

*The List:*

- **Interactive Role Game Platform - between and in companies.**

  - To share ideas
  - To create empathy/build strong empathy field
  - Meet new people
  - Try yourself in a different role (vocational guidance)
  - Work on an extraordinary project together/capability of gathering a team of professionals from different fields of knowledge - that improves the efficiency of design thinking

- **Global Techniques of "Fingers Magic"**
- There is a new concept of a special brain activity test to be introduced, which would support monitoring of activating the capabilities of the worker's brain.

  - To improve the brain abilities of workers and students.

- **VR Program Equipment of Communication** (VR: presentation, meetings, design thinking, and other company activities which to be found of extra use - modern distant company/team pattern)

- The equipment can be easily integrated into the company's infrastructure
- For general use in innovation areas among development teams

• **Experience-Empathy Exchange System/Platform** (based on Empathy Mirror Concept)

  • Sharing experience/ideas from your side of view using VR technologies - an easy way to transfer empathy (with the help of special equipment, record your experience in the first person and send it to your partner so that he could watch this experience on his VR headset as if it were in the first person)

• **Professional Industry Cosplay Parties**

# References

1. Taratukhin, V., Pulyavina, N.: The future of project-based learning for engineering and management students: towards an advanced design thinking approach. In: The American Society for Engineering Education Annual Conference & Exposition Proceedings, ASEE (2018)
2. Taratukhin, V., Baryshnikova, A., Kupriyanov, Y., Becker, J.: Digital business framework: shaping engineering education for next-gen in the era of digital economy. Paper presented at 2016 ASEE annual conference & exposition, New Orleans, Louisiana (2016). https://doi.org/10.18260/p.26840
3. Taratukhin, V., Pulyavina, N., Becker, J.: Next-gen design thinking. using project-based and game-oriented approaches to support creativity and innovation. In: ICID 2019 - Proceedings of the 1st International Conference of Information Systems and Design, CEUR (2020)
4. Brown, T., Katz, B.: Change by Design: How Design Thinking Transforms Organizations, 272 p. Harper Publisher, New York (2009)
5. Skibina, V., Taratukhin, V.: Virtual reality usage for innovations generation. In: Proceedings of the International Scientific and Technical Congress "Intelligent Systems and Information Technologies, vol. 1, 362 p. (2021)
6. Rizzolatti, G., Craighero, L.: The mirror-neuron system. Ann. Rev. Neurosci. **27**(1), 169–192. (2004)
7. Keysers, C.: Mirror Neurons. Curr. Biol. **19**(21), 971–973 (2010)
8. Rizzolatti, G., Fadiga, L.: Resonance behaviors and mirror neurons. Italiennes de Biologie **137**(2–3), 85–100 (1999)
9. Hameroff, S.: How quantum brain biology can rescue conscious free will. Front. Integrat. Neurosci. **6**, 93 (2012)
10. Hameroff, S., Penrose, R.: Reply to seven commentaries on consciousness in the universe: review of the 'Orch OR' theory. Phys. Life Rev. **11**(1), 94–100 (2014)
11. Church, E.: Fingerplay Fun! Early Childh. Today **16**(1), 60–61 (2001)
12. Curtis, K., DeCelle-Newman, P.: The PTA handbook: keys to success in school and Career for the Physical Therapist Assistant (2004)
13. Korgel, B.: Innovation Arts and the Conception of Rapid Design Pivot - performance (2016)
14. https://play.google.com/store/apps/details?id=com.sleepmonitor.aio&hl=ru&gl=US
15. https://www.design-science.org/empathy-mirror
16. https://apphaus.sap.com/resource/scenes#downloadscenes
17. https://www.interaction-design.org/literature/article/design-thinking-getting-started-with-empathy

# The Use of Virtual Reality to Drive Innovations. VRE-IP Experiment

Yulia Skrupskaya[1]($\boxtimes$) (iD), Valeriia Skibina[2] (iD), Victor Taratukhin[3] (iD), and Elvira Kozlova[1] (iD)

[1] National Research University Higher School of Economics, Moscow 101000, Russia
yubskrupskaya@edu.hse.ru
[2] Lomonosov Moscow State University, Moscow 119991, Russia
[3] University of Muenster, 48149 Muenster, Germany

**Abstract.** In the context of widespread digitalization of businesses and academia due to COVID-19, the new effective tool has become necessary to drive innovations. This paper observes the international scientific experiment VRE-IP that tested the hypothesis of the advantages of using virtual reality for teamwork, creativity, and innovation. VRE-IP experiment was based on design thinking methodology and supported SAP innovation process. The analysis of the trends of using virtual reality for educational and business development goals, and VRE-IP experiment revealed that employing virtual reality technology to organize workshops, distant meetings, plenary reports, and other tasks is exceptionally efficient, improves the engagement rate and creativity and contributes to the process of innovative ideas stimulation. The experiment also demonstrated the need for building a hybrid model for teamwork. The combination of traditional teamwork instruments and virtual technologies is proposed as the most sustainable and systematic model for online collaboration at the current stage of technological development.

**Keywords:** Virtual reality · Digital transformation · Design thinking

## 1 Introduction

The COVID-19 pandemic 2020 had a strong impact on different spheres of life especially on education since universities were forced to adjust to new reality with remote education. In most cases, the learning process was based on the use of platforms for virtual meetings, such as Zoom and Microsoft Teams (Navleen 2020). These platforms were originally created for video conferencing and work calls, and therefore they were not specialized for teaching and gaining knowledge. Interaction between students and teachers has become a routine process consisting only of presentations, questions and answers. The necessity to adapt to new learning conditions lead to the search for innovative solutions that would contribute to the improvement of the student productivity. Distance learning opens the new opportunities for educational processes. However, now there are no methodologies and studies demonstrating new options for obtaining and submitting educational material and their feasibility of implementation. The scientific novelty and practical significance

© Springer Nature Switzerland AG 2022
V. Taratukhin et al. (Eds.): ICID 2021, CCIS 1539, pp. 336–345, 2022.
https://doi.org/10.1007/978-3-030-95494-9_28

of the work is the hypothesis that the use of virtual reality in the process of remote learning will positively affect the speed and quality of learning and increase motivation to learn.

## 1.1 Application of Virtual Reality Technologies in Education

The development of virtual reality has led to various experiments of its practical application, including integration into the educational process. There are a few reasons for the research on the impact of virtual reality technologies on education. Firstly, the experience of practical application in other industries shows positive dynamics and optimization of processes. Virtual reality is used in the oil and gas industry, metallurgy, telecommunications, and advertising, gradually moving away from being used only in the gaming industry (Krayushkin 2021). Secondly, over the past few years, the average cost of technical equipment has decreased (The Economist 2020). Increasing availability of virtual reality devices has a positive effect on the growth of the audience interested in the technology.

## 1.2 VR Programs

There are many programs on the virtual reality market that are suitable for educational purposes (Smith 2021). Most of them have the same minimal functionality: the presentation demonstration, whiteboards, conference halls. However, there are also specialized programs for training in narrow professional or subject areas. At the first step of the experiment, we analyzed the most popular communication virtual reality applications for the further use during the experiment.

Spatial: a virtual reality program created for video conferences (Spatial 2021). Spatial allows users create realistic avatars by transferring their appearance to a virtual environment. It provides an opportunity to receive and exchange information (images, pdf files) inside a virtual environment. The application monitors the position of the hands using a controller during immersion in a virtual environment, users can see the gestures during a dialogue. The conference limit of participants equals 40 individuals. The tools include a whiteboard for writing and drawing using controller, speech notes, and a virtual keyboard.

Meetinvr: a platform that provides a virtual space for meetings and conferences, allowing users to work together on presentations, notes and mind map diagrams (Meetinvr 2021). Maximum 32 users can be present in the presentation hall at the same time. The advantages of Meetinvr: the ability to provide access to pre-uploaded files, visualize PowerPoint files, whereas in most virtual conference applications only PDF format is available. The Meetinvr application is not free but it has a trial version for a month of free use with some restrictions on functionality. VR headsets that could be used with Meetinvr are limited to: Oculus Quest, Oculus Quest 2, Pico Neo 2 and Pico Neo 2 Eye.

Rumii: the most accessible and free program for various types of VR headsets (Rumii 2021). In every meeting hall there is a large multifunctional central screen on one of the walls of the three-dimensional room. The user can activate the screen: a whiteboard, access to the Internet or a demonstration of a PDF file or images. The user's avatar is more animated compared to other applications and is created manually by the user. The maximum capacity of the virtual hall in the application Rumii is 20 active users, which is suitable for group work.

All the three programs are the examples of widely applicable applications in the learning process. The essential characteristics for us were the capability to interact with 10 users at the same time, make smaller rooms, make and then later present projects inside virtual reality, record the interaction process, compatibility with the Oculus Go virtual reality headset. Thus, the application Rumii was chosen for the further experiment.

### 1.3 Methodology of Design Thinking

The first paragraphs that follows a table, figure, equation etc. does not have an indent, either. All the three programs are the examples of widely applicable applications in the learning process. The essential characteristics for us were that the Design thinking is based on a creative approach to solving tasks, it differs a lot from analytical techniques that include critical analysis (Leifer 2016). One of the main parts of this methodology is teamwork. That why is why there could be an assumption that design thinking in team projects would have a positive impact on the results of students (Taratukhin et al. 2020). The methodology of design thinking in the approach to problem solving that includes five stages (Leifer 2016).

1. Empathize. This stage includes observation, engagement, and immersion. The team studies the intended users of the final solution, meets, and communicates with them, learns what kind of experience users get and what problems they go through.
2. Define. At the second stage of design thinking, it is necessary to focus on and define a specific problem. This can be done by achieving two goals: understanding the target audience and the user of the final product and determining the POV (point- of-view) and further tasks based on the understanding of the user.
3. Ideate. This stage usually begins with brainstorming – a way to find a large number of ideas through teamwork and collective reflection on the task in a short time. Then the ideas can be selected by voting, thereby narrowing the range of solutions for further prototyping.
4. Prototype. Prototyping is a part of implementation process. The selected ideas are implemented in the real world using improvised means. Anything can become a prototype, for example, a wall in paper stickers or a rough version of the application interface. Design thinking does not limit the freedom of actions and manifestations of imagination at all stages.
5. Test. The testing stage is aimed at improving solutions to the problem. Testing is usually carried out repeatedly, interactively and in real conditions. Testing usually involves demonstration of prototypes and getting feedback from users.

Design thinking is applied in various fields and has no limitations (Leifer et al. 2019). According to our hypothesis, the application of design thinking methodology in distance learning processes can have a positive impact on the interaction of students in a team.

## 2 Description of the Experiment

The experiment was based on the hypothesis that the introduction of virtual reality technologies into the learning processes will help to increase the productivity of students and reduce the routine of processes, which, in turn, will lead to more successful and high results compared to traditional distance learning processes (Dzardanova et al. 2021). VRE-IP experiment was organized jointly with the University of Muenster (ERCIS - University of Muenster) and SAP University Alliances. Students from different cities and universities such as Lomonosov Moscow State University, National Research University Higher School of Economics (HSE), Wilhelm University of Westphalia (University of Muenster), Norilsk State Industrial Institute, Arctic and Antarctic Research Institute, Stanford Janus Project (Stanford Project Janus) participated in the project. The students were divided into 4 groups of 2 people. Three groups used virtual reality technologies – glasses and the Oculus Go controller, Rumii program. The remaining group was a focus group and used the traditional methods of remote interaction – Zoom, Telegram and Google Slides. Group division was necessary for the further comparison and analysis of the results at the end of the experiment to find a connection between the results of the participants and the use of virtual reality technology. Surveys were conducted among students before and after the experiment.

The experiment was an accelerated version of the hackathon – the event when the teams solve a task with a time limit (Taratukhin et al. 2021).

The duration of the experiment was limited to 2 days. On the first day there was the introductory lecture and task announcement. On the second day the teams demonstrated their work. Intra-team interaction was carried out throughout the experiment. The introductory lecture was held in Zoom. The interaction of the participants within the three teams was carried out using virtual reality technologies. This allows us to assess the presence of the impact of VR implementation on student productivity and improving the quality of team interaction at a distance.

The final team presentations at the end of the experiment were shown in the environment in which they interacted during the experiment. The completion of the hackathon and the announcement of the results of the participants' speeches took place in Zoom.

### 2.1 Preparation for the Experiment

During the preparation of the experiment, instructions were developed for quick and easy cooperation. The instructions described the sequence of actions required during the installation and use of the program Rumii. The survey was also created for the participants of the experiment to obtain information about their expectations from virtual reality technologies, as well as their experience in distance education. The main task for the participants during the experiment was to offer solutions to the environmental problems: recycling plastic waste and reducing $CO_2$ emissions. A lecture on the methodology of design thinking was prepared and presented before the hackathon to all participants of the experiment. It contained all the necessary information about the approach of design

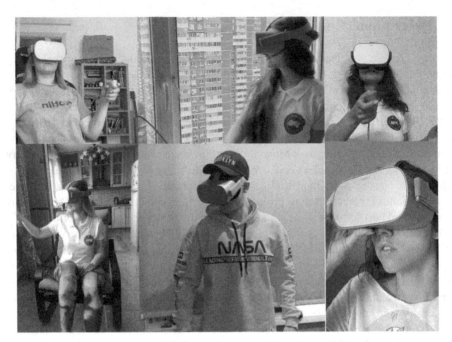

**Fig. 1.** Participants of the experiment wearing Oculus Go Headsets.

thinking in solving problems and described the stages of design thinking on which it was necessary to rely during the solution of the problem.

In the Rumii application, three separate halls for each team were created. The team members could stay in this hall for an unlimited amount of time during the experiment. The participants were not limited in the possibility of using any programs and communication methods convenient for interaction.

Participants were offered a survey after the first day of using the technology in order to form an idea of the first impression of virtual reality technology, another survey at the end of the experiment contained questions aimed at obtaining detailed feedback from the use of the new technology.

## 2.2 Experiment

The initial purpose of the experiment was to receive feedback on remote interaction through the use of virtual reality technologies, and to compare it with conventional distance education processes by involving a focus group in the experiment. Given that the Oculus Go virtual reality headsets have a limited operating time without connecting to batteries. Teamwork according to the experiment schedule was limited to 1–2 h. It was followed by a break for charging/recharging devices. Thus, the participants totally spent an average of 4.5 h in virtual reality. The rest of the team interaction took place in Telegram messenger and Zoom application.

After the beginning of the practical part of the experiment, some participants had a problem connecting to Rumii, they had black screen when trying to enter previously created conference halls and returning to the main screen of the program. The occurrence of this problem might have been due to both the instability of the Internet connection and the technical characteristics of the selected devices - Oculus Go has weaker characteristics compared to other Oculus devices. Due to the technical characteristics of virtual reality headset, there is also a problem of overheating after an hour and a half of the use. The problem was observed in 100% of the participants of the experiment.

During the experiment, after the initial setup of virtual reality devices and gaining access to the halls for team interaction, there were no massively encountered problems. During the preparation of presentations of their projects to solve the problem of ecology, most of the participants used a hybrid format of interaction, using virtual reality technologies and other methods of remote interaction. For example, the program Rumii does not allow you to create and edit presentations, so participants used Google Slides to create presentations. The participants preferred virtual reality to be used at such stages of design thinking as empathy, focusing, generating ideas, and choosing the best one. In other words, all the teams preferred team interaction with the help of the virtual reality program Rumii and used it during the need for intensive interaction and dialogue. Also, Rumii allows to draw inside a virtual environment, so the participants used this function on a white board inside the virtual hall to mark the main points of discussion and ideas. At the prototyping stage, the participants used programs for design and prototyping: Figma, Miro, Wireframepro.

**Fig. 2.** Modeling in VR.

Increase in concentration was found among the participants when virtual reality technologies were introduced into the interaction process. For example, according to the results of the survey the pronounced change in the distribution of self-concentration is noticeable. For the lecture, which was held at Zoom, half of the participants chose a score below or equal to 5 out of 10, and the average concentration score among all participants was 8.25 out of 10. If we talk about the evaluation of team interaction, then all scores exceed or equal to 6, and the average score among all participants takes the value of 8.25. In connection with the results obtained, it is possible to confirm the assumption of increased concentration and attention when using virtual reality technology in distance learning processes. One of the reasons explaining such a result is the impossibility of losing attention due to surrounding distractions: a smartphone or a book. Being in virtual reality, the user is completely detached from the real world around him, unable to pick up a smartphone or simultaneously engage in third-party business. Virtual reality requires the user to be completely immersed in the process, thereby reducing the impact of third-party stimuli. The second reason is the novelty of the participants' feelings from immersion in a virtual environment. The participants had not previously used virtual reality technology. The use of virtual reality during the experiment reduced the routine of interaction processes, thereby increased the interest of participants in what was happening.

Moreover, it is worth noting that the participants spoke positively about team interaction. According to the participants' feedback on the comfort of interacting with each other, the average score of the participants takes the value of 9, which is a good indicator, close to the maximum possible value 10. From this it can be assumed that most of the participants received positive impressions from the use of virtual reality technology in order to solve the task as a team.

**Fig. 3.** Three-dimensional representation of virtual reality in the program Rumii.

At the end of the experiment, 66% of participants rated the virtual reality technology as promising and are looking forward to unlocking the potential in the field of distance education.

The results of the experiment demonstrate the success of integrating virtual reality technologies into the process of remote team interaction, the ultimate goal of which is to obtain a solution to the task. The participants of the experiment speak positively about the technology used and note an increase in motivation due to the new sensations of immersion in a virtual environment.

The practical use of virtual reality technology in the educational process confirms the hypothesis of improving the built remote process. There is a clear connection between an increase in interest in receiving and working with information, as well as an increase in direct motivation motives among the participants of the experiment, and the integration of virtual reality technology.

Some of the disadvantages were also revealed such as poor Wi-Fi connection which delayed some steps of the design thinking, overheating, inability to arrange and take surveys inside virtual reality, the participants also noted that it was quite difficult to draw using the remote control on the board, the absence of the eraser. The last major disadvantage was fatigue. Even though VR glasses were quickly discharged, during this short period of time, the participants became very tired.

The experiment demonstrated the feasibility of building a hybrid model of distance learning using virtual reality technologies. The participants of the experiment were not ready to use virtual reality throughout all stages of the experiment, and, in addition to the virtual environment, they used other technologies previously used in distance learning.

### 2.3   Recommendations for Integrating Reality into Learning Processes

It was revealed that the optimal solution for improving distance education processes is the integration of a hybrid learning model. The list of recommendations might be useful for teachers or employees of educational institutions interested in integrating virtual reality technology into distance learning processes.

- During the initial integration of virtual reality technologies into the process of distance learning, a virtual environment should be used as a substitute for some types of receiving and presenting information in training. In other words, to create a hybrid model of the distance learning process, affecting both virtual reality technologies and other applications and programs for remote interaction.
- Virtual conference programs implemented in virtual reality are recommended for use during the explanation and repetition of new material - lectures, as well as during interaction in teams at the initial stages of solving a problem or seminars that require discussions and close interaction (McShaneh 2021).
- The requirements of the means for the virtual reality technology - headset and controller, include the limited time of use until the moment of complete shutdown due to the rarefaction of the battery. In this regard, when building distance learning using virtual reality technology, it is necessary to consider the operating time of the headset and controller used and provide additional time for charging.
- When integrating virtual reality technology, it is recommended to use the methodology of design thinking during the learning process. This methodology is closely related to the quality of team interaction and implies carrying out most of the work on the task using communication and discussion in teams. Due to the isolation that led to the mandatory use of distance education, students began to interact less with each other. Coupled with the use of design thinking, virtual reality will make it possible to fill in the gaps in communication and interaction among students.
- For the successful initial use of virtual reality programs, it is recommended to develop and provide detailed instructions for the teacher and students. These instructions

should help you install the program and use all the functions necessary for the learning process. It is also recommended to instruct the teacher inside the virtual environment for a more detailed and clear explanation of the principles of interaction with the environment and the use of functionality.

The use of the recommendations presented above will help to optimize the implementation process and will also help to avoid possible failures during the initial use of programs.

## 3 Conclusion

Virtual reality technologies represent a promising direction for development in new areas, including distance learning. The results of the experiment demonstrated an increase in the concentration of participants when using virtual reality technology, as well as a positive assessment of team interaction. Moreover, the development of virtual reality technologies leads to wider use and implementation of new programs, including those created for educational purposes. Hybrid type of interaction is the most practical way to study since it leaves some tasks to be performed using traditional means of remote interaction, which will improve and diversify the learning process, thereby improving the quality of education. The outcomes of this work can be useful for educational purposes in academic institutions and businesses.

## References

Dzardanova, E., Kasapakis, V., Gavalas, D., et al.: Virtual reality as a communication medium: a comparative study of forced compliance in virtual reality versus physical world. Virtual Reality (2021)

Krayushkin, N.: Virtual reality in education, Competence Development Center in Business Informatics of the Higher School of Business (2021), https://hsbi.hse.ru/articles/virtualnaya-rea lnost-v-obrazovanii/. Accessed 15 Oct 2021

Leifer, L., Dym, C., Agogino, A., Eris, O., Frey, D., Engineering design thinking, teaching, and learning, J. Eng. Educ. **94**(1), 103–120 (2016)

Leifer, L., Sonalkar, N., Mabogunje, A., Miller, M.R., Bailenson, J.N.: Augmenting learning of design teamwork using immersive virtual reality. In: Meinel, C., Leifer, L. (eds.) Design Thinking Research. UI, pp. 67–76. Springer, Cham (2019). https://doi.org/10.1007/978-3-030-28960-7_5

McShaneh, M.: Classical Education Meets Virtual Reality, Forbes (2021). https://www.forbes.com/sites/mikemcshane/2021/10/19/classical-education-meets-virtual-reality/?sh=89a404 94eb53. Accessed 15 Dec 2021

Meetinvr, the program for holding virtual conferences (2021). https://www.meetinvr.com/. Accessed 15 Oct 2021

Navleen, K., Supriya, S.: Fighting COVID-19 with technologyand innovation, evolving and advancing with technological possibilities. Int. J. Adv. Res. Eng. Technol. **11**, 395–405 (2020)

Rumii, the program for virtual conferences (2021). https://www.oculus.com/experiences/go/131 2559188845495/?locale=ru_RU. Accessed 15 Oct 2021

Smith, S.: Virtual Reality or Education in 2021, AR VR EDTECH (2021). https://arvredtech.com/blogs/news/virtual-reality-for-education-in-2021-how-to-utilize-the-benefits-of-vr-in-education. Accessed 15 Dec 2021
Spatial, the program for virtual conferences (2021). https://spatial.io/about. Accessed 15 Oct 2021
Taratukhin, V., Pulyavina, N., Becker, J.: The Future of Design Thinking for Management Education. Project-based and Game-oriented methods are critical ingredients of success, Issue Vol. 47: Developments in business simulation and experiential learning. Volume: Innovations and Future Directions in Education (2020)
Taratukhin, V., Skrupskaya, Y., Kozlova, E., Pulyavina, N., Yudina, V.: Bringing together engineering and management students for a project-based global idea-thon: towards next-gen design thinking methodology. In: 2021 ASEE Virtual Annual Conference Content Access, Virtual Conference (2021)
The Economist, Headsettechnology is cheaper and better than ever (2020). https://www.economist.com/technology-quarterly/2020/10/01/headset-technology-ischeaper-and-better-than-ever. Accessed 15 Oct 2021

# Author Index

Printed in the United States
by Baker & Taylor Publisher Services